Proof and System-Reliability

T0235711

NATO Science Series

A Series presenting the results of scientific meetings supported under the NATO Science Programme.

The Series is published by IOS Press, Amsterdam, and Kluwer Academic Publishers in conjunction with the NATO Scientific Affairs Division

Sub-Series

I. Life and Behavioural Sciences	IOS Press
II. Mathematics, Physics and Chemistry	Kluwer Academic Publishers
III. Computer and Systems Science	IOS Press
IV. Earth and Environmental Sciences	Kluwer Academic Publishers
V. Science and Technology Policy	IOS Press

The NATO Science Series continues the series of books published formerly as the NATO ASI Series.

The NATO Science Programme offers support for collaboration in civil science between scientists of countries of the Euro-Atlantic Partnership Council. The types of scientific meeting generally supported are "Advanced Study Institutes" and "Advanced Research Workshops", although other types of meeting are supported from time to time. The NATO Science Series collects together the results of these meetings. The meetings are co-organized bij scientists from NATO countries and scientists from NATO's Partner countries – countries of the CIS and Central and Eastern Europe.

Advanced Study Institutes are high-level tutorial courses offering in-depth study of latest advances in a field.
Advanced Research Workshops are expert meetings aimed at critical assessment of a field, and identification of directions for future action.

As a consequence of the restructuring of the NATO Science Programme in 1999, the NATO Science Series has been re-organised and there are currently Five Sub-series as noted above. Please consult the following web sites for information on previous volumes published in the Series, as well as details of earlier Sub-series.

http://www.nato.int/science
http://www.wkap.nl
http://www.iospress.nl
http://www.wtv-books.de/nato-pco.htm

Series II: Mathematics, Physics and Chemistry – Vol. 62

Proof and System-Reliability

edited by

Helmut Schwichtenberg

Mathematisches Institut,
Ludwig-Maximilians-Universität,
München, Germany

and

Ralf Steinbrüggen

Institut für Informatik,
Technische Universität,
München, Germany

Kluwer Academic Publishers

Dordrecht / Boston / London

Published in cooperation with NATO Scientific Affairs Division

Proceedings of the NATO Advanced Study Institute on
Proof and System-Reliability
Marktoberdorf, Germany
24 July–5 August 2001

A C.I.P. Catalogue record for this book is available from the Library of Congress.

ISBN 1-4020-0607-1 (HB)
ISBN 1-4020-0608-X (PB)

Published by Kluwer Academic Publishers,
P.O. Box 17, 3300 AA Dordrecht, The Netherlands.

Sold and distributed in North, Central and South America
by Kluwer Academic Publishers,
101 Philip Drive, Norwell, MA 02061, U.S.A.

In all other countries, sold and distributed
by Kluwer Academic Publishers,
P.O. Box 322, 3300 AH Dordrecht, The Netherlands.

Printed on acid-free paper

All Rights Reserved
© 2002 Kluwer Academic Publishers
No part of this work may be reproduced, stored in a retrieval system, or transmitted
in any form or by any means, electronic, mechanical, photocopying, microfilming,
recording or otherwise, without written permission from the Publisher, with the exception
of any material supplied specifically for the purpose of being entered
and executed on a computer system, for exclusive use by the purchaser of the work.

Printed in the Netherlands.

TABLE OF CONTENTS

PREFACE

As society relies more heavily on software for its welfare and prosperity, the need to create systems in which it can trust is a major concern. Experience has shown that confidence can only come from sharper understanding and can only be based on logically sound foundations.

The Marktoberdorf Summer School 2001 on *Proof and System-Reliability* brought together leading researchers and promising scientists in the critical disciplines of computing and information science, mathematics, logic and complexity. The elaborated results are joined together here. All contributions are self-contained aiming at comprehensibility as well as comprehensiveness. It should be pointed out that the volume also comprises introductory hints to technical issues, for instance on

> *CTL* by O. GRUMBERG,
> *LF* by G. NECULA,
> *Isabelle/HOL* by T. NIPKOW.

Furthermore, concise surveys or even introductions are worked out by the authors of monographs, books, and handbook articles that recently appeared like

> *Ludics* by J.-Y. GIRARD,
> *Dynamic Logic* by D. HAREL, D. KOZEN. J. TIURYN,
> *Computability and Complexity* by N. D. JONES,
> *Logical Frameworks* by F. PFENNING,
> *Network Algebra* by G. STEFANESCU.

The strength of the volume will be the wide spectrum of in-depths tutorials and elaborated surveys. Moreover, the volume comprises various fresh results and new perspectives and is devided into the following topics:

Domains and Semantics

Algorithmic Game Semantics: A Tutorial Introduction
by SAMSON ABRAMSKY

Game semantics models computation as the playing of a certain kind of game with two participants, called Player and Opponent. Player is to be thought of as representing the system under consideration, while Opponent represents the environment. In this way, the distinction between actions of the system and those of its environment is made explicit. Recently Game Semantics is being developed in a new algorithmic direction, carying over the methods of model checking, which are used efficiently for analysing circuits and protocols, to verifying assertions on programs.

Algebra of Networks
by GHEORGHE STEFANESCU

If the term network is used in the wide sense of a collection of interconnecting cells, general results on abstract networks apply to flowchart schemes, finite automata, process algebra, data-flow networks, or Petri nets. Network algebra models for control, space, and time can be mixed and we obtain models of finite systems of agents, their behaviour and their interaction.

Cartesian Closed Categories of Effective Domains
by GÖRAN HAMRIN and VIGGO STOLTENBERG-HANSEN

Domain theory is by now a mathematical theory with applications in central fields of computer science. Type structures of effective domains carry a notion of computability which corresponds to notions of recursive function theory, recursive analysis, and semantics of programming languages.

Logical Foundations

Computability and Complexity from a Programming Perspective
by NEIL D. JONES

Regarding programs as data instead of Gödel numbers accomodates computer scientists to reason about models of computation like imperative or applicative languages, Turing machines or counter machines. Results from complexity theory can be enlightened as evaluation, partial evaluation, compilation, and compiler generation. Whereas computability theory concerns only what is computable, and pays no attention at all to how much time or space is required to carry out a computation, the central computational concepts of complexity theory all involve resources, especially time, space, and nondeterminism. Various complexity classes can be represented by specific complete problems.

Logical Frameworks: A Brief Introduction
by FRANK PFENNING

Formal deduction systems require a coprising framework in which objects, rules, and derivtions can be represented and manipulated. Generic frameworks for logical calculi have successfully been based on type theory. Design isues are that the calculi are represented adequate and that the correctness of derivations can be checked efficiently. If derivations are represented as objects, checking the correctness of a derivation is reduced to type checking in the meta-language.

Ludics: An Introduction
by JEAN-YVES GIRARD

Ludics is a new approach to semantics for various logics that uses the behaviour of derivations under normalization to characterize the meaning of logical connectives from a game-theoretic perspective. In particular, it provides a thorough understanding of linear logic that in recent years has found many applications in modelling resource-bounded computations. Due diligence to the motivation results in clear abstractions. Both foundational implications and concrete applications are important aspects of the theory.

Proof and Security

Naïve Computational Type Theory
by ROBERT CONSTABLE

This is a beginners' course on types with bibliographic notes. What is new in this account is the treatment of classes and computational complexity along lines that seem very natural. It is shown how types for reasoning about algorithmic complexity are introduced into the world of algebraic specifications.

Proof-Carrying Code. Design and Implementation
by GEORGE NECULA

A new approach is being introduced for the safe execution of untrusted mobile code which is an interesting alternative to traditional approaches such as cryptography, interpretation by virtual machines or protection via hardware address spaces. The idea behind "proof carrying code" is to package each piece of mobile code together with a proof that the code does not violate specific security policies. This proof can then easily be checked by a client before executing the code.

Abstractions and Reductions in Model Checking
by ORNA GRUMBERG

The task of checking a model given by a finite state transition system with regard to a certain property specified in propositional temporal logic can be solved efficiently in time but it suffers from the state explosion problem: The number of states in the model grows exponentially with the number of system variables and components. Given a high level description of the system, abstraction is a method of hiding some of the system details that might be irrelevant for the checked property. Abstraction is supplemented by counterexample-guided refinement, on demand. Abstraction can be performed on the data domain and alternatively on the state space, that is within in the framework of abstract interpretation.

Programming Logic

Hoare Logic: From First Order to Propositional Formalism
by JERZY TIURYN

Properties of programs have mostly been investigated in the setting of first-order logic. It follows from complexity considerations that there is no sound and complete proof system capable of deriving all valid partial correctness assertions. The calculi are relatively complete if they are bound to sufficiently expressive structures, such that all weakest liberal preconditions are first-order definable. The propositional approach is concerned with iteration and nondeterministic binary choice generalizing the conditional and while constructs. Kripke frames are used for interpretation. A completeness result holds in this setting.

Hoare Logics in Isabelle/HOL
by TOBIAS NIPKOW

In practice, Hoare's assertion calculus, after beeing generally accepted for proving imperative programs correct, relies on sound proof systems. Investigations consider typical laguage constructs seperately. Hoare logics for while-loops, even with exceptions and side effects, as well as mutual recurvive procedures combined with unbounded nondeterminism can be formalized and shown to be sound and complete in the theorem prover Isabelle/HOL. In this higher-order setting, completeness results for complicated language constructs require various advanced proof techniques.

Logic and Feasibility

Proof Theory and Complexity
by GEOFFREY E. OSTRIN and STANLEY S. WAINER

Feasibly computable functions are a major concern of the recursion theoretic characterization of provably terminating functions. A reformulation of elementary arithmetic imposing a separation of variables into 'induction variables' and 'quantifier variables', leads to complexity hierarchy which is closely related to the increasing levels of induction complexity of termination proofs.

Feasible Computation with Higher Types
by HELMUT SCHWICHTENBERG and STEPHEN J. BELLANTONI.

A characterization of polynomial time functions can be obtained by restricting recursion in finite types: Besides separating variables into 'incomplete' and 'complete' ones, two forms of function types are used. The intuition is that functions of one type may use their argument several times, whereas the others are allowed to use their argument only once.

The authors envolved, elaborating their Marktoberdorf manuscripts, were aiming at a state-of-the-art vorlume. They had been joined with a group of talented young scientists living and working together for eleven days. The comfortable conditions at the Gymnasium Marktoberdorf made the Summer School a laboratory for Proof and System-Reliability. It is our privilege to thank all those who helped to make it a success. These are the participants, the lecturers, our hosts in Marktoberdorf, people helping in the staff, and especially our engaged secretary Mrs. INGRID LUHN. We also thank the publishing assistants from Kluwer, who have exercised great care in management of the printing.

The Marktoberdorf Summer School was arranged as an Advanced Study Institute of the NATO Science Committee with significant support by the Commission of the European Communities, and with financial aid from the town and district of Marktoberdorf. We thank all authorities involved.

Munich in January, 2002   The editors

CARTESIAN CLOSED CATEGORIES OF EFFECTIVE DOMAINS

G. HAMRIN and V. STOLTENBERG-HANSEN
Department of Mathematics, Uppsala University,
Box 480, S-75106 Uppsala, Sweden

1. Introduction

Perhaps the most important and striking fact of domain theory is that important categories of domains are cartesian closed. This means that the category has a terminal object, finite products, and exponents. The only problematic part for domains is the exponent, which in this setting means the space of continuous functions. Cartesian closed categories of domains are well understood and the understanding is in some sense essentially complete by the work of Jung [5], Smyth [11], and others.

In this paper we consider categories of *effective* domains. Again the function space is the crucial construction in order to obtain a cartesian closed category and we therefore concentrate on that. The case for effective algebraic domains is satisfactory. We introduce a natural notion of effective bifinite domains and show that the category of such is cartesian closed. This generalises the well-known construction for effective consistently complete algebraic cpos. The situation for continuous cpos is more problematic. One way to study effectivity on continuous cpos is to note that each continuous cpo is a projection of an algebraic cpo. Thus the more satisfactory theory of effectivity on algebraic cpos can be induced onto the continuous cpos via the projections. This was first done in Smyth [10] and consequently we use the term Smyth effective. Using the result for effective bifinite domains we prove that we can build Smyth effective type structures over Smyth effective continuous domains as long as these are projections of effective bifinite domains.

The ability to build type structures of effective domains is important for several reasons. Such type structures carry a notion of effectiveness or computability which is given externally by recursive function theory via

1

H. Schwichtenberg and R. Steinbrüggen (eds.), Proof and System-Reliability, 1–20.
© 2002 *Kluwer Academic Publishers. Printed in the Netherlands.*

numberings. These computations can in principle be performed on a digital computer. External computations over type structures have applications, for example, in recursive analysis.

There are typed programming languages with semantics on certain effective type structures. Examples of such languages are PCF (Plotkin [8, 9]) and Real PCF (Escardó [4]). These induce an internal notion of effectivity on the type structures in question. Thus we have a framework to compare the strength of the internal and external notions of effectivity. An interesting example of such comparison, which also involves the notion of totality, is carried out in Normann [6] for PCF.

Section 2 contains a brief review of continuous and algebraic cpos including a notion of weak effectivity. Section 3 develops the theory of bifinite domains and Section 4 gives a treatment of effective bifinite domains showing these to be closed under the function space construction. Then in Section 5 the theory of Smyth effectivity is developed. We show that projections of effective bifinite domains that are Smyth effective are closed under the function space construction.

2. Continuous and algebraic cpos

In this section we recall some basic notions and results of continuous and algebraic cpos. General references for this material are [2] and [13].

2.1. BASIC NOTIONS

Let $D = (D; \sqsubseteq, \bot)$ be a partially ordered set with least element $\bot$. A nonempty set $A \subseteq D$ is *directed* if for each $x, y \in A$ there is $z \in A$ such that $x, y \sqsubseteq z$. D is a *complete partial order* (abbreviated *cpo*) if whenever $A \subseteq D$ is directed then $\bigsqcup A$ (the least upper bound or supremum of A) exists in D. A function $f: D \to E$ between cpos is *continuous* if f is monotone and for each directed set $A \subseteq D$

$$f(\bigsqcup_D A) = \bigsqcup_E \{f(x) : x \in A\}.$$

For cpos D and E we define the *function space* $[D \to E]$ of D and E by

$$[D \to E] = \{f: D \to E \mid f \text{ continuous}\}.$$

We order $[D \to E]$ by

$$f \sqsubseteq g \Longleftrightarrow (\forall x \in D)(f(x) \sqsubseteq g(x)).$$

Then $[D \to E]$ is a cpo where, for a directed set $\mathcal{F} \subseteq [D \to E]$ and $x \in D$,

$$(\bigsqcup \mathcal{F})(x) = \bigsqcup_E \{f(x) : f \in \mathcal{F}\}.$$

We form a category whose objects are cpos and whose morphisms are continuous functions between cpos. It is well-known and easy to prove that this category is cartesian closed, where the exponent is the function space and the product of cpos D and E is given by

$$D \times E = \{(x,y): x \in D, y \in E\}$$

and ordered by

$$(x,y) \sqsubseteq (z,w) \Longleftrightarrow x \sqsubseteq_D z \text{ and } y \sqsubseteq_E w.$$

Definition 2.1. Let $D = (D; \sqsubseteq, \perp)$ be a cpo.

(i) For $x, y \in D$ we say x is *way below* y, denoted $x \ll y$, if for each directed set $A \subseteq D$,

$$y \sqsubseteq \bigsqcup A \Longrightarrow (\exists z \in A)(x \sqsubseteq z).$$

(ii) $a \in D$ is said to be *compact* if $a \ll a$. The set of compact elements in D is denoted D_c.

It is easily verified that $x \ll y \implies x \sqsubseteq y$, and that $z \sqsubseteq x \ll y \sqsubseteq w \implies z \ll w$.

Definition 2.2. Let $D = (D; \sqsubseteq, \perp)$ be a cpo. Then D is *continuous* if

(i) the set $\{y \in D: y \ll x\}$ is directed (w.r.t. $\sqsubseteq$); and
(ii) $x = \bigsqcup\{y \in D: y \ll x\}$.

We use the notation $\downarrow x = \{y \in D: y \ll x\}$ and $\uparrow x = \{y \in D: x \ll y\}$. Similarly we let $\downarrow x = \{y \in D : y \sqsubseteq x\}$ and $\uparrow x = \{y \in D : x \sqsubseteq y\}$.

As observed above, the way below relation $\ll$ is reflexive only for compact elements. However, for continuous cpos it satisfies the following crucial *interpolation property*.

Lemma 2.3. *Let D be a continuous cpo. Let $M \subseteq D$ be a finite set and suppose $M \ll y$, i.e., $(\forall z \in M)(z \ll y)$. Then there is $x \in D$ such that $M \ll x \ll y$.*

It follows that if D is a continuous cpo then $\downarrow y = \{x \in D: x \ll y\}$ is directed with respect to $\ll$ for each $y \in D$.

Let $D = (D; \sqsubseteq, \perp)$ be a cpo. A subset $B \subseteq D$ is a *base* for D if for each $x \in D$,

$$\mathrm{approx}_B(x) = \{a \in B: a \ll x\}$$

is directed and $\bigsqcup \mathrm{approx}_B(x) = x$. Thus all the information about the cpo D is contained in a base.

Proposition 2.4. *A cpo is continuous if, and only if, it has a base.*

Also continuous functions between continuous cpos are determined by their behaviour on the bases.

Proposition 2.5. *Let D and E be continuous cpos with bases B_D and B_E respectively. A function $f: D \to E$ is continuous if, and only if, f is monotone and for each $x \in D$,*

$$(\forall b \in \mathrm{approx}_{B_E}(f(x)))(\exists a \in \mathrm{approx}_{B_D}(x))(b \ll f(a)).$$

Definition 2.6. A cpo D is *algebraic* if the set D_c of compact elements is a base for D.

Thus the algebraic cpos make up a subclass of the continuous cpos. An algebraic cpo is in general a simpler structure to deal with than a continuous cpo since the way below relation $\ll$ coincides with $\sqsubseteq$ on its canonical base of compact elements. This is particularly useful when dealing with effectivity. Nonetheless, for each continuous cpo D there is an algebraic cpo E such that D is a projection of E.

Let D and E be cpos. Then a pair of functions $e: D \to E$ and $p: E \to D$ is a *projection pair* from D to E if they are continuous and

$$p \circ e = \mathrm{id}_D \qquad \text{and} \qquad e \circ p \sqsubseteq \mathrm{id}_E$$

where id is the identity function.

Let $P = (P, \leq)$ be a preorder. A set $I \subseteq P$ is an *ideal* if directed and if $x \in I$ and $y \leq x$ then $y \in I$. Let $\mathrm{Idl}(P, \leq)$ be the set of ideals ordered under inclusion. It is easily verified that $\mathrm{Idl}(P, \leq)$ is an algebraic cpo.

Let D be a continuous cpo with a base B and let $E = \mathrm{Idl}(B; \sqsubseteq)$. Define $e: D \to E$ and $p: E \to D$ by

$$e(x) = \mathrm{approx}_B(x) = \{a \in B: a \ll x\} \qquad \text{and} \qquad p(I) = \bigsqcup_D I.$$

Proposition 2.7. *The pair (e, p) is a projection pair from D to E.*

The proof is straightforward, noting that the continuity of e depends on the interpolation property.

2.2. THE FUNCTION SPACE

It is well-known that the categories of continuous cpos and algebraic cpos are not cartesian closed. Here we briefly review the fact that the categories of consistently complete continuous cpos and consistently complete algebraic cpos are cartesian closed.

Definition 2.8. A cpo $D = (D; \sqsubseteq, \bot)$ is *consistently complete* if whenever $x, y \in D$ is bounded above (or consistent) then $x \sqcup y$, the supremum of x and y, exists in D.

Given cpos D and E with bases B_D and B_E we want to construct a base for the function space $[D \to E]$. It turns out that such a base, under appropriate conditions, can be taken as finite suprema of step functions determined from B_D and B_E. Here is the definition of a step function.

Definition 2.9. Let $D = (D; \sqsubseteq, \bot)$ and $E = (E; \sqsubseteq, \bot)$ be cpos. For $a \in D$ and $b \in E$ define $\langle a; b \rangle : D \to E$ by

$$\langle a; b \rangle(x) = \begin{cases} b & \text{if } a \ll x \\ \bot & \text{otherwise.} \end{cases}$$

It is easily verified that each step function is continuous. Recall that if a is compact then $a \ll x \iff a \sqsubseteq x$.

Proposition 2.10. *Let D and E be cpos and let $a \in D$ and $b \in E$.*

(i) *Suppose $f : D \to E$ is continuous. Then*

$$b \ll f(a) \implies \langle a; b \rangle \ll f.$$

(ii) *If D and E are continuous cpos with bases B_D and B_E and $f : D \to E$ is continuous then*

$$f = \bigsqcup \{\langle a; b \rangle : a \in B_D, \ b \in B_E, \ \langle a; b \rangle \ll f\}.$$

Proof. We prove (ii). Suppose $\{\langle a; b \rangle : a \in B_D, \ b \in B_E, \ \langle a; b \rangle \ll f\}$ has g as an upper bound and let $x \in D$. Then $x = \bigsqcup \{a \in B_D : a \ll x\}$ and hence $f(x) = \bigsqcup \{f(a) : a \ll x\}$. For $b \ll f(x)$ we obtain (by interpolation) $a \ll x$ such that $b \ll f(a)$ and hence, by (i), $\langle a; b \rangle \ll f$. Thus $\langle a; b \rangle \sqsubseteq g$. In particular, $b = \langle a; b \rangle(x) \sqsubseteq g(x)$ so $f(x) \sqsubseteq g(x)$. $\qquad \square$

In important cases (i) is an equivalence. For example, if a and b are compact then $\langle a; b \rangle$ is compact and $\langle a; b \rangle \sqsubseteq f \iff b \sqsubseteq f(a)$.

It is clearly not the case that the set in (ii) is directed in general. In order to obtain a directed set we expand it by including suprema of finite subsets of step functions way below f. Consistent completeness of E suffices to obtain such suprema. (In Section 3 we consider a weaker sufficient condition.)

The following characterisation is important when considering the effectivity of the function space construction.

Proposition 2.11. *Let D be a continuous cpo, E a consistently complete cpo, and let $a_1, \ldots, a_n \in D$ and $b_1, \ldots, b_n \in E$. Then*

$$\{\langle a_1; b_1 \rangle, \ldots, \langle a_n; b_n \rangle\} \quad \text{is consistent in} \quad [D \to E]$$

if, and only if,

$$\forall I \subseteq \{1, \ldots, n\} (\bigcap_{i \in I} \uparrow a_i \neq \emptyset \implies \{b_i : i \in I\} \quad \text{consistent}).$$

Proof. For the non-trivial direction define $h : D \to E$ by

$$h(x) = \bigsqcup \{b_i : a_i \ll x\}.$$

Then h is well-defined by consistent completeness and h is monotone. Suppose $A \subseteq D$ is directed and $a_i \ll \bigsqcup A$. By the continuity of D there is $d_i \in A$ such that $a_i \ll d_i$ and hence $b_i \sqsubseteq h(d_i)$. Thus

$$h(\bigsqcup A) = \bigsqcup \{b_i : a_i \ll \bigsqcup A\} \sqsubseteq \bigsqcup h[A].$$

$\square$

Note that if $\{\langle a_1; b_1 \rangle, \ldots, \langle a_n; b_n \rangle\}$ is consistent then the function h in the proof is $\bigsqcup_{i=1}^{n} \langle a_i; b_i \rangle$.

Using Proposition 2.10 (ii) it is now straightforward to prove that the categories of consistently complete continuous and algebraic cpos are cartesian closed.

Theorem 2.12. *Let D and E be continuous cpos with bases B_D and B_E. If E is consistently complete then $[D \to E]$ is continuous and consistently complete. A base $B_{[D \to E]}$ for $[D \to E]$ is*

$$\{\bigsqcup_{i=1}^{n} \langle a_i; b_i \rangle : a_i \in B_D, \ b_i \in B_E, \ \{\langle a_1; b_1 \rangle, \ldots, \langle a_n; b_n \rangle\} \text{ consistent}\}.$$

For consistently complete algebraic cpos we let the bases be D_c and E_c. For $a \in D_c$ and $b \in E_c$ the step function $\langle a; b \rangle$ is compact. It follows that $B_{[D \to E]}$ is a base for $[D \to E]$ consisting only of compact elements. This shows that $[D \to E]$ is a consistently complete algebraic cpo.

2.3. EFFECTIVITY

In this section we initiate the study of computability or effectivity on domains. Here we only give some basic definitions and results. The main results will appear in later sections.

The type of effectivity we consider is based on what may be described as "concrete" computability, i.e., based on computations which may in principle be performed on a digital computer. Put differently, our computability theory is driven by the partial recursive functions. We use the Mal'cev-Ershov-Rabin theory of numberings in order to extend computability from the natural numbers to other structures, such as domains.

We assume some very basic knowledge of recursion theory. This can be found in any text on recursion theory and also in [13]. We choose a primitive recursive pairing function $\langle \cdot, \cdot \rangle : \omega \times \omega \to \omega$ along with its primitive recursive projections, π_1 and π_2.

Let A be a set. A *numbering* of A is a surjective function $\alpha : \omega \to A$. It should be thought of as a coding of A by natural numbers. A subset $S \subseteq A$ is α-*semidecidable* if $\alpha^{-1}(S)$ is r.e. and S is α-*decidable* if $\alpha^{-1}(S)$ is recursive.

Suppose $\alpha : \omega \to A$ is a numbering of a set A. Then let $\alpha^* : \omega \to \mathcal{P}_f(A)$ be the numbering defined by $\alpha^*(e) = \alpha[K_e]$, where $K_e \subseteq \omega$ is the finite subset with canonical index e. ($\mathcal{P}_f(A)$ denotes the set of finite subsets of A.) If, in addition, β is a numbering of B, then $\alpha \times \beta : \omega \to A \times B$ is the numbering defined by

$$\alpha \times \beta(n) = (\alpha(\pi_1(n)), \beta(\pi_2(n))).$$

A function $f : A \to B$ is said to be (α, β)-*computable* if there is a recursive function $\bar{f} : \omega \to \omega$ such that for each n,

$$f(\alpha(n)) = \beta(\bar{f}(n)).$$

We say that $\bar{f}$ *tracks* f.

Definition 2.13. A continuous cpo $D = (D; \sqsubseteq, \bot)$ is *weakly effective* if D has a base B for which there is a numbering

$$\alpha : \omega \to B$$

such that the relation $\alpha(n) \ll \alpha(m)$ is a recursively enumerable relation on ω.

We denote a continuous cpo weakly effective under a numbering α by (D, α). Implicit in this notation is a fixed base $B = \alpha[\omega]$. We will use the notation B for such a base. Thus we let $\operatorname{approx}_\alpha(x) = \{a \in B : a \ll x\}$, the set of approximations of x with respect to the base B determined by α. The condition on the numbering in Definition 2.13 is usually stated as $\ll$ is α-*semidecidable*.

Computable elements are those that can be effectively approximated. Similarly, a function that can be effectively approximated is said to be effective. Here are the precise definitions. (Note the distinction between effective and computable in regard to functions.)

Definition 2.14. Let (D, α) and (E, β) be weakly effective domains.

(i) An element $x \in D$ is *α-computable* if the set

$$\{n \in \omega : \alpha(n) \ll x\} = \alpha^{-1}(\mathrm{approx}_\alpha(x))$$

is r.e. An r.e. index for the set $\alpha^{-1}(\mathrm{approx}_\alpha(x))$ is an *index* for x. The set of computable elements is denoted by $D_{k,\alpha}$.

(ii) A continuous function $f: D \to E$ is *(α, β)-effective* if the relation

$$\beta(m) \ll f(\alpha(n))$$

is r.e. An r.e. index for the set $\{(m, n): \beta(m) \ll f(\alpha(n))\}$ is an *index* for f.

Proposition 2.15. *Let* (D, α), (E, β) *and* (F, γ) *be weakly effective continuous cpos and suppose* $f: D \to E$ *and* $g: E \to F$ *are* (α, β)*-effective and* (β, γ)*-effective, respectively.*

(i) *If* $x \in D$ *is α-computable then* $f(x)$ *is β-computable.*

(ii) $h = g \circ f$ *is* (α, γ)*-effective.*

Proof. (i) Let $W = \{n: \alpha(n) \ll x\}$. Then W is r.e. and

$$\beta(m) \ll f(x) \iff (\exists n \in W)(\beta(m) \ll f(\alpha(n)))$$

by Proposition 2.5. (ii) is proved similarly. $\square$

Observe that the proof is *uniform*. This means there is a total recursive function which given indices for x and f computes an index for $f(x)$. Similarly, there is a total recursive function which given indices for f and g computes an index for $g \circ f$.

There is a natural numbering $\bar{\alpha}$ of $D_{k,\alpha}$ such that $\bar{\alpha}$ is recursively complete and such that for weakly effective (D, α) and (E, β), if $f: D \to E$ is (α, β)-effective then $f|_{D_{k,\alpha}}: D_{k,\alpha} \to E_{k,\beta}$ is $(\bar{\alpha}, \bar{\beta})$-computable. The theory for this weak form of effectivity can be developed further, including a generalisation of the Myhill-Shepherdson theorem (see Stoltenberg-Hansen [12]).

We now restrict ourselves to consistently complete algebraic cpos. Continuous cpos are considered in Section 5.

Definition 2.16. A consistently complete algebraic cpo $D = (D; \sqsubseteq, \bot)$ is *effective* if there is a numbering $\alpha: \omega \to D_c$ such that the following relations are recursive:

(i) $\alpha(m) \sqsubseteq \alpha(n)$;

(ii) $\exists k(\alpha(m), \alpha(n) \sqsubseteq \alpha(k))$; and

(iii) $\alpha(m) \sqcup \alpha(n) = \alpha(k)$.

Let (D, α) and (E, β) be effective consistently complete algebraic cpos. By Theorem 2.12, $[D \to E]_c$ is the set

$$\{\bigsqcup_{i=1}^{n} \langle a_i; b_i \rangle : a_i \in D_c, \ b_i \in E_c, \ \{\langle a_1; b_1 \rangle, \ldots, \langle a_n; b_n \rangle\} \text{ consistent}\}.$$

Furthermore,

$$\bigsqcup_{i=1}^{n} \langle a_i, b_i \rangle \sqsubseteq \bigsqcup_{j=1}^{m} \langle c_j, d_j \rangle \iff (\forall i)(\langle a_i, b_i \rangle \sqsubseteq \bigsqcup_{j=1}^{m} \langle c_j, d_j \rangle)$$

and

$$(\bigsqcup_{j=1}^{m} \langle c_j, d_j \rangle)(x) = \bigsqcup \{d_j : c_j \sqsubseteq x\}.$$

The characterisation in Proposition 2.11 shows that $D_c \times E_c$ is $(\alpha \times \beta)^*$-decidable. Thus we obtain a numbering γ of $[D \to E]_c$ such that the relations in Definition 2.16 are recursive.

Theorem 2.17. *Let (D, α) and (E, β) be effective consistently complete algebraic cpos. Then $[D \to E]$ is an effective consistently complete algebraic cpo with a numbering obtained uniformly from α and β.*

3. Bifinite domains

The category of consistently complete algebraic cpos is not closed under the important Plotkin power domain construction (see Plotkin [7]). Fortunately there is a larger subcategory of the algebraic cpos, the bifinite domains, which is cartesian closed and which is also closed under the Plotkin power domain construction. It turns out, as shown by Jung [5], that the category of bifinite domains is a maximal cartesian closed full subcategory of the algebraic cpos. When restricting to the category of countably based algebraic cpos (i. e. the set of compact elements is countable), the countably based bifinite domains make up the largest cartesian closed full subcategory, i.e. it contains all other cartesian closed full subcategories (see Smyth [11]). A countably based bifinite domain is also called an *sfp*-object, indicating that it is the limit of a sequence of finite partial orders.

To motivate our definition let $(P; \sqsubseteq)$ be a partial order and let us consider $\sqsubseteq$ as an information ordering. Suppose $A \subseteq P$. When is A sufficiently well-structured so that A contains witnesses to all the consistent pieces of information in A?

Definition 3.1. Let $(P; \sqsubseteq, \bot)$ be a partial order with a least element.

(i) $B \subseteq P$ is a *complete set* (in P) if

$$(\forall C \subseteq B)(\forall x \sqsupseteq C)(\exists b \in B)(C \sqsubseteq b \sqsubseteq x).$$

(ii) A family $\mathcal{F} = \{B_i : i \in I\}$ of finite subsets of P is a *complete cover* of P if each B_i is complete and for each $A \subseteq_f P$ there is $i \in I$ such that $A \subseteq B_i$.

As usual, $C \sqsubseteq b$ means $(\forall c \in C)(c \sqsubseteq b)$ and $A \subseteq_f B$ means that A is a finite subset of B. Note that if B is complete then $\bot \in B$.

The following immediate observation is crucial.

Lemma 3.2. *If $A \subseteq P$ is a finite complete set then $\max(A \cap \downarrow x)$ exists for each $x \in P$.*

What we require of a bifinite domain is that each finite subset of compact elements be *covered* by a finite complete set of compact elements.

Definition 3.3. D is a *bifinite domain* if D is an algebraic cpo and D_c has a complete cover.

The covering property holds for each $x \in D$ since D is algebraic.

For a partial order $(P; \sqsubseteq)$ and $A \in \mathcal{P}_f(P)$ we let $\text{mub}(A)$ be the set of minimal upper bounds of A.

Suppose D is a bifinite domain. Then for each $A \in \mathcal{P}_f(D_c)$ there is a finite complete set $B \supseteq A$. Clearly this set contains $\text{mub}(A)$ so $\text{mub}(A)$ is finite. Furthermore, if $A \sqsubseteq x$ then there is $a \in \text{mub}(A)$ such that $A \sqsubseteq a \sqsubseteq x$, again since B is complete. Now we iterate the mub operation as follows:

(i) $\text{mub}^0(A) = \text{mub}(A)$.
(ii) $\text{mub}^{n+1}(A) = \bigcup\{\text{mub}(C) : C \in \mathcal{P}_f(\text{mub}^n(A))\}$.

And then we set $\text{mc}(A) = \bigcup_{n \in \omega} \text{mub}^n(A)$.

Note that each $\text{mub}^n(A) \subseteq B$ and hence $\text{mc}(A)$ is finite. In addition $\text{mc}(A)$ is complete. For if $C \subseteq \text{mc}(A)$ then $C \subseteq \text{mub}^n(A)$ for some n and hence for each $x \sqsupseteq C$ there is $a \in \text{mub}^{n+1}(A)$ which witnesses the completeness.

Proposition 3.4. *Let D be a bifinite domain. Then $\mathcal{F} = \{\text{mc}(A) : A \in \mathcal{P}_f(D_c)\}$ is a complete cover of D_c. Furthermore $\text{mc}(A)$ is the least complete set containing A, so $\mathcal{F}$ is the finest complete cover of D_c.*

Remark 3.5. A consistently complete algebraic cpo D is bifinite. For if $A \in \mathcal{P}_f(D_c)$ then $\mathrm{mc}(A) = \mathrm{mub}^1(A) = \{\bigsqcup B : B \subseteq A \text{ and } B \text{ consistent}\}$.

There is an equivalent characterisation of bifinite domains in terms of finite projections.

Definition 3.6. Let D be a cpo. A continuous function $p: D \to D$ is a *projection* if $p \sqsubseteq \mathrm{id}_D$ and p is idempotent. A projection p is *finite* if its image $\mathrm{im}(p)$ is finite.

Lemma 3.7. *Let p and q be finite projections on a cpo D. Then*

(i) $p \sqsubseteq_{[D \to D]} q \iff \mathrm{im}(p) \subseteq \mathrm{im}(q)$;
(ii) $\mathrm{im}(p) \subseteq D_c$;
(iii) $p \in [D \to D]_c$.

Here is the characterisation in terms of finite projections.

Proposition 3.8. *Let D be a cpo. Then D is a bifinite domain if, and only if, there exists a directed family $(p_i)_{i \in I}$ of finite projections such that*

$$\bigsqcup_{i \in I} p_i = \mathrm{id}_D.$$

Proof. For the only if direction let $\mathcal{F}$ be a complete cover of D_c and define, for $A \in \mathcal{F}$,

$$p_A(x) = \bigsqcup \{a \in A : a \sqsubseteq x\}.$$

Then the conclusion holds for $(p_A)_{A \in \mathcal{F}}$.

For the converse let $\mathcal{F} = \{\mathrm{im}(p_i) : i \in I\}$. The algebraicity of D follows from Lemma 3.7. $\square$

We now show that the function space between bifinite domains is bifinite. There are several ways to do this. Let us first sketch the result using finite projections.

Theorem 3.9. *Let D and E be bifinite domains. Then $[D \to E]$ is a bifinite domain.*

Proof. Let $(p_i)_{i \in I}$ and $(q_j)_{j \in J}$ be families of finite projections witnessing that D and E are bifinite. For $(i, j) \in I \times J$ define $F_{ij}: [D \to E] \to [D \to E]$ by $F_{ij}(h) = q_j \circ h \circ p_i$. It is routine to verify that F_{ij} is continuous and idempotent and that $\bigsqcup_{ij} F_{ij} = \mathrm{id}_{[D \to E]}$. The image $\mathrm{im}(F_{ij})$ is finite since it is determined by $\mathrm{im}(q_j)$ and $\mathrm{im}(p_i)$, which are finite by assumption. $\square$

In order to consider the effectivity of the function space $[D \to E]$ of bifinite domains D and E we need a finitary characterisation of the compact elements in $[D \to E]$ in terms of the compact elements in D and E. The following definition is a variant of one given by Abramsky [1].

Definition 3.10. Let D and E be bifinite domains.

(i) A non-empty finite set $\{(a_i, b_i) : i \in I\} \subseteq D_c \times E_c$ is said to be *joinable* if the set $\{a_i : i \in I\}$ is complete and $a_i \sqsubseteq a_j \implies b_i \sqsubseteq b_j$.

(ii) Suppose $u = \{(a_i, b_i) : i \in I\}$ is joinable and let $A = \{a_i : i \in I\}$. Then $s_u \colon [D \to E] \to [D \to E]$ is defined by

$$s_u(x) = b_i \iff a_i = \max(A \cap \downarrow x).$$

Note that s_u is well defined by Lemma 3.2.

Lemma 3.11. *Let D and E be bifinite domains and let u be joinable.*

(i) s_u *is continuous.*
(ii) $s_u = \bigsqcup\{\langle a; b \rangle : (a, b) \in u\}$.
(iii) $[D \to E]_c = \{s_u : u \subseteq D_c \times E_c,\ u \text{ joinable}\}$.

Proof. (i) and (ii) are routine given Lemma 3.2. To show (iii) we first note that each step function $\langle a; b \rangle$ is compact and hence by (ii) that s_u is compact when u is joinable. Now suppose $f \in [D \to E]_c$. By the argument of Proposition 2.10,

$$f = \bigsqcup\{\langle a; b \rangle : \langle a, b \rangle \sqsubseteq f, a \in D_c, b \in E_c\}.$$

Consider a finite set $\{(a_i, b_i) : i \in I\}$ such that each $\langle a_i, b_i \rangle \sqsubseteq f$. Let $A = \{a_i : i \in I\}$ and $B = \{b_i : i \in I\}$ and let $\bar{A} \supseteq A$ and $\bar{B} \supseteq B$ be finite complete sets of compact elements. For each $a \in \bar{A}$ let $b_a = \max(\bar{B} \cap \downarrow f(a))$ and let $u = \{(a, b_a) : a \in \bar{A}\}$.

Clearly s_u is joinable. For suppose $a_1, a_2 \in \bar{A}$ and $a_1 \sqsubseteq a_2$. Then $f(a_1) \sqsubseteq f(a_2)$ and hence $b_{a_1} \sqsubseteq b_{a_2}$. Furthermore $\langle a; b \rangle \sqsubseteq f$ for $(a, b) \in u$, since then $b = b_a \sqsubseteq f(a)$. By (ii) we then have $s_u \sqsubseteq f$.

By an analogous argument we see that the set

$$\{s_u : s_u \sqsubseteq f \text{ and } s_u \text{ joinable}\}$$

is directed and its supremum is f. It follows that $f = s_u$ for some joinable u since f is compact. $\square$

Remark 3.12. It is easily seen from the above proof that it suffices to consider the joinable sets obtained from a complete cover of D_c in order to obtain all of $[D \to E]_c$.

4. Effective bifinite domains

In this section we consider a notion of effective bifinite domains, a generalisation of effective consistently complete algebraic cpos, and show that the function space of effective bifinite domains is again an effective bifinite domain. It follows that the category of effective bifinite domains with effective continuous functions is cartesian closed.

Definition 4.1. A bifinite domain D is an *effective bifinite domain* if there is a numbering $\alpha: \omega \to D_c$ such that

(i) the relation $\alpha(n) \sqsubseteq \alpha(m)$ is recursive, i.e. $\sqsubseteq$ is α-decidable; and
(ii) there is a complete cover $\mathcal{F}$ of D_c such that $\mathcal{F}$ is α^*-decidable.

We leave the following as an exercise.

Proposition 4.2. *Let D be a bifinite domain and let α be a numbering of D_c such that $\sqsubseteq$ is α-decidable. Then (D, α) is an effective bifinite domain if, and only if,* mub: $\mathcal{P}_f(D_c) \to \mathcal{P}_f(D_c)$ *is (α^*, α^*)-computable.*

Let (D, α) be an effective bifinite domain. Then by the above we get that mub^n is (α^*, α^*)-computable for each n, uniformly in n. It follows that mc is (α^*, α^*)-computable since $\mathrm{mc}(A) = \mathrm{mub}^n(A)$ where n is such that $\mathrm{mub}^n(A) = \mathrm{mub}^{n+1}(A)$. Thus we can also conclude that the relation "A is complete" for $A \in \mathcal{P}_f(D_c)$ is α^*-decidable, since

$$A \text{ is complete} \iff \mathrm{mc}(A) = A.$$

Proposition 4.3. *Let (D, α) be an effective bifinite domain. Then*

(i) mc: $\mathcal{P}_f(D_c) \to \mathcal{P}_f(D_c)$ *is (α^*, α^*)-computable;*
(ii) *The relation "A is complete" is α^*-decidable.*

We now prove that the category of effective bifinite domains is cartesian closed.

Theorem 4.4. *Let (D, α) and (E, β) be effective bifinite domains. Then $[D \to E]$ is an effective bifinite domain with a numbering obtained uniformly from α and β.*

Proof. From an α^*-decidable complete cover $\mathcal{F}$ of D_c we obtain (by Remark 3.12) an $(\alpha \times \beta)^*$-decidable family $\mathcal{U} \subseteq D_c \times E_c$ of joinable sets such that

$$[D \to E]_c = \{s_u : u \in \mathcal{U}\}.$$

Thus we obtain in a standard way a numbering $\gamma: \omega \to [D \to E]_c$, uniformly from α, β and $\mathcal{F}$. The relation $\sqsubseteq$ on $[D \to E]_c$ is γ-decidable since

$$s_u \sqsubseteq s_v \iff (\forall(a,b) \in u)(b = \bot_E \vee (\exists(c,d) \in v)(c \sqsubseteq a \wedge b \sqsubseteq d)).$$

We are to construct a γ^*-decidable complete cover of $[D \to E]_c$. Let $\mathcal{G}$ be a β^*-decidable complete cover of E_c. Then, using the notation in the proof of Proposition 3.8, $(p_A)_{A \in \mathcal{F}}$ and $(q_B)_{B \in \mathcal{G}}$ are directed families of projections witnessing that D and E are bifinite. Using the notation and proof of Theorem 3.9 it suffices, by Proposition 3.8, to show that the family

$$\{\text{im}(F_{AB}) : A \in \mathcal{F}, B \in \mathcal{G}\}$$

is γ^*-decidable. For this it suffices to show that for all u joinable with respect to $\mathcal{F}$,

$$s_u \in \text{im}(F_{AB}) \iff (\exists v \subseteq A \times B)(v \text{ joinable } \wedge \pi_1(v) = A \wedge s_u = s_v),$$

where π_1 is the first projection function.

We leave the straight forward proof of the "if" direction as an exercise. To prove the "only if" direction let $s_u \in \text{im}(F_{AB})$ where $u = \{(c, d_c) : c \in C\}$ is joinable and $C \in \mathcal{F}$. For $a \in A$ let $c_a = \max(C \cap \downarrow a) \in C$. Note that $a \sqsubseteq a' \implies c_a \sqsubseteq c_{a'}$ and hence $d_{c_a} \sqsubseteq d_{c_{a'}}$. Thus the set $v = \{(a, d_{c_a}) : a \in A\}$ is joinable. For $(a, d_{c_a}) \in v$ we have

$$d_{c_a} = \langle c_a; d_{c_a} \rangle(a) \sqsubseteq s_u(a)$$

and hence $s_v \sqsubseteq s_u$.

For $c \in C$ let $c_A = \max(C \cap \downarrow p_A(c))$. We have for $c \in C$,

$$\begin{aligned}
F_{AB}(s_u)(c) &= q_B s_u p_A(c) \\
&= q_B \langle c_A; d_{c_A} \rangle p_A(c) \\
&= q_B(d_{c_A}).
\end{aligned}$$

But $F_{AB}(s_u)(c) = s_u(c) = d_c$ and hence $d_c = q_B(d_{c_A}) \sqsubseteq d_{c_A}$. But $c_A \sqsubseteq c$ so $d_{c_A} \sqsubseteq d_c$, that is equality holds. In particular $d_c \in B$ and hence $v \subseteq A \times B$.

To prove $s_u \sqsubseteq s_v$ let $(c, d_c) \in u$. Then $c_A \sqsubseteq p_A(c) \sqsubseteq c$ and, in fact, $c_A = c_{p_A(c)}$. From the above, $d_c = d_{c_A}$, and hence

$$d_c \sqsubseteq \langle p_A(c); d_{c_A} \rangle(c) \sqsubseteq s_v(c)$$

which proves that $s_u \sqsubseteq s_v$. $\qquad\qquad\square$

Note that for joinable $u \subseteq D_c \times E_c$ and continuous $f : D \to E$,

$$\begin{aligned}
s_u \sqsubseteq f &\iff (\forall (a, b) \in u)(\langle a; b \rangle \sqsubseteq f) \\
&\iff (\forall (a, b) \in u)(b \sqsubseteq f(a)).
\end{aligned}$$

It follows that if f is effective then f is a computable element in $[D \to E]$. The converse is immediate since $b \sqsubseteq f(a) \iff \langle a; b \rangle \sqsubseteq f$.

One can also show that if (D, α) is an effective bifinite domain then the Plotkin power domain $\mathcal{P}_P(D)$ is an effective bifinite domain with a numbering obtained uniformly from α (see Blanck [3]).

5. Smyth effective domains

In this section we consider effectivity of continuous cpos induced by effectivity on algebraic cpos via projection pairs. The simple idea is to work on algebraic cpos since these are easier and more well behaved, in particular concerning effectivity, than cpos that are merely continuous.

We show that if we start with a cartesian closed category of effective algebraic cpos then we obtain in this way a cartesian closed category of effective continuous cpos. This was first done in Smyth [10], where effective consistently complete algebraic cpos were considered. Using the results from the previous section we extend his result to projections of effective bifinite domains.

The setting is as follows. Let E be an algebraic cpo, let D be a cpo, and let (e, p) be a projection pair from D to E. Note that $(ep)^2 = ep$, that is, ep is a retraction.

Define a relation $\prec$ on $E_c \times E$ by

$$a \prec x \iff a \sqsubseteq ep(x).$$

Note that, since $ep \sqsubseteq \mathrm{id}_E$, $a \prec x \implies a \sqsubseteq x$ and $a' \sqsubseteq a \prec x \sqsubseteq x' \implies a' \prec x'$.

Lemma 5.1. (i) $a \prec x \implies p(a) \ll p(x)$.
(ii) $a \prec x \implies (\exists b \in E_c)(a \prec b \prec x)$.
(iii) $\{a \in E_c : a \prec x\}$ is a directed set under $\prec$.
(iv) For each $x \in E$, $p(x) = \bigsqcup_D \{p(a) : a \prec x\}$.

Proof. (i) Let $a \prec x$ and suppose $A \subseteq D$ is directed and $p(x) \sqsubseteq \bigsqcup_D A$. Then $a \sqsubseteq ep(x) \sqsubseteq \bigsqcup_E \{e(y) : y \in A\}$ so $a \sqsubseteq e(y)$ for some $y \in A$. But then $p(a) \sqsubseteq pe(y) = y \in A$.
(ii) Again let $a \prec x$. Then, since ep is idempotent,

$$\begin{aligned}
a \sqsubseteq ep(x) &= epep(x) \\
&= ep(\bigsqcup_E \{b \in E_c : b \sqsubseteq ep(x)\}) \\
&= \bigsqcup_E \{ep(b) : b \in E_c \text{ and } b \sqsubseteq ep(x)\}.
\end{aligned}$$

Thus $a \prec b$ for some $b \prec x$.
(iii) Suppose $a_1, a_2 \prec x$. From (ii) we obtain b_1 and b_2 such that $a_i \prec b_i$ and $b_i \sqsubseteq ep(x)$. Let $b \in E_c$ be such that $b_i \sqsubseteq b \sqsubseteq ep(x)$. Then $a_1, a_2 \prec b \prec x$.

(iv) For $x \in E$ we get

$$ep(x) = \bigsqcup_E \{a \in E_c : a \sqsubseteq ep(x)\}$$
$$= \bigsqcup_E \{ep(a) : a \in E_c \text{ and } a \sqsubseteq ep(x)\},$$

again since ep is idempotent. But then $p(x) = \bigsqcup_D \{p(a) : a \prec x\}$.

$\square$

The lemma says, essentially, that $p(E_c)$ is a base for D and that the relation $\prec$ induces a sufficiently large part of the way below relation $\ll$ on D. We have the following characterisation of $\ll$ on $p(E_c) \times D$.

Lemma 5.2. (i) *Let* $a \in E_c$ *and* $x \in E$. *Then*

$$p(a) \ll p(x) \iff (\exists b \in E_c)(p(a) \sqsubseteq p(b) \wedge b \prec x)$$
$$\iff (\exists b \in E_c)(p(a) \ll p(b) \wedge b \prec x).$$

(ii) $p(E_c)$ *is a base for* D.

Proof. (i) The if direction follows from Lemma 5.1 (i). For the converse, suppose $p(a) \ll p(x)$. By Lemma 5.1 (iv), $p(x) = \bigsqcup \{p(b) : b \prec x\}$ so there is $b \prec x$ such that $p(a) \sqsubseteq p(b)$. The second equivalence follows from Lemma 5.1 (iii).

(ii) It suffices to show that the set $\{p(a) : a \in E_c \wedge p(a) \ll p(x)\}$ is directed. This follows from Lemma 5.1 (iii).

$\square$

Remark 5.3. One can show that D is obtained as the ideal completion of E_c with respect to $\prec$, that is

$$D \cong \mathrm{Idl}(E_c, \prec).$$

We now consider function spaces.

Definition 5.4. Let (e_i, p_i) be a projection pair from D_i to E_i for $i = 1, 2$. Define $\mathcal{E}: [D_1 \to D_2] \to [E_1 \to E_2]$ and $\mathcal{P}: [E_1 \to E_2] \to [D_1 \to D_2]$ by

$$\mathcal{E}(g) = e_2 \circ g \circ p_1 \text{ and } \mathcal{P}(f) = p_2 \circ f \circ e_1.$$

The following is easily verified.

Lemma 5.5. $(\mathcal{E}, \mathcal{P})$ *is a projection pair from* $[D_1 \to D_2]$ *to* $[E_1 \to E_2]$.

Recall that $[E_1 \to E_2]$ may not be algebraic even when E_1 and E_2 are algebraic. To guarantee algebraicity, E_1 and E_2 must be bifinite (when countably based).

Continuous functions can in this setting be characterised in the usual way using $\prec$. Here we assume E_1 and E_2 are algebraic.

Lemma 5.6. *A function* $f: D_1 \to D_2$ *is continuous if, and only if, f is monotone, and for each* $x \in E_1$,

$$(\forall b \prec \mathcal{E}(f)(x))(\exists a \prec x)(b \prec \mathcal{E}(f)(a)).$$

Proof. Suppose f is continuous and let $x \in E_1$. Then $p_1(x) = \bigsqcup_D\{p_1(a) : a \prec x\}$ so $\mathcal{E}(f)(x) = \bigsqcup_{E_2}\{e_2 f p_1(a) : a \prec x\}$. If $b \prec e_2 f p_1(x) = e_2 p_2 e_2 f p_1(x)$ then $b \sqsubseteq e_2 f p_1(x) = \mathcal{E}(f)(x)$ so there is $a \prec x$ such that $b \sqsubseteq e_2 f p_1(a) = e_2 p_2 e_2 f p_1(a)$, that is $b \prec \mathcal{E}(f)(a)$.

For the converse we need only show (by Proposition 2.5)

$$p_2(b) \ll f p_1(x) \implies (\exists a \in E_{2,c})(p_1(a) \ll p_1(x) \text{ and } p_2(b) \ll f p_1(a)).$$

So suppose the condition holds and $p_2(b) \ll f p_1(x)$. By Lemma 5.2 there is $b' \in E_{2,c}$ such that $p_2(b) \sqsubseteq p_2(b')$ and $b' \prec e_2 f p_1(x)$. The condition now provides an $a \prec x$ (and hence $p_1(a) \ll p_1(x)$) such that $b' \prec e_2 f p_1(a)$. But then $p(b) \sqsubseteq p(b') \ll p_2 e_2 f p_1(a) = f p_1(a)$. $\qquad\square$

Let $D = (D; \sqsubseteq, \bot)$ be a continuous cpo with a base B and let $E = \mathrm{Idl}(B, \sqsubseteq)$, the ideal completion of B with respect to $\sqsubseteq$. Let (e, p) be the projection pair from D to E given by Proposition 2.7. The compact elements E_c in E consists of the principal ideals $[a]$ for $a \in B$. It is easily verified that for $a \in B$ and $I \in E$,

$$[a] \prec I \iff [a] \subseteq ep(I) \iff a \ll \bigsqcup_D I \iff a \ll p(I).$$

In particular for $b \in B$, $[a] \prec [b] \iff a \ll b$. Thus in this canonical construction there is a precise connection between $\prec$ on E_c and $\ll$ on B.

Recall that one element in a projection pair (e, p) from D to E determines the other. Thus we let (E, p, D) denote that $p: E \to D$ is a projection onto D and we then denote the corresponding embedding by e. We say that (E, p, D) is an *AP-domain* if E is algebraic.

Definition 5.7. (i) Let (E, p, D) be an AP-domain. Then $((E, p, D), \alpha)$ is *Smyth effective* if $\alpha: \omega \to E_c$ is a numbering such that the relation $\prec$ on E_c is α-semidecidable.

(ii) Let $((E, p, D), \alpha)$ be Smyth effective. Then $x \in D$ is α-*Smyth computable* if the relation $a \prec e(x)$ is α-semidecidable.

(iii) Let $((E_1, p_1, D_1), \alpha)$ and $((E_2, p_2, D_2), \beta)$ be Smyth effective. Then a continuous function $f: D_1 \to D_2$ is (α, β)-*Smyth effective* if the relation $b \prec \mathcal{E}(f)(a)$ is (α, β)-semidecidable.

If D is a weakly effective continuous cpo in the sense of Definition 2.13 and E is the canonical algebraic cpo constructed from D (with the base determined by the numbering) then (E, p, D) is Smyth effective.

Now we can develop the theory of Smyth effective cpos just as is done for weakly effective continuous cpos in Stoltenberg-Hansen [12]. In this paper, however, we are concerned with cartesian closed categories and hence, by necessity, we need stronger effectivity assumptions. Therefore we just note the following basic results.

Lemma 5.8. *Let $((E_1, p_1, D_1), \alpha)$, $((E_2, p_2, D_2), \beta)$, and $((E_3, p_3, D_3), \gamma)$ be Smyth effective and let $f\colon D_1 \to D_2$ and $g\colon D_2 \to D_3$ be (α, β)-effective and (β, γ)-effective, respectively.*

(i) *If $x \in D_1$ is α-Smyth computable then $f(x)$ is β-Smyth computable.*
(ii) *$h = g \circ f$ is (α, γ)-Smyth effective.*

Proof. Let $W = \{n : \alpha(n) \prec e_1(x)\}$. Then W is r.e. and

$$\beta(m) \prec e_2 f(x) = e_2 f p_1 e_1(x) \iff (\exists n \in W)(\beta(m) \prec \mathcal{E}(f)(\alpha(n)))$$

by Lemma 5.6. Part (ii) is proved similarly. $\qquad\square$

From the proof we see that the lemma holds uniformly.

The Smyth effective functions between the continuous domains correspond to the effective functions between the algebraic domains in the expected fashion.

Lemma 5.9. *Let $((E_1, p_1, D_1), \alpha)$ and $((E_2, p_2, D_2), \beta)$ be Smyth effective AP-domains and assume (E_1, α) and (E_2, β) are weakly effective.*

(i) *If $f\colon D_1 \to D_2$ is (α, β)-Smyth effective then $\mathcal{E}(f)\colon E_1 \to E_2$ is (α, β)-effective.*
(ii) *If $g\colon E_1 \to E_2$ is (α, β)-effective then $\mathcal{P}(g)\colon D_1 \to D_2$ is (α, β)-Smyth effective.*

Proof. (i) For $a \in E_{1,c}$ and $b \in E_{2,c}$ we have

$$b \sqsubseteq \mathcal{E}(f)(a) \iff b \sqsubseteq e_2 p_2 e_2 f p_1(a) \iff b \prec \mathcal{E}(f)(a).$$

(ii) Similarly we have

$$b \prec \mathcal{EP}(g)(a) = e_2 p_2 g e_1 p_1(a) \iff b \sqsubseteq e_2 p_2 e_2 p_2 g e_1 p_1(a) = \mathcal{EP}(g)(a)$$

and the latter relation is semidecidable. $\qquad\square$

We now prove the main theorem.

Theorem 5.10. *Let $((E_1, p_1, D_1), \alpha)$ and $((E_2, p_2, D_2), \beta)$ be Smyth effective AP-domains and assume (E_1, α) and (E_2, β) are effective bifinite domains. Then there is a numbering γ of $[E_1 \to E_2]$, obtained uniformly from α and β, such that $(([E_1 \to E_2], \mathcal{P}, [D_1 \to D_2]), \gamma)$ is a Smyth effective AP-domain and $([E_1 \to E_2], \gamma)$ is an effective bifinite domain.*

Proof. Let γ be the numbering making $([E_1 \to E_2], \gamma)$ into an effective bifinite domain obtained from Theorem 4.4. Thus it remains to show that $(([E_1 \to E_2], \mathcal{P}, [D_1 \to D_2]), \gamma)$ is Smyth effective. For this it suffices to show $\langle a; b \rangle \sqsubseteq \mathcal{EP}(s_u)$ is γ-semidecidable where $a \in E_{1,c}$, $b \in E_{2,c}$, and u a joinable set. But

$$\langle a; b \rangle \sqsubseteq \mathcal{EP}(s_u) \iff b \sqsubseteq \mathcal{EP}(s_u)(a) \iff b \sqsubseteq e_2 p_2 s_u e_1 p_1(a).$$

s_u is (α, β)-effective uniformly in a γ-index for u. Furthermore $e_1 p_1$ is α-effective and $e_2 p_2$ is β-effective by the assumed Smyth effectivities. Thus the relation is γ-semidecidable by the uniformity of composition. $\square$

Let $D = (D; \sqsubseteq, \bot)$ be a consistently complete continuous cpo and suppose α is a numbering of a base B of D such that the following holds:

(i) $\alpha(m) \ll \alpha(n)$ is r.e.
(ii) $\exists k (\alpha(m), \alpha(n) \sqsubseteq \alpha(k))$ is recursive.
(iii) $\alpha(m) \sqcup \alpha(n) = \alpha(k)$ is recursive.

Note that it follows that $\sqsubseteq$ is α-decidable on B.

Let $E = \mathrm{Idl}(B, \sqsubseteq)$ be the canonical algebraic domain constructed from D. From α we obtain a numbering $\bar{\alpha} \colon \omega \to E_c$ defined by $\bar{\alpha}(n) = [\alpha(n)]$, where $[a]$ is the principal ideal generated by $a \in B$. Clearly E is consistently complete since D is consistently complete. Furthermore, as already noted, $[a] \prec [b] \iff a \ll b$, $[a] \sqsubseteq [b] \iff a \sqsubseteq b$, and a and b are consistent $\iff$ $[a]$ and $[b]$ are consistent in which case $[a] \sqcup [b] = [a \sqcup b]$. It follows that $(E, \bar{\alpha})$ is an effective consistently complete algebraic cpo and $((E, P, D), \bar{\alpha})$ is Smyth effective.

It now follows from Theorem 5.10 that we can build a Smyth effective type structure over D.

An important example that fits into this framework is the continuous interval domain $\mathbb{CI}$. Recall that $\mathbb{CI} = \{[x, y] : x \le y; \ x, y \in \mathbb{R}\} \cup \mathbb{R}$ ordered under reverse inclusion. A base for $\mathbb{CI}$ is the set of intervals with rational end points. It is easy to see that $[a, b] \ll [c, d] \iff a < c \le d < b$, and $[a, b] \sqsubseteq [c, d] \iff a \le c \le d \le b$. It is therefore straight forward to define a numbering satisfying the conditions above and hence build a Smyth effective type structure over $\mathbb{CI}$.

References

1. S. ABRAMSKY, Domain theory in logical form, *Annals of Pure and Applied Logic* 51 (1991), 1 – 77.
2. S. ABRAMSKY AND A. JUNG, Domain theory, in *Handbook of Logic in Computer Science*, volume 3, (S. Abramsky, D. Gabbay, and T. S. E. Maibaum, editors), Oxford University Press, Oxford, 1995, 1 – 168.
3. J. BLANCK, Effective domain representation of $\mathcal{H}(X)$, the space of compact subsets, *Theoretical Computer Science* 219 (1999), 19 – 48.
4. M. ESCARDÓ, PCF extended with real numbers, *Theoretical Computer Science* 162 (1996), 79 – 115.
5. A. JUNG, Cartesian closed categories of algebraic cpo's, *Theoretical Computer Science* 70 (1990), 233 – 250.
6. D. NORMANN, Computability of the partial continuous functionals, *Journal of Symbolic Logic* 65 (2000), 1133 – 1142.
7. G. PLOTKIN, A power domain construction, *SIAM Journal on Computing* 5 (1976), 452 – 488.
8. G. PLOTKIN, LCF considered as a programming language, *Theoretical Computer Science* 5 (1977), 223 – 255.
9. G. PLOTKIN, Full abstraction, totality and PCF, *Mathematical Structures in Computer Science* 9 (1999), 1 – 20.
10. M. B. SMYTH, Effectively given domains, *Theoretical Computer Science* 5 (1977), 257 – 274.
11. M. B. SMYTH, The largest cartesian closed category of domains, *Theoretical Computer Science* 27 (1983), 109 – 119.
12. V. STOLTENBERG-HANSEN Effective domains and concrete computability: a survey, in F. L. Bauer and R. Steinbrüggen (editors) *Foundations of Secure Computation*, IOS Press, 2000.
13. V. STOLTENBERG-HANSEN, I. LINDSTRÖM AND E. R. GRIFFOR, *Mathematical Theory of Domains*, Cambridge University Press, 1994.

ALGORITHMIC GAME SEMANTICS

A Tutorial Introduction

SAMSON ABRAMSKY
Oxford University Computing Laboratory

samson@comlab.ox.ac.uk

1. Introduction

Game Semantics has emerged as a powerful paradigm for giving semantics to a variety of programming languages and logical systems. It has been used to construct the first syntax-independent fully abstract models for a spectrum of programming languages ranging from purely functional languages to languages with non-functional features such as control operators and locally-scoped references [4, 21, 5, 19, 2, 22, 17, 11]. A substantial survey of the state of the art of Game Semantics *circa* 1997 was given in a previous Marktoberdorf volume [6].

Our aim in this tutorial presentation is to give a first indication of how Game Semantics can be developed in a new, algorithmic direction, with a view to applications in computer-assisted verification and program analysis. Some promising steps have already been taken in this direction. Hankin and Malacaria have applied Game Semantics to program analysis, e.g. to certifying secure information flows in programs [25]. A particularly striking development was the work by Ghica and McCusker [15] which captures the game semantics of a fragment of Idealized Algol in a remarkably simple form as regular expressions. This leads to a decision procedure for observation equivalence on this fragment. Ghica has subsequently extended the approach to a call-by-value language with arrays [14], and to model checking Hoare-style program correctness assertions [13].

We believe the time is ripe for a systematic development of this algorithmic approach to game semantics. Game Semantics has several features which make it very promising from this point of view. It provides a very *concrete* way of building *fully abstract* models. It has a clear operational content, while admitting *compositional methods* in the style of denotational semantics. The basic objects studied in Game Semantics are games, and strategies on games. Strategies can be seen as certain kinds of highly-constrained processes, hence they admit the same kind of automata-theoretic representations central to model checking and allied

21

H. Schwichtenberg and R. Steinbrüggen (eds.), Proof and System-Reliability, 21–47.
© 2002 *Kluwer Academic Publishers. Printed in the Netherlands.*

methods in computer-assisted verification. At the same time, games and strategies naturally form themselves into rich mathematical structures which yield very accurate models of advanced high-level programming languages, as the various full abstraction results show. Thus the promise of this approach is to carry over the methods of model checking, which has been so effective in the analysis of circuit designs and communications protocols, to much more *structured* programming situations, in which data-types as well as control flow are important.

A further benefit of the algorithmic approach is that by embodying game semantics in tools, and making it concrete and algorithmic, it should become more accessible and meaningful to practitioners. We see Game Semantics as having the potential to fill the role of a "Popular Formal Semantics" called for in an eloquent paper by David Schmidt [31], which can help to bridge the gap between the semantics and programming language communities. Game Semantics has been successful in its own terms as a semantic theory; we aim to make it useful to and usable by a wider community.

In relation to the extensive current activity in software model checking and computer assisted verification (see e.g. [8, 12]), our approach is distinctive, being founded on a highly-structured *compositional* semantic model. This means that we can directly apply our methods to *program phrases* (i.e. terms-in-context with free variables) in a high-level language with procedures, local variables and data types; moreover, the soundness of our methods is guaranteed by the properties of the semantic models on which they are based. By contrast, most current model checking applies to relatively "flat" unstructured situations, in which the system being analyzed is presented as a transition system or automaton. Our aim is to build on the tools and methods which have been developed in the verification community, while exploring the advantages offered by our semantics-based approach.

1.1. OVERVIEW

In the following section, we shall begin with an informal overview of game semantics, followed by a step-by-step development of how constructs in a procedural programming language can be modelled in this approach. We formalize these descriptions using elementary tools from formal language theory; this will guarantee that the semantic descriptions are themselves effective, and can serve as the basis for model-checking and program analysis.

A more systematic account is then given in section 3, while section 4 discusses model-checking.

2. Game semantics for a procedural language

2.1. INFORMAL INTRODUCTION

Before proceeding to a precise formalization, we will give an informal presentation of the main ideas through examples, with the aim of conveying how close to programming intuitions the formal model is.

As the name suggests, game semantics models computation as the playing of a certain kind of game, with two participants, called Player (P) and Opponent (O). P is to be thought of as representing the system under consideration, while O represents the environment. In the case of programming languages, the system corresponds to a term (a piece of program text) and the environment to the context in which the term is used. This is a key point at which games models differ from other process models: the distinction between the actions of the system and those of its environment is made explicit from the very beginning. (For a fuller discussion of the ramifications of this distinction, see [1]).

In the games we shall consider, O always moves first—the environment sets the system going—and thereafter the two players make moves alternately. What these moves are, and when they may be played, are determined by the rules of each particular game. Since in a programming language a *type* determines the kind of computation which may take place, types will be modelled as games; a *program* of type A determines how the system behaves, so programs will be represented as *strategies* for P, that is, by the specification of responses by P to the moves O may make.

2.2. MODELLING VALUES

In standard denotational semantics, values are *atomic*: a natural number is represented simply as $n \in \omega$. In game semantics, each number is modelled as a simple interaction: the environment starts the computation with an initial move q (a *question*: "What is the number?"), and P may respond by playing a natural number (an *answer* to the question). So the game $I\!N$ of natural numbers looks like this:

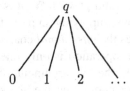

and the strategy for 3 is "When O plays q, I will play 3":

$$\begin{array}{cc} \textit{IN} & \\ q & \text{O} \\ 3 & \text{P} \end{array}$$

In diagrams such as the above, time flows downwards: here O has begun by playing q, and at the next step P has responded with 3, as the strategy dictates.

2.3. EXPRESSIONS

The interactions required to model expressions are a little more complex. The view taken in game semantics is that the environment of an expression consumes the output and provides the input, while the expression itself consumes the input and produces the output. Thus the game involved in evaluating an expression such as

$$x : I\!N, y : I\!N \vdash x + y : I\!N$$

is formed from "three copies of $I\!N$", one for each of the inputs x and y, and one for the output—the result of evaluating $x + y$. In the output copy, O may demand output by playing the move q and P may provide it. In the input copies, the situation is reversed: P may demand input with the move q. Thus the O/P role of moves in the input copy is reversed. A typical computation of a natural strategy interpreting this expression has the following form.

$$\begin{array}{ccc} I\!N & , & I\!N \vdash I\!N \\ & & q \quad \text{O} \\ q & & \text{P} \\ 3 & & \text{O} \\ & q & \text{P} \\ & 2 & \text{O} \\ & & 5 \quad \text{P} \end{array}$$

The play above is a particular run of the strategy modelling the addition operation:

"When O asks for output, I will ask for my first input x; when O provides input m for x, I will ask for my second input y; when O provides the input n for y, I will give output $m + n$."

It is important to notice that the play in each copy of $I\!N$ (that is, each column of the above diagram) is indeed a valid play of $I\!N$: it is not possible for O to begin with the third move shown above, supplying an input to the function immediately.

This example also illustrates the *intensional* character of game semantics. The above strategy for the addition operation is only one possibility; another would be the strategy which evaluated the two arguments in the opposite order. These would be *distinct* strategies for computing the *same* operation (function). This distinction may appear otiose in the purely functional setting; but the ability to make such distinctions becomes crucial when we come to model non-functional features such as state or control.

2.4. INTERACTION: COMPOSITION OF STRATEGIES

Game semantics is intended to provide a *compositional* interpretation of programs: just as small programs can be put together to form large ones, so strategies can be combined to form new strategies. The fundamental "glue" in traditional denotational semantics is function application; for game semantics it is *interaction* of strategies which gives us a notion of composition.

Consider the example of addition again, with the type

$$x : I\!N, y : I\!N \vdash x + y : I\!N$$

We would like to combine this with $\vdash 3 : I\!N$:

$$\frac{\vdash 3 : I\!N \quad x : I\!N, y : I\!N \vdash x + y : I\!N}{y : I\!N \vdash 3 + y : I\!N}$$

This is just *substitution* of 3 for x, or in logical terms, the Cut rule [16]. In order to represent this composition, we let the two strategies interact with one another. When add plays a move in the first copy of $I\!N$ (corresponding to asking for the value of x), we feed it as an O-move to the strategy for 3; conversely, when this strategy responds with 3 in $I\!N$, we feed this move as an O-move back to add.

$$
\begin{array}{cccc}
I\!N & , \ I\!N & \vdash & I\!N \\
 & & & q \\
q & & & \\
3 & & & \\
 & & q & \\
 & & 5 & \\
 & & & 8
\end{array}
$$

By hiding the action in the first copy of $I\!N$, we obtain the strategy

$$
\begin{array}{ccc}
I\!N & \vdash & I\!N \\
 & & q \\
q & & \\
n & & \\
 & & n + 3
\end{array}
$$

representing the (unary) operation which adds 3 to its argument, as expected. So in game semantics, composition of functions is modelled by CSP-style "parallel composition + hiding" [18, 3].

2.5. VARIABLES AND COPY-CAT STRATEGIES

To interpret a variable

$$x : I\!N \vdash x : I\!N$$

we play as follows:

$$
\begin{array}{ccc}
I\!N & \vdash & I\!N \\
& & q \\
q & & \\
n & & \\
& & n
\end{array}
$$

This is a basic example of a *copy-cat strategy*. Note that at each stage the strategy simply copies the preceding move by O from one copy of $I\!N$ to the other. This is clearly not specific to $I\!N$—the same idea can be applied to any game. Copy-cat strategies have a fundamental importance in game semantics, as first recognized in [3]. Note that they provide *identities* with respect to composition. For example, if we form the composition

$$
\frac{z : I\!N \vdash z : I\!N \qquad x : I\!N, y : I\!N \vdash x + y : I\!N}{z : I\!N, y : I\!N \vdash z + y : I\!N}
$$

then the corresponding strategy is *the same strategy for addition*.

By repeatedly applying composition to the strategies for constants, variables, and operations such as addition, we can build up interpretations for arbitrary expressions

$$x_1 : B_1, \ldots, x_k : B_k \vdash e : B$$

where $B_1, \ldots, B_n, B$ are basic data types such as $I\!N$ and **bool**, the latter interpreted by the game

$$
\begin{array}{cc}
& q \\
\diagup & \diagdown \\
\textbf{true} & \textbf{false}
\end{array}
$$

An important additional example of an operation is the *conditional*

$$b : \textbf{bool}, x : B, y : B \vdash \text{cond}(b, x, y) : B$$

which plays as follows

$$\mathbf{bool} , B , B \vdash B \qquad\qquad \mathbf{bool} , B , B \vdash B$$

$$q \qquad\qquad\qquad\qquad q$$

$$q \qquad\qquad\qquad\qquad\qquad\qquad q$$
$$\mathbf{true} \qquad\qquad\qquad\qquad\quad \mathbf{false}$$

$$q \qquad\qquad\qquad\qquad\qquad\qquad q$$
$$a \qquad\qquad\qquad\qquad\qquad\qquad b$$

$$a \qquad\qquad\qquad\qquad\qquad\qquad b$$

The strategy for conditional is:

"Ask for the boolean argument; if I get **true**, I will play copy-cat between the second argument and the result; if I get **false**, I will play copy-cat between the third argument and the result".

2.6. FORMALIZATION

Before we continue with the development of the semantics, we will set up a simple framework in which we can make the definitions formal and precise. We will interpret each type of the programming language by an *alphabet* of "moves". A "play" of the game will then be interpreted as a *word* (string, sequence) over this alphabet. A strategy will be represented by the set of complete plays following that strategy, *i.e.* as a *language* over the alphabet of moves. For a significant fragment of the programming language, the strategies arising as the interpretations of terms will turn out to be *regular languages*, which means that they can be represented by finite automata [20].

In fact, our preferred means of specification of strategies will be by using a certain class of *extended regular expressions*.

We recall firstly the standard syntax of regular expressions:

$$R \cdot S \quad R + S \quad R^* \quad a \quad \epsilon \quad 0$$

where a ranges over some alphabet Σ. These expressions are then interpreted as languages over Σ, i.e. as subsets of Σ^*, the set of all strings over Σ; we shall recall the basic definitions, while for more extensive background, we refer to standard texts such as [20].

Notation When we need to refer explicitly to the language denoted by a regular expression R, we shall write $\mathcal{L}(R)$.

We briefly recall the definitions: given $L, M \subseteq \Sigma^*$,

$$L \cdot M = \{st \mid s \in L \wedge t \in M\}$$

$$L^* = \bigcup_{i \in \mathbb{N}} L^i$$

where

$$L^0 = \{\epsilon\}, \quad L^{i+1} = L \cdot L^i.$$

Then

$$
\begin{aligned}
\mathcal{L}(R \cdot S) &= \mathcal{L}(R) \cdot \mathcal{L}(S) \\
\mathcal{L}(R) + \mathcal{L}(S) &= \mathcal{L}(R) \cup \mathcal{L}(S) \\
\mathcal{L}(R^*) &= \mathcal{L}(R)^* \\
\mathcal{L}(a) &= \{a\} \\
\mathcal{L}(\epsilon) &= \{\epsilon\} \\
\mathcal{L}(0) &= \varnothing
\end{aligned}
$$

The extended regular expressions we will consider have the additional constructs

$$R \cap S \qquad \phi(R) \qquad \phi^{-1}(R)$$

where $\phi : \Sigma_1 \longrightarrow \Sigma_2^*$ is a homomorphism (more precisely, such a map uniquely induces a homomorphism between the free monoids Σ_1^* and Σ_2^*; note that, if Σ_1 is finite, such a map is itself a finite object). The interpretation of these constructs is the obvious one: $R \cap S$ is intersection of languages, $\phi(R)$ is direct image of a language under a homomorphism, and $\phi^{-1}(R)$ is inverse image.

$$
\begin{aligned}
\mathcal{L}(R \cap S) &= \mathcal{L}(R) \cap \mathcal{L}(S) \\
\mathcal{L}(\phi(R)) &= \{\phi(s) \mid s \in \mathcal{L}(R)\} \\
\mathcal{L}(\phi^{-1}(R)) &= \{s \in \Sigma_1^* \mid \phi(s) \in \mathcal{L}(R)\}
\end{aligned}
$$

The following is standard [20].

Proposition 2.1 *If we restrict to finite alphabets (so that in particular all homomorphisms map between finite alphabets) then every extended regular expression denotes a regular language, which can be recognized by a finite automaton which can be effectively constructed from the extended regular expression.*

We briefly indicate the arguments showing that regularity is preserved by the extended regular operations.

- $R \cap S$: a product automaton construction is used.
- $\phi(R)$: homomorphisms commute with regular expression operations.
- $\phi^{-1}(R)$: if (Q, q_0, F, δ) recognizes $\mathcal{L}(R)$, (Q, q_0, F, δ') recognizes $\mathcal{L}(\phi^{-1}(R))$, where:

$$\delta'(q, a) = q' \equiv \delta^*(q, \phi(a)) = q'.$$

Alphabet transformations Since we will interpret types as alphabets, getting the types right in our interpretation of terms as strategies will require transforming appropriately between different alphabets, and it is in this restricted form that homomorphisms will be used. The ideas are simple, but will be used repeatedly, and must be mastered at this stage.

We will "glue" types together using disjoint union:

$$X + Y = \{x^1 \mid x \in X\} \cup \{y^2 \mid y \in Y\}$$

We then have canonical maps arising from these disjoint unions.

$$X \underset{\text{outl}}{\overset{\text{inl}}{\rightleftarrows}} X + Y \underset{\text{outr}}{\overset{\text{inr}}{\rightleftarrows}} Y$$

$$\text{inl}(x) = x^1 \quad \text{inr}(y) = y^2$$
$$\text{outl}(x^1) = x \quad \text{outr}(x^1) = \epsilon$$
$$\text{outl}(y^2) = \epsilon \quad \text{outr}(y^2) = y$$

Exercise Verify that

$$\text{outl} \circ \text{inl} = \text{id}_X \qquad \text{outr} \circ \text{inr} = \text{id}_Y.$$

What can you say about outr ∘ inl? About inl ∘ outl?

Example Shuffle Product: for regular expressions R and S over the alphabet Σ,

$$R \parallel S = \nabla(\text{outl}^{-1}(R) \cap \text{outr}^{-1}(S))$$

$$\Sigma \underset{\text{outl}}{\overset{\text{inl}}{\rightleftarrows}} \Sigma + \Sigma \underset{\text{outr}}{\overset{\text{inr}}{\rightleftarrows}} \Sigma$$

$$\nabla : \Sigma + \Sigma \longrightarrow \Sigma$$

$$\nabla(a^1) = a = \nabla(a^2)$$

Exercise Verify that this expression does yield the shuffle product of the languages denoted by R and S, *i.e.* the set of "shuffles" or interleavings of strings drawn from $\mathcal{L}(R)$ and $\mathcal{L}(S)$.

The reader may wonder why we did not immediately stipulate that all alphabets are finite. This is because infinite alphabets do arise naturally in defining game semantics (for example, our game for the basic type $I\!N$ has infinitely many moves), so it is useful to consider this more general situation, even though we will lose effectiveness in general. (Note that the definitions of the meanings of extended regular expressions as formal languages still make sense even if the alphabet is infinite). For the same reason, we consider a further extension to the syntax of "regular expressions", even though it certainly does not preserve regularity in general. Namely, we consider infinite summation $\sum_{i \in I} R_i$, which of course is to be interpreted as union of a family of languages. (This extension is also used in Conway's classic treatise [9]).

What will be important is that for a significant fragment of the language whose semantics we will describe, the extended regular expressions (excluding infinite summations) over finite alphabets suffice, and hence Proposition 2.1 will apply.

2.7. DENOTATIONAL SEMANTICS À LA HOPCROFT AND ULLMAN

We now proceed to the formal description of the semantics.

Firstly, for each basic data type with set of data values D, we specify the following alphabet of moves:

$$M_D = \{q\} \cup D.$$

Note that $I\!N$ and **bool** follow this pattern. In each case, the idea is the same: O can initially use q to request a value, and the possible responses by P are drawn from the set D.

Note that $M_{I\!N}$ is an infinite alphabet. We will also consider finite truncations $M_{I\!N_k}$ where the set of data values is $\{0, \ldots, k-1\}$.

To interpret

$$x_1 : B_1, \ldots, x_k : B_k \vdash t : B$$

we will use the disjoint union of the alphabets:

$$M_{B_1} + \cdots + M_{B_k} + M_B$$

The "tagging" of the elements of the disjoint union to indicate which of the games $B_1, \ldots, B_k, B$ each move occurs in makes precise the "alignment by columns" of the moves in our informal displays of plays such as

$$I\!N \vdash I\!N$$

$$q$$

$$q$$

$$n$$

$$n$$

The corresponding play, as a word over the alphabet $M_{I\!N} + M_{I\!N}$, will now be written as

$$q^2 \cdot q^1 \cdot n^1 \cdot n^2.$$

Our general procedure is then as follows. Given a term in context

$$x_1 : B_1, \ldots, x_k : B_k \vdash t : B$$

we will give an extended regular expression

$$R = [\![x_1 : B_1, \ldots, x_k : B_k \vdash t : B]\!]$$

such that the language denoted by R is the strategy interpreting the term.

As a first example, for a constant $\vdash c : B$,

$$[\![\vdash c : B]\!] = q \cdot c.$$

The operation of addition, of type[1]

$$I\!N^1, \ I\!N^2 \vdash I\!N^3$$

is interpreted by

$$q^3 \cdot q^1 \cdot \sum_{n \in I\!N} (n^1 \cdot q^2 \cdot \sum_{m \in I\!N} (m^2 \cdot (n+m)^3))$$

Note that this extended regular expression is over an infinite alphabet, and involves an infinite summation.

Exercise Give a definition of addition for the truncated natural numbers $I\!N_k$. Note that this is not just a matter of restricting the infinite summations to finite ones. Verify that the resulting expression does denote a regular language.

Exercise Write down the regular expression giving the interpretation of the conditional **bool**, $B, B \vdash B$ for $B =$ **bool**. Now use infinite summation to do the same for $B = I\!N$.

A variable

$$x : B^1 \vdash x : B^2$$

is interpreted by

$$q^2 \cdot q^1 \cdot \sum_{b \in V(B)} b^1 \cdot b^2$$

(here we use $V(B)$ for the set of data values in the basic type B). Next, we show how to interpret composition.

$$\frac{\Gamma \vdash t : A \quad x : A, \Delta \vdash u : B}{\Gamma, \Delta \vdash u[t/x] : B}$$

Our interpretation (of composition in this instance!) is, of course, *compositional*. That is, we assume we have already defined

$$R = [\![\Gamma \vdash t : A]\!] \qquad S = [\![x : A, \Delta \vdash u : B]\!]$$

as the intepretations of the premises in the rule for composition.

Next, we assemble the alphabets of the premises and the conclusion, and assign names to the canonical alphabet transformations relating them.

---

[1] Here and in subsequent examples, we tag the copies of the type to make it easier to track which component of the disjoint union—*i.e.* which "column"—each move occurs in.

$$M_\Gamma + M_A \xrightarrow[\text{out}_1]{\text{in}_1} M_\Gamma + M_A + M_\Delta + M_B \xrightarrow[\text{out}_2]{\text{in}_2} M_A + M_\Delta + M_B$$

$$\text{out} \Big\downarrow \Big\uparrow \text{in}$$

$$M_\Gamma + M_\Delta + M_B$$

The central type in this diagram combines the types of both the premises, including the type A, which will form the "locus of interaction" between t and u. The type of the conclusion arises from this type by "hiding" or erasing this locus of interaction, which thereby is made "internal" to the compound system formed by the composition. Thus the interpretation of composition is by "parallel composition plus hiding". Formally, we write

$$[\![\Gamma, \Delta \vdash u[t/x] : B]\!] = \text{out}(\text{out}_1^{-1}(R^*) \cap \text{out}_2^{-1}(S)).$$

This algebraic expression may seem somewhat opaque: note that it is equivalent to the more intuitive expression

$$\{s/M_A \mid s/(M_\Delta + M_B) \in \mathcal{L}(R^*) \wedge s/M_\Gamma \in \mathcal{L}(S)\}$$

Here s/X means the string s with all symbols from X erased. This set expression will be recognised as essentially the definition of parallel composition plus hiding in the trace semantics for CSP [18]. Although less intuitive, the algebraic expression we gave as the "official" definition has the advantage of making it apparent that regular languages are closed under composition, and indeed of immediately yielding a construction of an automaton to recognise the resulting language.

2.8. COMMANDS

At this point, we have formal descriptions of all the programming constructs we have considered thus far. We shall now go on to complete our description of the semantics of a procedural language. What may come as a pleasant surprise is that the tools we have already developed can be easily extended to cover commands, local variables, and procedures.

Firstly, we consider commands. In our language, we have a basic type **com**, with the following operations:

$$
\begin{array}{ll}
\textbf{skip} & : \textbf{com} \\
\textbf{seq} & : \textbf{com} \times \textbf{com} \rightarrow \textbf{com} \\
\textbf{cond} & : \textbf{bexp} \times \textbf{com} \times \textbf{com} \rightarrow \textbf{com} \\
\textbf{while} & : \textbf{bexp} \times \textbf{com} \rightarrow \textbf{com}
\end{array}
$$

with the following, more colloquial equivalents:

$$\begin{aligned}
\mathbf{seq}(c_1, c_2) &\equiv c_1; c_2 \\
\mathbf{cond}(b, c_1, c_2) &\equiv \mathbf{if}\ b\ \mathbf{then}\ c_1\ \mathbf{else}\ c_2 \\
\mathbf{while}(b, c) &\equiv \mathbf{while}\ b\ \mathbf{do}\ c
\end{aligned}$$

For each operation

$$\omega : B_1 \times \cdots \times B_k \longrightarrow B$$

there is a typing rule

$$x_1 : B_1, \ldots, x_k : B_k \vdash \omega(x_1, \ldots, x_k) : B$$

which can be combined with the use of the composition rule to build up complex commands, just as we did for expressions.

The game interpreting the type **com** is extremely simple:

run

done

This can be thought of as a kind of "scheduler interface": the environment of a command has the opportunity to schedule it by playing the move **run**. When the command is finished, it returns control to the environment by playing **done**.

Formally, note that **com** is just another example of a basic type—in fact the *unit type*, with just one data value. Thus its alphabet of moves is

$$M_{\mathbf{com}} = \{\mathbf{run}, \mathbf{done}\}.$$

(We use **run** rather than q for descriptive effect).
The strategies interpreting the operations are also disarmingly simple. Firstly, note that **skip** is the unique constant for the unit type:

$$[\![\vdash \mathbf{skip} : \mathbf{com}]\!] = \mathbf{run} \cdot \mathbf{done}.$$

Now the following strategy interprets sequential composition.

$$\mathbf{seq}:\ \mathbf{com}\ \Rightarrow\ \mathbf{com}\ \Rightarrow\ \mathbf{com}$$

run

run
done

run
done

done

Formally, this is just

$$seq : com^1 \times com^2 \to com^3$$

$$[\![seq]\!] = run^3 \cdot run^1 \cdot done^1 \cdot run^2 \cdot done^2 \cdot done^3.$$

Thus we think of sequential composition as working as follows. When asked to run by the environment, it begins by asking its first argument to run; when this argument responds, signalling that it has run to completion, it asks its second argument to run; when this has completed, it signals completion to its environment.

Exercise Give strategies to interpret **cond** and **while**. These should be expressible as (non-extended) regular expressions.

Note that this way of modelling commands and their sequential composition is radically different to the traditional approach in denotational semantics (see e.g. [33, 32, 35, 34]), in which commands are modelled as *state transformers*, *i.e.* (roughly speaking) functions from states to states, and sequential composition as function composition. To fully understand the difference in our point of view, we must see how imperative variables are modelled in our approach; this is our next topic.

2.9. IMPERATIVE VARIABLES

The most distinctive part of an imperative language is of course the store upon which the commands operate. To interpret mutable variables, we will take an "object-oriented view" as advocated by John Reynolds [30]. In this view, a variable (say for example being used to store values of type $I\!N$) is seen as an object with two methods:

- the "read method", for dereferencing, giving rise to an operation of type $var \Rightarrow I\!N$;
- the "write method", for assignment, giving an operation of type $var \Rightarrow I\!N \Rightarrow com$.

We *identify* the type of variables with the product of the types of these methods, setting

$$var = (I\!N \Rightarrow com) \times I\!N.$$

Now assignment and dereferencing are just the two projections, and we can interpret a command $x := !x+1$ as the strategy

$$(I\!N \Rightarrow \textbf{com}) \times I\!N \Longrightarrow \textbf{com}$$

$$\begin{array}{l} \hspace{9cm} \texttt{run} \\[4pt] \hspace{4.5cm} \texttt{read} \\[2pt] \hspace{4.7cm} n \\[4pt] \hspace{3cm} \texttt{write} \\[2pt] \hspace{2cm} q \\[2pt] \hspace{1.8cm} n+1 \\[4pt] \hspace{3.4cm} \texttt{ok} \\[4pt] \hspace{6cm} \texttt{ok} \end{array}$$

(We use $\texttt{write}$ and $\texttt{ok}$ in place of $\texttt{run}$ and $\texttt{done}$ in the assignment part, and $\texttt{read}$ in place of q in the dereferencing part, to emphasize that these moves initiate assignments and dereferencing rather than arbitrary commands or natural number expressions.)

In fact, we shall slightly simplify this description. We shall elide the opening two moves in a write operation, and simply have moves of the form $\textbf{write}(d)$, for each possible value d of the data-type, and $\texttt{ok}$ as the only possible reponse.

Thus our alphabet for the type $\textbf{var}[D]$ of imperative variables which can have values from the set D stored in them, is given by

$$M_{\textbf{var}[D]} = \{\texttt{read}\} \cup D \cup \{\textbf{write}(d) \mid d \in D\} \cup \{\texttt{ok}\}$$

The operations for reading and writing have the form

$$\begin{array}{ll} \textbf{assign} : \textbf{var}[D] \times \textbf{exp}[D] \to \textbf{com} & \textbf{assign}(v, e) \equiv v := e \\[4pt] \textbf{deref} : \textbf{var}[D] \to \textbf{exp}[D] & \textbf{deref}(v) \equiv !v \end{array}$$

The strategy for assign is:

$$\textbf{assign} : \textbf{var}[D]^1 \times \textbf{exp}[D]^2 \to \textbf{com}^3$$

$$[\![\textbf{assign}]\!] = \texttt{run}^3 \cdot q^2 \cdot \sum_{d \in D} (d^2 \cdot \textbf{write}(d)^1) \cdot \texttt{ok}^1 \cdot \texttt{done}^3$$

Exercise Give the strategy for $\textbf{deref}$. Compute the strategy obtained for $x := !x+1$.

2.9.1. *Block structure*

The key point is to interpret the *allocation* of variables correctly, so that if the variable x in the above example has been bound to a genuine storage cell, the various reads and writes made to it have the expected relationship. In general, a term M with a free variable x will be interpreted as a strategy for $\textbf{var} \Rightarrow A$, where A is the type of M. We must interpret **new** x **in** M as a strategy for A by "binding x to a memory cell". With game semantics, this is easy! The strategy for M will play some moves in A, and may also make repeated use of the **var** part. The play in the **var** part will look something like this.

var

write(d_1)
ok
write(d_2)
ok
read
d_3
read
d_4

$\vdots$

Of course there is nothing constraining the reads and writes to have the expected relationship. However, there is an obvious strategy

$$\textbf{cell} : \textbf{var}$$

which plays like a storage cell, always responding to a **read** with the last value written. Once we have this strategy, we can interpret **new** by composition with **cell**, so

$$[\![\textbf{new } x \textbf{ in } M]\!] = [\![M]\!] \circ \textbf{cell}.$$

Two important properties of local variables are immediately captured by this interpretation:

Locality Since the action in **var** is hidden by the composition, the environment is unaware of the existence and use of the local variable.

Irreversibility As M interacts with **cell**, there is no way for M to undo any writes which it makes. Of course M can return the value stored in the cell to be the same as it has been previously, but only by performing a new **write**.

2.9.2. *Commands revisited*

We can now get a better perspective on how command combinators work. Suppose that we have a sequential composition

$$\frac{x : \textbf{var} \vdash c_1 : \textbf{com} \qquad x : \textbf{var} \vdash c_2 : \textbf{com}}{x : \textbf{var} \vdash c_1 ; c_2 : \textbf{com}}$$

The game semantics of this command, formed by the composition of the **seq** combinator with the commands c_1 and c_2, can be pictured as follows.

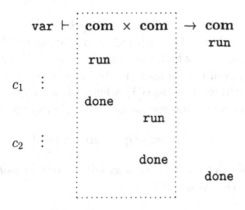

Note that the simple behaviour of the sequential composition combinator can be used, via composition of strategies, to sequence arbitrarily complex commands c_1 and c_2.

Exercise Compute the semantics of: x:=0; x:=!x+1.

This should be contrasted with the traditional denotational semantics of imperative state

$$\text{State} = \text{Val}^{\text{Loc}}$$

in which states are modelled as mappings from locations to values, i.e. as "state snapshots". Programs are then modelled as state transformers, i.e. as functions or relations on these state snapshots, and sequential composition as function or relation composition. It turns out to be hard to give accurate models of locally scoped imperative state with these models; one has to introduce functor categories, and even then full abstraction is hard to achieve [27].

By contrast, we track the "stream of consciousness" of each variable as a separate, autonomous process. A similar approach to modelling variables has also been taken in other process models, for example by Milner in CCS [26]. However, the greater degree of structure provided by game semantics (the distinction between O and P, the constraints imposed by the rules of the game and so on) will enable us to obtain not merely a sound model but a fully abstract one.

3. The full language

We now give a more systematic account of the syntax and semantics of the full language we are considering, including (first-order) procedures.

3.1. SYNTAX

Firstly, we shall be more precise about the syntax of the language we are interpreting.

We assume given a family of basic data types D; these are just sets of data values, such as $I\!N$ and **bool**. For each such data type, we will have types **exp**$[D]$ and **var**$[D]$ of expressions which can produce values of type D, and variables which can store values of type D respectively. We will also have a data type **com** of commands; this will really be **exp**$[1]$, where **1** is a one-element set, but it will be convenient to distinguish this special case. Thus our syntax of *basic types* is

$$B \quad ::= \quad \textbf{exp}[D] \mid \textbf{com} \mid \textbf{var}[D].$$

The general class of types we shall consider will contain *first-order procedures* as well as basic types. The syntax is

$$T \quad ::= \quad B \mid B \to T.$$

The reason for the restriction to first-order is that it allows much simpler representations of the game semantics, and in particular makes effectivity much easier to achieve. The effective representation of the game semantics of programs at higher types is a topic of current research.

The language is an applied typed λ-calculus over this set of types. Typing judgements have the form

$$\Gamma \vdash t : T \qquad \Gamma = x_1 : T_1, \ldots, x_k : T_k$$

The type system is as follows.
Typed λ-calculus (I): Variables and Abstraction.

$$\frac{}{\Gamma, x : T \vdash x : T} \qquad \frac{\Gamma, x : B \vdash t : T}{\Gamma \vdash \lambda x.t : B \to T}$$

Typed λ-calculus (II): Application.

$$\frac{\Gamma \vdash t : B \to T \quad \Gamma \vdash u : B}{\Gamma \vdash tu : T}$$

Block structure:

$$\frac{\Gamma, x : \textbf{var}[D] \vdash t : T}{\Gamma \vdash \textbf{new } x \textbf{ in } t : B}$$

There is also a set of constants

$$\kappa : B_1 \to \cdots \to B_k \to B$$

as already described.

For the purposes of giving a semantics, it will be convenient to use a variant of the above system, in which the application rule is replaced by the following two rules:

Linear Application:

$$\frac{\Gamma \vdash t : B \to T \quad \Delta \vdash u : B}{\Gamma, \Delta \vdash tu : T}$$

Contraction:

$$\frac{\Gamma, x : U, y : U \vdash t : T}{\Gamma, z : U \vdash t[z/x, z/y] : T}$$

It is well know that the resulting system is equivalent, in the sense that exactly the same typing judgements can be derived.

Exercise Show how to derive (arbitrary instances of) the two new rules in the original system, and conversely how to derive any instance of the original rule for Application in the new system.

3.2. TYPES AS ALPHABETS

The first step in the game semantics is to interpret each type T by an alphabet of moves M_T. In fact, for basic types B it will be convenient to define sets Q_B and A_B (*questions* and *answers* of type B respectively), such that

$$Q_B \cap A_B = \varnothing, \qquad Q_B \cup A_B = M_B.$$

$$
\begin{aligned}
Q_{\exp[D]} &= \{q\} & A_{\exp[D]} &= D \\
Q_{\text{com}} &= \{\text{run}\} & A_{\text{com}} &= \{\text{done}\} \\
Q_{\text{var}[D]} &= \{\text{read}\} \cup \{\text{write}(d) \mid d \in D\} & A_{\text{var}[D]} &= D \cup \{\text{ok}\}
\end{aligned}
$$

We extend this to general types by

$$M_{B \to T} = M_B + M_T.$$

3.3. THE INTERPRETATION OF TERMS

The general format is as follows. A typing judgement

$$x_1 : T_1, \ldots, x_k : T_k \vdash t : T$$

is interpreted by a regular expression

$$R = [\![x_1 : T_1, \ldots, x_k : T_k \vdash t : T]\!]$$

over the alphabet

$$M_{T_1} + \cdots + M_{T_k} + M_T.$$

3.3.1. *Variables*

Let $T = B_1 \to \cdots B_k \to B$. The corresponding alphabet is $M_\Gamma + M_T^1 + M_T^2$. Then $[\![\Gamma, x : T \vdash x : T]\!]$ is given by the following regular expression.

$$\sum_{q \in Q_B} q^2 \cdot q^1 \cdot \left(\sum_{i=1}^{k} R_i\right)^* \cdot \sum_{a \in A_B} (a^1 \cdot a^2)$$

where

$$R_i = \sum_{q' \in Q_{B_i}} q'^1 \cdot q'^2 \cdot \sum_{a' \in A_{B_i}} (a'^2 \cdot a'^1), \qquad 1 \leq i \leq k.$$

This is the copy-cat strategy for first-order procedure types. At such types the "generic behaviour" of a procedure is, on being called by its environment, to perform some sequence of calls of its arguments, and then to return a result. The behaviour of higher-order procedures can be much more complicated. See [28] for a description of the behaviours of second-order functions as deterministic context-free languages.

Exercise Try to give an interpretation of a second-order function variable, of type $(\exp[I\!N] \to \exp[I\!N]) \to \exp[I\!N]$. To get an idea of some of the complications which arise, it may be useful to consider the strategies corresponding to the following two terms:

$$M_1 = \lambda f. \, f(\lambda x. \, f(\lambda y. \, y))$$
$$M_2 = \lambda f. \, f(\lambda x. \, f(\lambda y. \, x))$$

For a discussion of this example, see [7].

3.3.2. *Abstraction*

$$\frac{\Gamma, x : B \vdash t : T}{\Gamma \vdash \lambda x. t : B \to T}$$

If

$$[\![\Gamma, x : B \vdash t : T]\!] = R$$

then

$$[\![\Gamma \vdash \lambda x. t : B \to T]\!] = \phi(R)$$

where ϕ is the trivial associativity isomorphism for disjoint union:

$$\phi : (M_\Gamma + M_B) + M_T \xrightarrow{\cong} M_\Gamma + (M_B + M_T)$$

3.3.3. *Linear Application*

This follows very similar lines to our previous treatment of composition (as one would expect).

$$\frac{\Gamma \vdash t : B \rightarrow T \quad \Delta \vdash u : B}{\Gamma, \Delta \vdash tu : T}$$

$$R = [\![\Gamma \vdash t : B \rightarrow T]\!] \qquad S = [\![\Delta \vdash u : B]\!]$$

$$M_\Gamma + M_B + M_T \overset{\text{in}_1}{\underset{\text{out}_1}{\rightleftarrows}} M_\Gamma + M_\Delta + M_B + M_T \overset{\text{in}_2}{\underset{\text{out}_2}{\rightleftarrows}} M_\Delta + M_B$$

$$\text{out} \Big\downarrow \Big\uparrow \text{in}$$

$$M_\Gamma + M_\Delta + M_T$$

$$[\![\Gamma, \Delta \vdash tu : T]\!] = \text{out}(\text{out}_1^{-1}(R) \cap \text{out}_2^{-1}(S^*)).$$

3.3.4. *Contraction*

If $R = [\![\Gamma, x : T, y : T \vdash u : U]\!]$, then

$$[\![\Gamma, z : T \vdash u[z/x, z/y] : U]\!] = \text{id}_{M_\Gamma} + \nabla + \text{id}_{M_U}(R)$$

where $\nabla : M_T + M_T \longrightarrow M_T$. This simply "de-tags" moves from the two occurrences x and y so that they both come from the same occurrence z. It is a property of first-order types—but *not* of higher-order ones—that this de-tagging results in an unambiguous description, in which the "threads" of interaction involving the two occurrences can still be disentangled.

3.3.5. *Local variables*

$$\frac{\Gamma, x : \textbf{var}[D] \vdash t : T}{\Gamma \vdash \textbf{new } x \textbf{ in } t : T}$$

Let

$$R \quad = \quad [\![\Gamma, x : \textbf{var}[D] \vdash t : T]\!],$$
$$\textbf{cell}_D \quad = \quad (\textstyle\sum_{d \in D}(\texttt{write}(d) \cdot \texttt{ok} \cdot (\texttt{read} \cdot d)^*))^*$$

Then

$$[\![\Gamma \vdash \textbf{new } x \textbf{ in } t : T]\!] = \text{out}_2(R \cap \text{out}_1^{-1}(\textbf{cell}_D))$$

where

$$M_{\textbf{var}[D]} \overset{\text{in}_1}{\underset{\text{out}_1}{\rightleftarrows}} M_\Gamma + M_{\textbf{var}[D]} + M_T \overset{\text{in}_2}{\underset{\text{out}_2}{\rightleftarrows}} M_\Gamma + M_T$$

3.4. THE FINITARY SUB-LANGUAGE

We define the finitary fragment of our procedural language to be that in which only *finite* sets of basic data values D are used. For example, we may consider only the basic types over **bool** and $\mathbb{N}_k$, omitting $\mathbb{N}$.

The following result is immediate from the above definitions.

Proposition 3.1 *The types in the finitary language are intepreted by finite alphabets, and the terms are interpreted by extended regular expressions without infinite summations. Hence terms in this fragment denote regular languages.*

4. Model checking

Terms M and N are defined to be *observationally equivalent*, written $M \equiv N$, just in case for any context $C[\cdot]$ such that both $C[M]$ and $C[N]$ are programs (i.e. closed terms of type **com**), $C[M]$ converges (i.e. evaluates to **skip**) if and only if $C[N]$ converges. (Note that the quantification over all *program* contexts takes side effects of M and N fully into account; see e.g. [5].) The theory of observational equivalence is rich; for example, here are some non-trivial observationally equivalences (Ω is the divergent program):

$$P : \mathbf{com} \vdash \mathbf{new}\ x\ \mathbf{in}\ P \equiv P \tag{1}$$

$$
\begin{aligned}
P : \mathbf{com} \to \mathbf{com} \vdash\ &\mathbf{new}\ x{:=}0\ \mathbf{in}\ P(x{:=}1);\\
&\mathbf{if}\ !x = 1\ \mathbf{then}\ \Omega\ \mathbf{else}\ \mathbf{skip}\\
&\equiv\\
&P(\Omega)
\end{aligned}
\tag{2}
$$

$$
\begin{aligned}
P : \mathbf{com} \to \mathbf{com} \vdash\ &\mathbf{new}\ x{:=}0\ \mathbf{in}\\
&P(x{:=}!x + 2);\ \mathbf{if}\ \mathbf{even}(x)\ \mathbf{then}\ \Omega\\
&\equiv\\
&\Omega
\end{aligned}
\tag{3}
$$

$$
\begin{aligned}
P : \mathbf{com} \to \mathbf{bexp} \to \mathbf{com} \vdash\ &\mathbf{new}[\mathbf{int}]\ x{:=}1\ \mathbf{in}\ P(x{:=} - x)(x > 0)\\
&\equiv\\
&\mathbf{new}[\mathbf{bool}]\ x{:=}\mathbf{true}\ \mathbf{in}\ P(x{:=}\neg x)\ x
\end{aligned}
\tag{4}
$$

Exercise Try to reason informally about the validity of these equivalences.

Theorem 1 ([5, 7, 15])

$$\Gamma \vdash t \equiv u \iff \mathcal{L}(R) = \mathcal{L}(S)$$

where $R = [\![\Gamma \vdash t : T]\!]$, $S = [\![\Gamma \vdash u : T]\!]$.

Moreover, $\mathcal{L}(R) = \mathcal{L}(S)$ *(equality of (languages denoted by) regular expressions) is decidable. Hence observation equivalence for the finitary sub-language is decidable.*

If $R \neq S$ we will obtain a witness $s \in [\![R]\!]\triangle[\![S]\!]$, which we can use to construct a separating context $C[\cdot]$, such that

$$\text{eval}(C[t]) \neq \text{eval}(C[u]).$$

4.1. MODEL CHECKING BEHAVIOURAL PROPERTIES

The same algorithmic representations of program meanings which are used in deciding observational equivalence can be put to use in verifying a wide range of program properties, and in practice this is where the interesting applications are most likely to lie. The basic idea is very simple. To verify that a term-in-context $\Gamma \vdash M : A$ satisfies behavioural property $\phi \subseteq M_{\Gamma,A}^*$ amounts to checking $[\![\Gamma \vdash M : A]\!] \subseteq \phi$, which is decidable if ϕ is, for example, regular. Such properties can be specified in temporal logic, or simply as regular expressions.

As a first example of such a property, consider the sequent

$$x : \mathbf{var}[D], p : \mathbf{com} \vdash \mathbf{com}.$$

Suppose we wish to express the property

"x is written before p is (first) called".

The alphabet of the sequent is

$$M = M_{\mathbf{var}[D]}^1 + M_{\mathbf{com}}^2 + M_{\mathbf{com}}^3.$$

The following regular expression captures the required property:

$$X^* \cdot \sum_{d \in D} \mathbf{write}(d)^1 \cdot X^* \cdot \mathbf{run}^2 \cdot M^*$$

where $X = M \setminus M_{\mathbf{com}}^2$.

As a more elaborate example, consider the sequent

$$p : \mathbf{exp}[D] \to \mathbf{exp}[D], x : \mathbf{var}[D] \vdash \mathbf{com}$$

and the property:

"whenever p is called, its argument is read from x, and its result is immediately written into x".

This time, the alphabet is

$$M = (M^1_{\exp[D]} + M^2_{\exp[D]}) + M^3_{\text{var}[D]} + M^4_{\text{com}}$$

and the property can be captured by the regular expression

$$(X^* \cdot (q^1 \cdot \text{read}^3 \cdot \sum_{d \in D} (d^3 \cdot d^1) \cdot Y^* \cdot \sum_{d \in D} (d^2 \cdot \text{write}(d)^3) \cdot \text{ok}^3 \cdot Z^*)^*)^*$$

for suitable choices of sets of moves X, Y, Z.

Exercise Find suitable choices for X, Y and Z.

Exercise Find more interesting properties!

This example illustrates the inherent compositionality of our approach, being based on a compositional semantics which assigns meanings to terms-in-context.

Our approach combines gracefully with the standard methods of *over-approximation* and *data-abstraction*. The idea of over-approximation is simple and general:

$$[\![\Gamma, \vdash M : A]\!] \subseteq S \wedge S \subseteq \phi \implies [\![\Gamma, \vdash M : A]\!] \subseteq \phi. \quad .$$

This means that we can "lose information" in over-approximating the semantics of a program while still inferring useful information about it. This combines usefully with the fact that all the regular expression constructions used in our semantics are *monotone*, which means that if we over-approximate the semantics of some sub-terms $t_1, \ldots, t_n$, and calculate the meaning of the context $C[\cdot, \ldots, \cdot]$ in which the sub-terms are embedded in the standard way, then the resulting interpretation of $t = C[t_1, \ldots, t_n]$ will over-approximate the "true" semantics $[\![t]\!]$.

An important and natural way in which over-approximation arises is from *data abstraction*. Suppose, for a simple example, that we divide the integer data type $\mathbb{Z}$ into "negative" and "non-negative". Since various operations (e.g. addition) will not be compatible with this equivalence relation, we must also add a set "negative *or* non-negative"—i.e. the whole of $\mathbb{Z}$. Now arithmetic operations can be defined to work on these three "abstract values". To define boolean-valued operations on these values, we must extend the type **bool** with "true *or* false", which we write as ?. These extended booleans must in turn be propagated through conditionals and loops, which we do using the *non-determinism* which is naturally present in our regular expression formalism. For example, the conditional of type

$$\exp[\text{bool}]^1 \to \exp[\text{bool}]^2 \to \exp[\text{bool}]^3 \to \exp[\text{bool}]^4$$

can be defined thus:

$$q^4 \cdot q^1 \cdot (\textbf{true}^1 \cdot R + \textbf{false}^1 \cdot S + ?^1 \cdot (R + S))$$

where

$$R = q^2 \cdot \sum_{b \in \textbf{bool}} (b^2 \cdot b^4), \qquad S = q^3 \cdot \sum_{b \in \textbf{bool}} (b^3 \cdot b^4).$$

This over-approximates the meaning in the obvious way (which is of course quite classical in flow analysis): if we don't know whether the boolean value used to control the conditional is really true or false, then *we take both branches*. We can then use the monotonicity properties of the semantics to compute the interpretations of the λ-calculus constructs as usual, and conclude that the meaning assigned to the whole term over-approximates the "true" meaning, and hence that properties inferred of the abstraction hold for the original program. This gives an attractive approach to many of the standard issues in program analysis, e.g. inter-procedural control-flow analysis and reachability analysis [29, 8].

Of course, all of this fits into the framework of *abstract interpretation* [10] in a very natural way.

Another extant technique which can be nicely adapted to our setting is *data independence*. A program is said to be data independent with respect to a data type T if the only operations involving T it can perform are to input, output, and assign values of type T, as well as to test pairs of such values for equality. The impressive results on data independence in [23, 24] include sufficient conditions for reducing the verification of properties that are universally quantified over all instantiations of the data types, to the verification of the same properties for a finite number of finite instantiations.

In conclusion, this area seems ripe for further development, and to have the potential to act as a very effective bridge between semantics and computer-assisted verification and program analysis.

References

1. S. Abramsky. Semantics of interaction: an introduction to game semantics. In *Semantics and Logics of Computation*, pages 1–32. Cambridge Univ. Press, 1997.
2. S. Abramsky, K. Honda, and G. McCusker. Fully abstract game semantics for general references. In *Proceedings of IEEE Symposium on Logic in Computer Science, 1998*, pages 334–344. Computer Society Press, 1998.
3. S. Abramsky and R. Jagadeesan. Games and full completeness for multiplicative linear logic. *J. Symb. Logic*, 59:543–574, 1994.
4. S. Abramsky, R. Jagadeesan, and P. Malacaria. Full abstraction for PCF. *Information and Computation*, 163:409–470, 2000.
5. S. Abramsky and G. McCusker. Linearity, sharing and state: a fully abstract game semantics for Idealized Algol with active expressions. In P. W. O'Hearn and R. D. Tennent, editors, *Algol-like languages*, pages 297–330. Birkhaüser, 1997.
6. S. Abramsky and G. McCusker. Call-by-value games. In *Proceedings of CSL '97*, number 1414 in Lecture Notes in Computer Science, pages 1–17. Springer-Verlag, 1998.

46

7. S. Abramsky and G. McCusker. Game semantics. In H. Schwichtenberg and U. Berger, editors, *Computational Logic: Proceedings of the 1997 Marktoberdorf Summer School*, pages 1–56. Springer-Verlag, 1998.

8. T. Ball and S. K. Rajamani. Boolean programs: A model and process for software analysis. Technical Report MSR-TR-2000-14, MicroSoft Research, 2000.

9. J. H. Conway. *Regular Algebra and Finite Machines*. Chapman and Hall, 1971.

10. P. Cousot and R. Cousot. Abstract interpretation: A unified lattice model for static analysis of programs by construction or approximation of fixpoints. In *Proceedings of 4th ACM Symp. POPL*, pages 238–252, 1977.

11. V. Danos and R. Harmer. Probabilistic game semantics. In *Proc. IEEE Symposium on Logic in Computer Science, Santa Barbara, June, 2000*. Computer Science Society, 2000.

12. J. Corbett *et al*. Bandera: Extracting finite-state models from java source code. In *Proceedings of the 2000 International Conference on Software Engineering*, 2000.

13. D. R. Ghica. A regular-language model for Hoare-style correctness statements. In *Proc. 2nd Int. Workshop on Verification and Computational Logic VCL'2001, Florence, Italy*, 2001. www.cs.queensu.ca/home/ghica/.

14. D. R. Ghica. Regular language semantics for a call-by-value programming language. In *Proc. 17th Conf. Mathematical Foundations of Programming Semantics, Aarhus, Denmark*, 2001. www.cs.queensu.ca/home/ghica/.

15. D. R. Ghica and G. McCusker. Reasoning about Idealized Algol using regular languages. In *Proceedings of 27th International Colloquium on Automata, Languages and Programming ICALP 2000*, pages 103–116. Springer-Verlag, 2000. LNCS Vol. 1853.

16. J.-Y. Girard, Y. Lafont, and P. Taylor. *Proofs and Types*. Cambridge University Press, 1989. Cambridge Tracts in Theoretical Computer Science 7.

17. R. Harmer and G. McCusker. A fully abstract game semantics for finite nondeterminism. In *Proceedings of Fourteenth Annual IEEE Symposium on Logic in Computer Science*. IEEE Computer Society Press, 1999.

18. C. A. R. Hoare. *Communicating Sequential Processes*. Prentice-Hall, 1985.

19. K. Honda and N. Yoshida. Game-theoretic analysis of call-by-value computation (extended abstract). In *Proc. of ICALP'97, Borogna, Italy, July, 1997*, LNCS. Springer-Verlag, 1997.

20. J. E. Hopcroft and J. D. Ullman. *Introduction to Automata Theory, Languages and Computation*. Addison-Wesley, 1979.

21. J. M. E. Hyland and C.-H. L. Ong. On Full Abstraction for PCF: I. Models, observables and the full abstraction problem, II. Dialogue games and innocent strategies, III. A fully abstract and universal game model. *Information and Computation*, 163:285–408, 2000.

22. J. Laird. *A semantic analysis of control*. PhD thesis, University of Edinburgh, 1998.

23. R. Lazic and D. Nowak. A unifying approach to data-independence. In *Proceedings of the 11th International Conference on Concurrency Theory (CONCUR 2000)*. Springer-Verlag, 2000. LNCS.

24. R. S. Lazic, T. C. Newcomb, and A. W. Roscoe. On model checking data-independent systems with arrays without reset (abstract). In *Proceedings of VCL 2001*, 2001.

25. P. Malacaria and C. Hankin. Non-deterministic games and program analysis: an application to security. In *Proceedings of 14th Annual IEEE Symp. Logic in Computer Science*, pages 443–452. IEEE Computer Society, 1999.

26. R. Milner. *Communication and Concurrency*. Prentice-Hall, 1989.

27. P. W. O'Hearn and R. D. Tennent. *Algol-like Languages: Volumes I and II*. Birkhäuser, 1997. Progress in Theoretical Computer Science.

28. C.-H. L. Ong. Equivalence of third order Idealized Algol is decidable. Technical report, Oxford University Computing Laboratory, 2001. In preparation.

29. T. Reps, S. Horowitz, and M. Sagiv. Precise interprocedural dataflow analysis via graph reachability. In *Proc. ACM Symp. POPL*, pages 49–61, 1995.

30. J. C. Reynolds. The essence of Algol. In J. W. de Bakker and J. C. van Vliet, editors, *Algorithmic Languages*, pages 345–372. North Holland, 1978.

31. D. A. Schmidt. On the need for a popular formal semantics. *ACM SIGPLAN Notices*, 32:115–116, 1997.

32. J. E. Stoy. *Denotational Semantics: The Scott-Strachey Approach to Programming Language Theory*. The MIT Press, 1979. The MIT Press Series in Computer Science.

33. C. Strachey. Fundamental concepts in programming languages. Lecture notes for the International Summer School in Computer Programming, Copenhagen, 1967.

34. R. D. Tennent. Denotational semantics. In S. Abramsky *et al*, editor, *Handbook of Logic in Computer Science*, pages 169–322. Oxford University Press, 1994.

35. G. Winskel. *The Formal Semantics of Programming Languages*. MIT Press, 1993. Foundations of Computing Series.

ALGEBRA OF NETWORKS

Modeling simple networks, as well as complex interactive systems

G. STEFANESCU
Department of Computer Science
National University of Singapore
gheorghe@comp.nus.edu.sg

Abstract. The first part of the paper contains an overview of *Network Algebra* (NA) book [35]. The second part introduces *finite interactive systems* as an abstract mathematical model of agents' behaviour and their interaction.

Network Algebra: This book is devoted to an algebraic study of networks and their behaviour. The kernel of the involved algebraic structures is BNA (Basic Network Algebra) structure. These axioms are sound and complete for (abstract) networks modulo graph isomorphism. Branching constants are added to the BNA signature, together with the corresponding weak axioms leading to an axiomatization for networks with branching constants, modulo graph isomorphism. Strong and enzymatic axioms for branching constants are used to depart from graph isomorphism models. Using them, Elgot theory may be introduced, as a sound and complete axiomatization for input behaviour (regular trees). Extensions to the input–output behaviour of deterministic networks (Park theories) and of nondeterministic networks (Kleene theories) are included.

These general results on abstract networks are then specialized on certain particular cases including flowchart schemes, finite automata, process algebra, data-flow networks, or Petri nets.

Finally, a brief presentation of the Mixed Network Algebra models is included in NA book. These models are obtained mixing network algebra models for control, space, and time. A few results are presented, including a challenging space-time duality thesis. Many interesting questions risen by this model are open.

Finite interactive systems: The second half of the paper introduces *finite interactive systems*, an abstract mathematical model of agents' behaviour and their interaction. Finite interactive systems may be seen as a version of the Mixed Network Algebra where the mixture is made at the Kleene theory level. Its aim is the fill in the gap between behaviour and class diagrams used in modern (UML) system design [4].

H. Schwichtenberg and R. Steinbrüggen (eds.), Proof and System-Reliability, 49–78.
© *2002 Kluwer Academic Publishers. Printed in the Netherlands.*

1. An overview of Network Algebra approach

1.1. KEY FEATURES

The term network is used in a broad sense here as consisting of a collection of interconnecting cells.

1.1.1. *Models*
Two radically different interpretations of this enlarged notion of network are particularly relevant. First, virtual networks are obtained using the Cantorian interpretation in which at most one cell is active at a given time. With this interpretation, Network Algebra (NA) covers the classical models of control, including finite automata or flowchart schemes. In a second Cartesian interpretation, each cell is always active. This implies models for reactive and concurrent systems such as Petri nets or data-flow networks may be covered as well.

These two interpretations may be mixed. A sketch of the resulting much more complicated Mixed Network Algebra (MixNA) models is included, as well.

1.1.2. *Algebraic structures*
The results are presented in the unified framework of the calculus of flownomials. This is an abstract calculus very similar to the classical calculus of polynomials.

The kernel structure is the *BNA* ("Basic Network Algebra"). After their introduction in the context of control flowcharts setting [33], the BNA axioms were rediscovered in various fields ranging from circuit theory to action calculi, from data-flow networks to knot theory (traced monoidal categories), from process graphs to functional programming.

In general, the involved algebraic structures are BNAs enriched with branching constants and additional specific axioms. The branching constants are specified by xy with $x \in \{a, b, c, d\}$ and $y \in \{\alpha, \beta, \gamma, \delta\}$; the axioms are divided into three groups: weak, strong, and enzymatic axioms.

As already said, an important feature of the NA approach is the uniform presentation of many apparently unrelated models which have appeared in the literature. A few such structures are described below.

(1) *BNA* (case $a\alpha$-weak, $a\alpha$-strong, $a\alpha$-enzymatic): any transformation which preserves the graph-isomorphism relation is valid (no duplication or removal of wires or cells is permitted);

(2) *xy-symocats with feedback* (case *xy*-weak, $a\alpha$-strong, $a\alpha$-enzymatic): in these cases duplication or removal is permitted for wires, but not for network cells;

(3) *Elgot theory* (case $b\delta$-weak, $a\delta$-strong, $a\delta$-enzymatic): in this setting, networks may be completely unfolded to get "regular" trees; Elgot theory axioms capture the transformation rules that are valid for these trees;

(4) *Kleene theory* (case $d\delta$-weak, $d\delta$-strong, $d\delta$-enzymatic): in this setting, the input-output sequences modeling network behaviour define "regular" languages; Kleene theory captures all the identities which are valid for regular languages (flownomial and regular expressions are equivalent in this context);

(5) *Park theory* (case $b\delta$-weak, $b\delta$-strong, $b\delta$-enzymatic): this is a setting in-between Elgot and Kleene theories; it deals with input-output behaviour, but in the deterministic case;

(6) *Căzănescu–Ungureanu theory* (case $b\delta$-weak, $a\delta$-strong, $a\alpha$-enzymatic): this structure is a BNA over a coalgebraic theory; Floyd–Hoare logic for program correctness may be developed in this setting;

(7) *Conway theory* (case $d\delta$-weak, $d\delta$-strong, $a\alpha$-enzymatic): this structure is a BNA over a matrix theory; this setting suffices for proving Kleene theorems;

(8) *Milner theory* (case $d\delta$-weak, $a\delta$-strong, $a\delta$-enzymatic): this setting may be seen as a lifting with weak constants of the Elgot theory to the nondeterministic case; this setting is used to build-up process algebra.

This ability of the calculus of flownomials to cover many particular algebras may be a good motivation for the reader to follow this, maybe too abstract, calculus.

1.1.3. Results

The main results presented in the book include: axiomatizations for isomorphic networks without or with branching constants; axiomatizations for several classes of relations; axiomatizations for regular trees or regular languages; the Universality Theorems; the Structural Theorems; correctness of Floyd–Hoare logic; the duality between flowcharts and circuits; correctness of ACP (Algebra of Communicating Processes) with respect to bisimulation semantics; axiomatization for input–output behaviour of deterministic data-flow networks; a Kleene-like theorem for Petri net languages; a construction of the free distributive category; the space-time duality principle.

1.2. A BRIEF INTRODUCTION TO NETWORK ALGEBRA

In this introduction we briefly present the *calculus of flownomials*, a unified algebraic theory for various types of networks. This approach is the result of an effort to integrate various algebraic approaches used in computer science.

We may agree: a variety of models is desirable. Then, one may use the most appropriate one for the task. But the models should not be completely unrelated. If so, there will be no transfer of information, results, techniques, etc. with the undesired result of duplication of effort and lost of time.

As a source of inspiration, one may look at the marvelous world of polynomials. There is a lot of freedom to accommodate various models, but still within a clear and unique (or coherent) algebraic framework.

Flownomials were designed with the explicit intention to inherit some of the qualities of polynomials. The motivation for their introduction comes from particular concrete models of computation, but the result is an abstract mathematical model that may be used whenever the underlying rules are valid. (A striking example is the coincidence of the kernel BNA structure of the calculus of flownomials [33] with the later introduced traced monoidal categories occurring in the study of quantum groups [21].)

Flownomials may be used as a mechanism for translating facts from one field to another, or, furthermore, may transform themselves into their own field from where translation to particular models is straightforward.

1.2.1. *Regular expressions*

A basic calculus for sequential computation is provided by Kleene's *calculus of regular expressions*, see, e.g., [23], [11]. It has a simple syntax and a simple semantics. The regular expressions over a set of atomic symbols X, denoted as $\mathsf{RegExp}(X)$, are given by the following grammar

$$E ::= E + E \mid E \cdot E \mid E^* \mid 0 \mid 1 \mid x(\in X)$$

In the standard semantics regular expressions are interpreted as languages, i.e., sets of strings. The operations are interpreted as "union," "catenation," and "iterated catenation," respectively, while the constants are interpreted as "empty set" and "empty string," respectively.

This calculus is a basic calculus for sequential computation and has a broad range of applications in the design and analysis of circuits and programming languages, study of automata and grammars, analysis of flowchart schemes, semantics of programming languages, etc.

However, the calculus has a strong, but less visible, restriction. In order to handle complex objects like automata, grammars, circuits, etc. one has to add one more operation to the above syntax, namely a *matrix building operator*: if $a_{ij}, i \in [m], j \in [n]$ ($[k]$ denotes the set $\{1, 2, \ldots, k\}$) are expressions, then the matrix (a_{ij}) is an (extended) expression. Such an expression is then used to specify the behaviour of a complex object with m inputs and n outputs.

By changing the point of view from "elements" to such "complex objects," one may see that the above construction is based on the following restriction:

Matrix theory rules: reducing m-to-n to 1-to-1 networks

• Let $f : m \to n$ denote the behaviour of a network F with m input ports and n output ports.
• For $i \in [m], j \in [n]$, let f_{ij} denote the behaviour of the network obtained by restricting F to its i-th input and j-th output only.
• By the matrix theory rules, f is uniquely determined by these f_{ij}.

This is a very strong restriction indeed and it makes the calculus unsuited for many computation models. In particular, as is well-known, in such a framework one may rather freely copy or discard network cells, a questionable property in certain cases.

1.2.2. *Iteration theories*

An algebraic setting in between regular expressions and the calculus of flownomials is provided by an algebraic theory modeling flowchart schemes or data-flow networks. In such a case a weaker hypothesis is used. [1]

Coalgebraic theory rules: from m-to-n to 1-to-n networks

• Let $f : m \to n$ denote the behaviour of a network F with m input ports and n output ports.
• For $i \in [m]$, let f_i denote the behaviour of the network obtained from F by considering only its i-th input port (all the output ports are preserved).
• By colgebraic theory rules, f is uniquely determined by these f_i.

The resulting algebraic calculus is more general than the calculus of regular expressions and it is useful to study certain semantic models, see, e.g., [12], [3], [26]. However, it is still restrictive and no simple syntactic model similar to Kleene's model of regular expressions exists.

1.2.3. *Flownomials*

The *calculus of flownomials* may be seen as an extension of the calculus of regular expressions to the case of many-input/many-output atoms. Its main ingredients were presented in a series of papers, including [31], [32], [33], [9], [10]; see [35] for more on this.

---

[1] In the given form the hypothesis fits with the semantic models of flowchart schemes. The case of data-flow models is dual, i.e., a many-output network is specified as a tuple of single-output ones.

> **Symocat rules: atomic m-to-n networks**
>
> Using symocats (shorthand for symmetric strict monoidal categories), no reductions of a network behaviour $f : m \to n$ to certain parts is a priori possible. The network is atomic; its behaviour cannot be decomposed into simpler parts.

The calculus of flownomials is an abstract calculus for networks (= labeled directed hyper-graphs) and their behaviours. It starts with two families of doubly-ranked elements:

- a family of variables $X = \{X(a,b)\}_{a,b \in M}$; these variables denote "black boxes," i.e., atomic elements with two types of connecting ports: input ports (described by a) and output ports (described by b).
- a connection structure $T = \{T(a,b)\}_{a,b \in M}$; the elements in T model known processes between the input/output interfaces; e.g., its elements may model the flow of messages that are transmitted via the connecting channels between the input/output pins; this is the "known" (or "interpreted") part of the model; the flownomial operations are supposed to be already defined on this part.

$M = (M, \star, \epsilon)$ is a monoid modeling network interfaces; if M is a free monoid, then an interface may be considered as a string of atomic ports/pins. We also use the functions $i : \ldots \to M$ and $o : \ldots \to M$ that give the corresponding interfaces for inputs and outputs, respectively.

Flownomial expressions over X and T are specified by: [2]

$$E ::= E \star E \mid E \cdot E \mid E \uparrow^c \mid \mathsf{I}_a \mid {}^a\mathsf{X}^b \mid \wedge_k^a \mid \vee_a^k \mid x(\in X) \mid f(\in T)$$

where $a, b, c \in M$ and $k \in \mathbb{N}$. Let $\mathsf{FlowExp}[X, T]$ denote their set.

The collection of *flownomial expressions* is obtained starting with the elements in X and T, certain constants (to be described below), and applying three operations:

juxtaposition "$\star$", *composition* "$\cdot$", and *feedback* "$\uparrow$"

Juxtaposition $f \star g$ is always defined and $i(f \star g) = i(f) \star i(g)$ and $o(f \star g) = o(f) \star o(g)$. The composite $f \cdot g$ is defined only in the case $o(f) = i(g)$ and, in such a case, $i(f \cdot g) = i(f)$ and $o(f \cdot g) = o(g)$. The feedback $f \uparrow^c$ is defined only in the case $i(f) = a \star c$ and $o(f) = b \star c$, for some $a, b \in M$ and, in such a case, $i(f \uparrow^c) = a$ and $o(f \uparrow^c) = b$. (Technically, a condition

---

[2] Notice the simple way regular expressions fit in this setting: this is the particular instance where $M = (\mathbb{N}, +, 0)$ and $0 = \wedge_0^1 \cdot \vee_1^0$, $1 := \mathsf{I}_1 (= \wedge_1^1 = \vee_1^1)$, $f + g := \wedge_2^1 \cdot (f \star g) \cdot \vee_1^2$, $f \cdot g := f \cdot g$, $f^* := [\vee_1^2 \cdot (1 + f) \cdot \wedge_2^1] \uparrow^1$.

$a \star c = a' \star c \Rightarrow a = a'$ is necessary for the feedback case; it is valid when M is a free monoid.)

The *acyclic case* refers to the case without feedback.

The support theory for connections T should "contain" certain finite binary relations. The main classes used below are: bijections $\mathbb{B}$i, functions $\mathbb{F}$n, partial functions $\mathbb{P}$fn and relations $\mathbb{R}$el. At the abstract level, this means that we should have some constants generating these relations. The collection of constants used for this purpose is:

identity I_a, *(block) transposition* ${}^a\mathsf{X}^b$, *(block) ramification* $\wedge_k^a$, and *(block) identification* $\vee_a^k$, where $a, b \in M$ and $k \in \mathbb{N}$

The types of constants are as follows: $i(\mathsf{I}_a) = o(\mathsf{I}_a) = a$; $i({}^a\mathsf{X}^b) = a \star b$, $o({}^a\mathsf{X}^b) = b \star a$; $i(\wedge_k^a) = a$, $o(\wedge_k^a) = a \star \ldots \star a(k$ times); $i(\vee_a^k) = a \star \ldots \star a(k$ times), $o(\vee_a^k) = a$.

Each of the above classes of relations (and 13 other classes) may be generated using certain sub-collections of these constants only. To specify them one has to *restrict* the using of the branching constants $\wedge_m$ and $\vee^n$ as follows:

Restrictions on branching degrees

• Restriction $x = a$ means the constants of the type $\wedge_m^a$ always have $m = 1$; restriction $x = b$ means $m \leq 1$; $x = c$, means $m \geq 1$; and $x = d$ means no restriction (arbitrary m).

• Restriction y is defined in a similar manner, using n of the constants $\vee_a^n$ and the corresponding Greek letters.

The notation xy, where $x \in \{a, b, c, d\}$ and $y \in \{\alpha, \beta, \gamma, \delta\}$, refers to the class of relations generated with branching constants satisfying restriction xy. E.g., $a\alpha$ refers to bijections, $a\delta$ to functions, $b\delta$ to partial functions, $d\delta$ to relations, etc.

For flownomials we may single out certain abstract algebraic structures to play the role the ring structure is playing for polynomials. They are extensions of a basic algebraic structure, called *BNA (Basic Network Algebra)*, or $a\alpha$-*flow*, which uses I and X in its definition only (no branching constants). The critical axioms used to define these extensions are:

(1) *commutation of the branching constants to the other elements*

$$f \cdot \wedge_k^b = \wedge_k^a \cdot (kf)$$
$$\vee_a^k \cdot f = (kf) \cdot \vee_b^k$$

where $f : a \to b$ and $kf = f \star \ldots \star f$, k-times.

This axiom scheme is used in two ways:

(1a) In the *weak* variant it is used only for certain ground terms f's.[3]
(1b) In the *strong* variant it applies to arbitrary f's.

(2) *enzymatic axiom for the looping operation*

$$f \cdot (I_b \star z) = (I_a \star z) \cdot g \quad \text{implies} \quad f \uparrow^c = g \uparrow^d$$

where $z : c \to d$ is an abstract relation and $f : a \star c \to b \star c$, $g : a \star d \to b \star d$ are arbitrary elements.

Roughly speaking (see Subsection 1.2.4 for more precise definitions):

- an *xy-symocat* (shorthand for *xy-symmetric strict monoidal category*) is the acyclic structure defined using the weak commutation for all *xy*-terms;
- in a *strong xy-symocat*, the strong commutation axioms hold for all *xy*-terms;
- finally, an *xy-flow* is: (1) a BNA over an *xy*-symocat, which is (2) a strong *xy*-symocat, and (3) obeys the enzymatic axiom whenever z is an abstract *xy*-relation.

A large number of algebraic structures may be obtained by choosing the branching constants and certain weak, strong or enzymatic axioms for them. This freedom gives flexibility to the calculus which is well-suited to model various kind of networks and equivalences.

Classification using weak/strong/enzymatic scheme

The main algebraic structures used in NA book are defined as

$$\text{BNA} + x_1 y_1 \text{-weak} + x_2 y_2 \text{-strong} + x_3 y_3 \text{-enzymatic}$$

The first restriction $x_1 y_1$ shows what branching constants may be used, the second $x_2 y_2$ which branching constants strongly commute with arbitrary arrows, and the third $x_3 y_3$ for which branching constants the enzymatic rule may be applied. (Usually $x_1 y_1 \geq x_2 y_2 \geq x_3 y_3$, where "$\geq$" means more constants, hence weaker restriction.)

Polynomials may be seen as classes of equivalent polynomial expressions. In a similar manner, *flownomials* are defined as classes of equivalent flownomial expressions under an appropriate relation of equivalence. (However, we have various classes of flownomials, for various restrictions *xy*.) Moreover, in both calculi there are two equivalent ways to introduce such equivalences:

---

[3] A *ground term*, or an *abstract relation*, is a term written with the flownomial operations and certain constants in $I, X, \wedge_k, \vee^k$. A *ground xy-term*, or an *abstract xy-relation*, is a term which may be written using only constants fulfilling restriction *xy*.

(1) by using the rules of the appropriate algebra, or (2) by using normal form representations.

For the first way, a standard equivalence $\sim_{xy}$ on flownomial expressions may be introduced using xy-flow rules.

The second way to define such an equivalence is to use *normal form flownomial expressions*. Such an expression over X and T is of the following type:

$$((\mathsf{I}_a \star x_1 \star \ldots \star x_k) \cdot f) \uparrow^{i(x_1)\star \ldots \star i(x_k)}$$

with $f : a \star o(x_1) \star \ldots \star o(x_k) \to b \star i(x_1) \star \ldots \star i(x_k)$ in T and $x_1, \ldots, x_k$ in X. Then, two normal form flownomial expressions $F = ((\mathsf{I}_a \star x_1 \star \ldots \star x_m) \cdot f) \uparrow^{i(x_1)\star \ldots \star i(x_m)}$ and $G = ((\mathsf{I}_a \star y_1 \star \ldots \star y_n) \cdot g) \uparrow^{i(y_1)\star \ldots \star i(y_n)}$ are *similar* via a relation $r : m \to n$ iff

   (i) $(i,j) \in r \Rightarrow x_i = y_j$;

   (ii) $f \cdot (\mathsf{I}_b \star i(r)) = (\mathsf{I}_a \star o(r)) \cdot g$.

Here $i(r)$ represents the "block" extension of r to inputs (e.g., for a free monoid M, if r relates two variables x_i and y_j, then $i(r)$ relates the 1st input of x_i to the 1st input of y_j, the 2nd to the 2nd, and so on); similarly $o(r)$ denotes the block extension of r to outputs.

Simulation is reflexive, transitive, and compatible with the flownomial operations, but not symmetric. The generated congruence, denoted $\overset{xy}{\Longleftrightarrow}$, is the equivalence relation generated by simulation via xy-relations.

It turns out that $\overset{xy}{\Longleftrightarrow}$ coincides with $\sim_{xy}$ in the basic cases studied in NA book: $a\alpha, a\delta, b\delta$, or $d\delta$ (sometimes the proofs require certain additional conditions). Generally speaking, $\sim_{xy}$ is coarser.

A basic model is $[T, X]_{xy}$, i.e., the algebra of normal form flownomial expressions modulo $\overset{xy}{\Longleftrightarrow}$ equivalence. Another one is $\mathsf{FlowExp}[X, T]_{xy}$ consisting of classes of $\sim_{xy}$-equivalent normal form flownomials over X and T. The above fact shows that these models are isomorphic in the basic cases $a\alpha, a\delta, b\delta$, and $d\delta$.

1.2.4. *Basic algebraic structures*

A few basic algebraic structures used in the calculus of flownomials are described below, with a brief description of the basic features of these models. The axioms are listed in Table I. We start with the graph-isomorphism models.

BNA Case ($a\alpha$-weak, $a\alpha$-strong, $a\alpha$-enzymatic): axioms I-II

   This is the basic algebraic structure we are working with. In this setting, one may represent cyclic networks. The cells in X may have multiple input or output pins. The signature allows us to use only bijective relations. More sophisticated connections may be used, but only

TABLE I.   The standard axioms for the calculus of flownomials

I. Axioms for symocats

B1 $f \star (g \star h) = (f \star g) \star h$

B2 $I_\epsilon \star f = f = f \star I_\epsilon$

B3 $f \cdot (g \cdot h) = (f \cdot g) \cdot h$

B4 $I_a \cdot f = f = f \cdot I_b$

B5 $(f \star f') \cdot (g \star g') = f \cdot g \star f' \cdot g'$

B6 $I_a \star I_b = I_{a \star b}$

B7 ${}^a X^b \cdot {}^b X^a = I_{a \star b}$

B8 ${}^a X^{b \star c} = ({}^a X^b \star I_c) \cdot (I_b \star {}^a X^c)$

B9 $(f \star g) \cdot {}^c X^d = {}^a X^b \cdot (g \star f)$
     for $f : a \to c, \quad g : b \to d$

II. Axioms for feedback

R1 $f \cdot (g \uparrow^c) \cdot h = ((f \star I_c) \cdot g \cdot (h \star I_c)) \uparrow^c$

R2 $f \star g \uparrow^c = (f \star g) \uparrow^c$

R3 $(f \cdot (I_b \star g)) \uparrow^c = ((I_a \star g) \cdot f) \uparrow^d$
     for $f : a \star c \to b \star d, \quad g : d \to c$

R4 $f \uparrow^\epsilon = f$

R5 $(f \uparrow^b) \uparrow^a = f \uparrow^{a \star b}$

R6 $I_a \uparrow^a = I_\epsilon$

R7 ${}^a X^a \uparrow^a = I_a$

III. Axioms for (angelic) branching constants, without feedback

A1a $\wedge^a \cdot (\wedge^a \star I_a) = \wedge^a \cdot (I_a \star \wedge^a)$

A1b $\wedge^a \cdot (\perp^a \star I_a) = I_a$

A1c $(V_a \star I_a) \cdot V_a = (I_a \star V_a) \cdot V_a$

A1d $(T_a \star I_a) \cdot V_a = I_a$

A2a $\wedge^a \cdot {}^a X^a = \wedge^a$

A2b ${}^a X^a \cdot V_a = V_a$

A3a $T_a \cdot \perp^a = I_\epsilon$

A3b $T_a \cdot \wedge^a = T_a \star T_a$

A3c $V_a \cdot \perp^a = \perp^a \star \perp^a$

A3d $V_a \cdot \wedge^a =$
      $(\wedge^a \star \wedge^a) \cdot (I_a \star {}^a X^a \star I_a) \cdot (V_a \star V_a)$

A4 $\wedge^a \cdot V_a = I_a$

A5a $\perp^{a \star b} = \perp^a \star \perp^b$

A5b $\wedge^{a \star b} = (\wedge^a \star \wedge^b) \cdot (I_a \star {}^a X^b \star I_b)$

A5c $T_{a \star b} = T_a \star T_b$

A5d $V_{a \star b} = (I_a \star {}^b X^a \star I_b) \cdot (V_a \star V_b)$

[A5e $\perp^\epsilon = I_\epsilon$

A5f $\wedge^\epsilon = I_\epsilon$

A5g $T_\epsilon = I_\epsilon$

A5h $V_\epsilon = I_\epsilon$]

IV. Feedback on (angelic) branching constants

R8 $\wedge^a \uparrow^a = T_a$

R9 $V_a \uparrow^a = \perp^a$

R10 $[(I_a \star \wedge^a) \cdot ({}^a X^a \star I_a)$
      $\cdot (I_a \star V_a)] \uparrow^a = I_a$

V. The strong axioms

Sa $f \cdot \perp^b = \perp^a$

Sb $f \cdot \wedge^b = \wedge^a \cdot (f \star f)$

Sc $T_a \cdot f = T_b$

Sd $V_a \cdot f = (f \star f) \cdot V_b$

VI. The enzymatic rule (E is a class of finite, abstract relations)

Enz$_E$:   if for $f : a \star c \to b \star c$ and $g : a \star d \to b \star d$ there exists $y : c \to d$ in E
          such that $f \cdot (I_b \star y) = (I_a \star y) \cdot g$, then $f \uparrow^c = g \uparrow^d$

within the connection theory T. Any transformation which preserves the graph-isomorphism relation is valid. This is a linear-like setting: no duplication or removal of network cells is permitted.

xy-symocats with feedback Case (xy-weak, $a\alpha$-strong, $a\alpha$-enzymatic):
I-II and III-IV(using xy constants)
Only 9 cases are allowed for xy, namely those with $x = b$, or $y = \beta$, or $xy \in \{a\alpha, d\delta\}$. In these cases branching constants are added to the syntactic part of the calculus. They come with appropriate algebraic rules, which were designed in such a way: (1) to be strong enough to allow usual relations with their (angelic) operations to be mapped in these structures, and (2) to preserve the graph-isomorphism setting. Hence, duplication or removing is permitted for connections/wires, but not for network cells.

Together with BNA, the next 3 cases enter into the category of xy-flows. In these cases, all three conditions (weak, strong, enzymatic) are applied for the same class of relations, with a slight exception for the weak case, where actually the closure of the restriction to feedback has to be used.

Elgot theory Case ($b\delta$-weak, $a\delta$-strong, $a\delta$-enzymatic):
I-II, III-IV(using $\top, \vee, \perp$), V(Sc-d), VI (for $E = \mathbb{F}n$)
In this case, by the $a\delta$-strong rules one can safely restrict to single-input networks, variables, etc. As a byproduct, such networks may be completely unfolded to get "regular" trees. Elgot theory axioms capture the transformation rules that are valid for these trees. Rather different networks may produce the same unfolding tree, but the enzymatic rule (together with the other ones) suffices to identify them.

Kleene theory Case ($d\delta$-weak, $d\delta$-strong, $d\delta$-enzymatic): I-V, VI($\mathbb{R}$el)
Kleene theory does the same as Elgot theory, but for input–output computation sequences (paths) for nondeterministic networks. In this case, due to the $d\delta$-strong rules, one can restrict to single-input/single-output networks, cells, etc. For these networks input–output sequences may be defined, producing the associated regular languages. Kleene theory captures all the identities which are valid in this model. (Actually, flownomials and regular expressions are equivalent in this context.) Again, the enzymatic rule proves to be of crucial importance in the proof of axioms' completeness.

Park theory Case ($b\delta$-weak, $b\delta$-strong, $b\delta$-enzymatic):
I-II, III-IV($\top, \vee, \perp$), V(Sa,Sc-d), VI($\mathbb{P}$fn)
This is a setting in-between Elgot and Kleene theories. It deals with

input–output behaviour, but in the deterministic case. Hence matrices cannot be used here. While this case has not (yet!) the widely recognized value of the two cases above, it has its own beauty. For instance, the Structural Theorem modeling deterministic minimization is an interesting and nontrivial extension of the one for Elgot theories, but still the proof is clean, with a rather natural additional "identification-free" hypothesis.

Three more cases deserve some special attention. In these structures, there is a mismatch between the restrictions used for the weak, strong, and enzymatic rules.

Căzănescu–Ungureanu theory Case ($b\delta$-weak, $a\delta$-strong, $a\alpha$-enzymatic): I-II, III-IV($\top, \vee, \bot$), V(Sc-d)
Căzănescu–Ungureanu theory is a BNA over a coalgebraic theory. Floyd–Hoare logic for program correctness may be developed in this setting.

Conway theory Case ($d\delta$-weak, $d\delta$-strong, $a\alpha$-enzymatic): I-V
This case is similar to the above one, so a Conway theory is a BNA over a matrix theory. This setting suffices for proving Kleene theorems.

Milner theory Case ($d\delta$-weak, $a\delta$-strong, $a\delta$-enzymatic): I-IV, V(Sc-d), VI($\mathbb{F}$n)
This may be seen either as a lifting (with weak constants) of the Elgot theory to the nondeterministic case, or as a relaxation of Kleene theory, preserving the strong and enzymatic rules just for functions. This setting is used to build-up process algebra.

1.2.5. *Basic results*
The main results on the calculus of flownomials presented in NA book are:

Expressiveness results: Various classes of networks are represented by flownomial expressions over appropriate connecting theories. E.g.: (a) nondeterministic networks are represented by flownomials over $\mathbb{R}$el; deterministic networks are represented by flownomials over $\mathbb{F}$n, etc. (b) networks with value passing channels are represented by flownomials over relational semantic models AddRel(D)/MultRel(D); (c) higher order flownomials are represented by flownomials over other flownomials.

Axiomatization results: Correct (i.e., sound and complete) axiomatizations for the classes of equivalent networks with respect to various natural equivalence relations are presented. E.g., BNA axioms are correct for graph isomorphism; xy-symocat with feedback axioms are correct for graph isomorphism with various constants; Elgot theory axioms are correct for deterministic input behaviour; Milner theory axioms are correct for nondeterministic input behaviour (bisimilar networks); Park theory axioms are correct for deterministic IO behaviour; Kleene theory axioms are correct for nondeterministic IO behaviour.

Uniformity: All the axiomatization results above are proved in an abstract setting, namely in the case the connecting wires are represented by morphisms in an appropriate abstract theory. Hence we have uniform proofs that may be used, for example, to the case of connections made by simple wiring relations, or by message passing channels, or by other flownomials. On the other hand, the proofs are ordered within a natural framework: from the simplest case of BNAs, to the most complicated case of Kleene theories.

Universality: In such an abstract setting the correctness problem consists of the preservation of the algebraic structure when one passes from connections to classes of equivalent networks. This problem is solved by theorems of the following type (in cases $b\delta$ and $d\delta$ the proofs are done under some mild additional conditions for the connecting theory T): If T is an xy-flow, then $[X,T]_{xy}$ is xy-flow; moreover, $[X,T]_{xy}$ forms the xy-flow freely generated by adding X to T.

If one replaces "flownomials" by "polynomials" and "xy-flow" by "ring," then one gets the classical universal property for polynomials. In particular, this result shows that one may use standard algebraic refinement methods.

Special networks: The calculus of flownomials is general enough to cover various types of networks, including flowchart schemes, automata, data-flow networks, Petri nets. (A process algebra version, re-cast in the timed-NA framework, may, perhaps, be included as well; in the current approach, process algebra has no "identities.")

Modularity: The approach is variable-free with respect to interface references. As a byproduct a high degree of modularity is achieved. One may freely shifts pieces of networks from one part to another.

1.2.6. *Mixed calculi*

Appropriate instances of the NA models may be found for control (flowcharts), space (data-flows) and time (processes). A very promising line for current research is to mix such models in order to have a unique calculus for all these features of the computing systems. This may lead towards an algebraic calculus for (concurrent, object-oriented) software systems. A few results are presented, including a challenging space-time duality thesis. Many interesting questions risen by this model are open.

The second part of this paper contains a detailed introduction to finite interactive system, which may be seen as a version of the Mixed Network Algebra where the mixture is made at the Kleene theory level.

2. Finite interactive systems

2.1. INTRODUCTION

This section gives a brief introduction to *finite interactive systems*, an abstract mathematical model of agents' behaviour and their interaction.

The agents we a talking about are those specified in standard *concurrent object-oriented languages (COOP)*. The key point is the observation that *planar words may be seen as interaction running patterns* for programs written in these COOP languages much in the same way as usual words/paths represent running paths of classical sequential programs. Then, finite interactive systems may be seen as a kind of two-dimensional finite automata melting together a state transforming automaton with a behaviour transforming one (this latter transformation is responsible for modeling the interaction of the threads generated by the first automaton).

2.2. PLANAR WORDS AND REGULAR EXPRESSIONS

2.2.1. *Planar words*

A *planar word* is a *rectangular* two-dimensional area filled in with atomic letters in a given alphabet V. Each letter in V is to be seen as a two-dimensional atom having its own *northern, southern, western* and *eastern* border. For the beginning we do not consider any typing mechanism, hence all atoms' borders are similar, denoted by 1; this border information is naturally extended to general planar words specifying the number of atoms laying on that border (rectangle's dimensions). For a planar word p, we let $n(p), s(p), w(p)$, and $e(p)$ denote the dimension of its northern, southern, western, and eastern border, respectively.

The term *picture* is often used as a substitute for two-dimensional or planar word, especially in the context of *picture languages*. Useful surveys on two-dimensional languages may be found in [20], [15], or [24].

It is known that the following are equivalent for a planar language L (called *recognizable two-dimensional language*):

1. L is recognized by a *on-line tessellation automaton*;
2. L is defined by a *simple regular expression with intersection and adding homomorphisms*;
3. L is defined by a *local lattice language plus homomorphisms (tile system)*;
4. L is defined by an *existential monadic second order formula*.

The proof of this and many other informations may be found in [16], [15], [25], etc.[4]

However, the situation is more complex in this two-dimensional case. For instance, deterministic, nondeterministic, and alternating (4-way) finite automata have distinct increasing power, and the class of languages recognized by the last one is incomparable with the above class of recognizable two-dimensional language; see [22].

2.2.2. Simple regular expressions

We present here a simple extension of classical regular expressions [23, 11] to cope with planar words.

Simple two-dimensional regular expressions. The *signature* of simple two-dimensional regular expressions consists of two collections of usual regular operators, sharing the additive part. Specifically, it is

$$+, \ 0, \ \cdot, \ ^\star, \ |, \ \triangleright, \ ^\dagger, \ -$$

where $(+, 0, \cdot, ^\star, |)$ is a usual Kleene signature to be used for the vertical dimension, while $(+, 0, \triangleright, ^\dagger, -)$ is another Kleene signature to be used now for the horizontal dimension. Notice that, "0", "|" and "−" are zero-ary operators (constants), "*" and "†" are unary, while the remaining ones are binary.

Our default rule is that if no dimension is specified, that the vertical one is considered. Hence, usually written Kleene operators refer to the vertical axis: "$\cdot$" is *(vertical) composition*, "*" is *iterated (vertical) composition or (vertical) star* and "|" is *(vertical) identity*. Similarly, "$\triangleright$" is *horizontal composition*, "†" is *iterated horizontal composition or horizontal star* and "−" is *horizontal identity*.

Simple two-dimensional regular expressions over an alphabet V (consisting of two-dimensional letters/atoms) are obtained starting from atoms

---

[4] Notice that regular expressions and tile systems do not directly cover the class of recognizable languages, but reach this using homomorphisms. This may be seen as an inconvenient of these mechanisms. On the other hand, this problem does not occur for on-line tessellation automata or for our interactive systems, as they are closed to homomorphisms.

and iteratively applying the operations in the described signature. Formally,

$$E ::= a \mid 0 \mid E + E \mid E \cdot E \mid E^\star \mid \shortmid \mid E \triangleright E \mid E^\dagger \mid -$$

Their set is denoted by 2RegExp(V).

From expressions to planar words. To each expression in 2RegExp(V) one may associate a language of planar words over V. To this end, we first describe how composition operations act on planar words and give the meaning of the identity operators.

If v, w are planar words, then their horizontal composition is defined only if $e(v) = w(w)$ and the result $v \triangleright w$ is the word obtained putting together v on the left and w on the right. Their vertical composition is defined only if $s(v) = n(w)$ and $v \cdot w$ is the word obtained putting v on top of w.

For each natural number k one may associate two "empty" planar words: the vertical identity ϵ_k having $w(\epsilon_k) = e(\epsilon_k) = 0$ and $n(\epsilon_k) = s(\epsilon_k) = k$ and the horizontal identity λ_k with $w(\lambda_k) = e(\lambda_k) = k$ and $n(\lambda_k) = s(\lambda_k) = 0$.

Now, the interpretation

$$\mid \ \mid \ : \ \text{2RegExp}(V) \quad \to \quad \text{LangPlanarWords}(V)$$

from expressions to languages of planar words is defined by:

- $|a| = \{a\}$; $|0| = \emptyset$; $|E + F| = |E| \cup |F|$;
- $|E \cdot F| = \{v \cdot w : v \in |E| \ \& \ w \in |F|\}$;
- $|E^\star| = \{v_1 \cdot \ldots \cdot v_k : k \in \mathbb{N} \ \& \ v_1, \ldots, v_k \in |E|\}$;
- $|\shortmid| = \{\epsilon_0, \ldots, \epsilon_k, \ldots\}$;
- $|E \triangleright F| = \{v \triangleright w : v \in |E| \ \& \ w \in |F|\}$;
- $|E^\dagger| = \{v_1 \triangleright \ldots \triangleright v_k : k \in \mathbb{N} \ \& \ v_1, \ldots, v_k \in |E|\}$;
- $|-| = \{\lambda_0, \ldots, \lambda_k, \ldots\}$.

Before giving some examples, we define a useful flattening operator mapping languages of planar words to languages of usual, one-dimensional words.

The flattening operator. The *flattening operator*

$$\flat \ : \ \text{LangPlanarWords}(V) \quad \to \quad \text{LangWords}(V)$$

maps sets of planar words to sets of strings representing their topological sorting. In more details it is defined as follows:

- Each planar word may be considered as an acyclic directed graph drawing: (1) horizontal edges from each letter to its right neighbor, if this exists, and (2) vertical edges from each letter to its bottom neighbor, if this exists.

- Being acyclic, the starting graph associated to a planar word and all of its subgraphs have *minimal atoms/vertices*, i.e., vertices without incoming arrows.
- The topological sorting procedure selects a minimal vertex, then deletes it and its outgoing edges and repeats this as long as possible. This way a sequence (i.e., a usual one-dimensional word) containing the atoms of the planar word is obtained.
- Varying the minimal vertex selection in the topological sorting procedure in all possible ways one gets a set of words that is the value of the flattening operator applied to the planar word.
- Finally, to define $\flat$ on a language L, take the union of $\flat(w)$, for $w \in L$.

To have an example, let us start with a planar word $\begin{smallmatrix}\text{abcd}\\\text{efgh}\end{smallmatrix}$; there is only one minimal element a and after its deletion we get $\begin{smallmatrix}\text{bcd}\\\text{efgh}\end{smallmatrix}$; now there are two minimal elements b and e to choose from and suppose we choose b; what remains is $\begin{smallmatrix}\text{cd}\\\text{efgh}\end{smallmatrix}$; next, from the minimal elements c and e, we choose e and get $\begin{smallmatrix}\text{cd}\\\text{fgh}\end{smallmatrix}$; and so on; finally a usual word, say abecfgdh, is obtained. Actually,

$$\flat(\begin{smallmatrix}\text{abcd}\\\text{efgh}\end{smallmatrix}) = \{\text{abcdefgh, abcedfgh, abcefdgh, abcefgdh, abecdfgh,}$$
$$\text{abecfdgh, abecfgdh, abefcdgh, abefcgdh, aebcdfgh,}$$
$$\text{aebcfdgh, aebcfgdh, aebfcdgh, aebfcgdh}\}$$

It is one of our main claims that this

> *flattening operator is responsible for the well-known state-explosion problem which occurs in the verification of concurrent object-oriented systems*

And one of our main hopes is that

> *the lifting of the verification techniques from usual words/paths to the planar version may avoid this problem*

Convention: we use terminal type letters a, b, ... in planar words; they should be identified with the corresponding italic version, used elsewhere.

Examples. 1. The expression $(a \cdot d^* \cdot g) \triangleright (b \cdot e^* \cdot h)^\dagger \triangleright (c \cdot f^* \cdot i)$ represents the language of planar words $\begin{smallmatrix}\text{a b . . b c}\\\text{d e . . e f}\\\text{. . . . . .}\\\text{. . . . . .}\\\text{d e . . e f}\\\text{g h . . h i}\end{smallmatrix}$. Notice that the same language may be represented by the expression $(a \triangleright b^\dagger \triangleright c) \cdot (d \triangleright e^\dagger \triangleright f)^* \cdot (g \triangleright h^\dagger \triangleright i)$.

2. The language associated to $a^\dagger \cdot b^\dagger$ consists of planar words $\begin{smallmatrix} \text{aa}\cdots\text{a} \\ \text{bb}\cdots\text{b} \end{smallmatrix}$. Its flattened version $\flat(a^\dagger \cdot b^\dagger) = \{w \in \{a,b\}^* : |w|_a = |w|_b \ \& \ \forall w = w'w'' : |w'|_a \geq |w'|_b\}$ is context-free, but not regular ($|w|_a$ denotes the number of the occurrences of a in w).

The slightly extended expression $a^\dagger \cdot b^\dagger \cdot c^\dagger$ represents the language of planar words $\begin{smallmatrix} \text{aa}\cdots\text{a} \\ \text{bb}\cdots\text{b} \\ \text{cc}\cdots\text{c} \end{smallmatrix}$. Notice that its flattened version is $\flat(a^\dagger \cdot b^\dagger \cdot c^\dagger) = \{w \in \{a,b,c\}^* : |w|_a = |w|_b = |w|_c \ \& \ \forall \ w = w'w'' : |w'|_a \geq |w'|_b \geq |w'|_c\}$. This latter language is even not context-free.

3. Notice that $(a+b)^{*\dagger} = (a+b)^{\dagger*}$. This shows that vertical and horizontal stars may commute, providing simple atoms are involved.

On the other hand, $\begin{smallmatrix} \text{aa} \\ \text{bb} \end{smallmatrix} \in (a + b \triangleright b)^{\dagger*} \setminus (a + b \triangleright b)^{*\dagger}$, showing that, in general, the stars do not commute.

4. Finally, notice that $a^\dagger \cdot b^\dagger \neq a^* \triangleright b^*$, but $\flat(a^\dagger \cdot b^\dagger) = \flat(a^* \triangleright b^*)$. This example shows that the flattening mechanism may loose some information when passing from planar to usual languages.

2.2.3. *Extended regular expressions*
More powerful iteration The regular expressions previously described are simple and natural, but they have a number of shortcomings. For instance, their generative power is quite limited, comparing with finite interactive systems. In this subsection we roughly describe a few possible extensions based on a more powerful iteration mechanism.

A first possibility is to use an alternating horizontal/vertical concatenation within the iteration process. The separate vertical and horizontal iteration operators are not strong enough to represent the languages L_{sq} consisting of squares of a's. But, starting with a and iterating a horizontal concatenation on the right with a vertical string a^* and then a vertical concatenation on the bottom with a horizontal string $a^\dagger \triangleright a$ one precisely gets the language of squares of a's. In other words, L_{sq} is the least solution of the equation $X = a + (X \triangleright a^*) \cdot (a^\dagger \triangleright a)$. (This type of iteration is studied in [27].)

Another example is given by the language L_{spir} of spiral words

```
x      2aa     2aaaa    ...
       2x1     22aa1
       bb1     22x11
               2bb11
               bbbb1
```

A corresponding equation is $X = x + (2 \triangleright a^\dagger) \cdot (2^\star \triangleright X \triangleright 1^\star) \cdot (b^\dagger \triangleright 1)$. While in the above example only concatenation at the right and the bottom parts were used within the iteration process, in the present case concatenation on all sites were used within the iteration process.

Space-time duality operator Space-time duality operator " $^\$$ " maps a planar word to its symmetric over the main diagonal.

The signature of Kleene algebra with space-time duality is:

$$(a, +, 0, \cdot, {}^\star, \mathsf{I}, {}^\$)$$

With space-time duality operator the horizontal operators are defined in terms of the vertical ones by:

$$
\begin{aligned}
v \triangleright w &= (v^\$ \cdot w^\$)^\$ \\
v^\dagger &= (v^{\$\star})^\$ \\
{-} &= \mathsf{I}^\$
\end{aligned}
$$

Using a parallel with boolean algebra case, the former regular expressions are somehow similar to lattice expressions, while with this space-time duality operator they become more similar to boolean expressions.

More constants One may depart from the restriction to have only rectangular planar words by allowing some new constants, i.e., "⊤", "⊥", "⊣", or "⊢". Their meaning may be obvious: they allow to initiate atoms from nothing ("⊤" and "⊢") or to block their continuation ("⊥" and "⊣"). One may be even more flexible and add a two-dimensional identity (crossing) "+", or even constants "⌐", " ⌐", " ∟", or "⌐ ". These latter constants may be used to switch data from time to space and vice-versa.

2.3. FINITE INTERACTIVE SYSTEMS

A finite interactive system may be seen as a kind of two-dimensional automaton mixing together a state-transforming machinery with a machinery used for the interaction of different threads of the first automaton.

While this way of presentation may be somehow useful, it is also misleading.

The first point is that there are not two different automata to be combined, but these two views are melted together to give this new concept of finite interactive systems.

A next critics is on using a term as 'automaton' for this device. While there may be some other interpretations, an automaton is generally considered as a state-transforming device. A state is a temporal section of a running system: it gives the information related to the current values of the involved variables at a given temporal point. This information is used to compose the systems, to get more complex behaviours out of simpler ones. Then, using a term as 'two-dimensional automaton' is more or less as considering automata, but with a more complicate, say two-dimensional, state structure.

The situation with finite interactive systems is by far different: one (currently, the vertical) dimension is used to model this state transformation, but the other dimension is considered to be a behaviour transforming device. Orthogonal to the previous case of states, a behaviour gives a spatial section of the running system: it consider information on all the actions a spatially located agent has made during its evolution. This information is used at the temporal interfaces by which the agents interact, to get more complicate intelligent systems out of simpler ones. An agent is considered as a job-transforming device: given a job-request at its input temporal interface, the agent acts, transforms it, and passes the transformed job-result at its output temporal interface.

One feature which is implicitly present in this view is that the systems we are talking about are open: to run such a system one has to give an appropriate initial state, but also an initial job-request, actually a place where the starting user interacts with the system. The result of the running is a pair consisting of a final state of the system and a job-result (or a stream of messages) the system passes to an output user.

2.3.1. *Finite interactive systems*
After the above general considerations let us give a precise definition of finite interactive systems and explain it on a concrete example.

Finite interactive systems. A *finite interactive system* is a finite hypergraph with two types of vertices and one type of (hyper) edges:
— the first type of vertices is distinguished using a labeling with numbers (or lower case letters); such a vertex denotes a *state* (memory of variables);
— the second type of vertices is distinguished using a labeling with capital letters; such a vertex is considered to be a *class* (of job requests);
— the actions/transitions are labeled by letters denoting atoms of planar words and are such that: (1) each transition has two incoming arrows: one from a class vertex, the other from a state vertex, and (2) each transition has two outgoing arrows: one to a class vertex, the other to a state vertex.

Some classes/states may be *initial* (in the graphical representation this situation is specified by using small incoming arrows) or *final* (in this case the double circle representation is used).

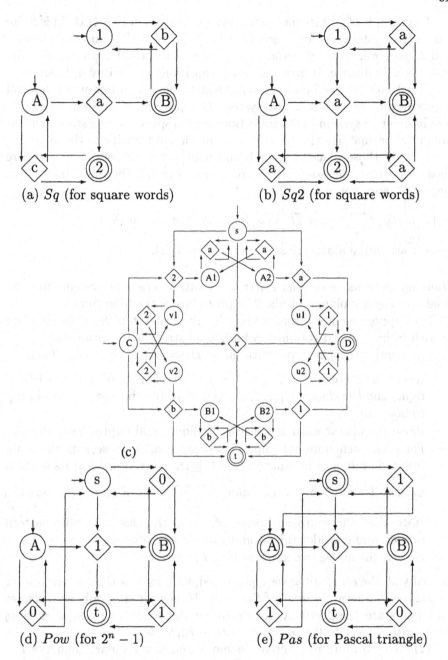

Figure 1.   Examples of finite interactive systems

An example of a finite interactive system is given in Fig. 1(a). This finite interactive system has four vertices: two classes A and B and two states 1 and 2; moreover, A is an initial class, B is a final class, 1 is an initial state, and 2 is a final state. It also has three transitions labeled by a, b, and c.

Sometimes it may be more convenient to have a simple non-graphical representation of an interactive system. One possibility is to represent a finite interactive system by its transitions and to specify which states/classes are initial or final. (In this 'crux' representation of transitions the incoming vertices are those placed at north and west, while the outgoing ones are those from east and south.) For the given example, the finite interactive system is represented by:

$$-\ \begin{array}{c} 1 \\ \boxed{A\ |\ a\ |\ B} \\ 2 \end{array},\quad \begin{array}{c} 2 \\ \boxed{A\ |\ c\ |\ A} \\ 2 \end{array},\ \text{and}\ \begin{array}{c} 1 \\ \boxed{B\ |\ b\ |\ B} \\ 1 \end{array}\ \text{and the information that}$$

$-$ $A, 1$ are initial states/classes and $B, 2$ are final.

Running patterns, accepting criteria. Finite interactive systems may be used to *recognize* planar words. The procedure is the following:

(1) Suppose we are given: a finite interactive system S; a planar word w with m lines and n columns; a horizontal string with initial states $b_n = s_1 \ldots s_n$; and a vertical string with initial classes $b_w = C_1 \ldots C_m$. Then:

- Insert the states $s_1, \ldots, s_n$ at the northern border of w (from left to right) and the classes $C_1, \ldots, C_m$ at the western border of w (from top to bottom).
- Parse the planar word w selecting minimal[5] still unprocessed atoms.
- For each such minimal unprocessed atom a, suppose s is the state inserted at its northern border and C is the class inserted at its western border. Then, choose a transition $\begin{array}{c} s \\ \boxed{C\ |\ a\ |\ C'} \\ s' \end{array}$ from S (if any), insert the state s' at the southern border of a and the class C' at its eastern border, and consider this atom to be already processed.
- Repeat the above two steps as long as possible.

If finally all the atoms of w were processed, then look at the eastern border b_e and the southern border b_s of the result. If they contains only final classes and final states, respectively, then consider this to be a *successful running for w with respect to the border conditions* b_n, b_w, b_e, b_s.

(2) Given a finite interactive system S and four regular languages B_n, B_w, B_e, and B_s for the border conditions, a planar word w is *recognized by*

---

[5] The notion of minimal atoms we are using here is the one described when the flattening operator has been introduced.

S with respect to the border conditions B_n, B_w, B_e, B_s if there exists border strings $b_n \in B_n$, $b_w \in B_w$, $b_e \in B_e$, $b_s \in B_s$ and a successful running for w with respect to the border conditions b_n, b_w, b_e, b_s; let us denote by $L(S; B_n, B_w, B_e, B_s)$ their set.

(3) Finally, a language of planar words L is *recognized by* S if there are some regular languages B_n, B_w, B_e, B_s for the border conditions such that $L = L(S; B_n, B_w, B_e, B_s)$.

(4) We simply denote the associated language by $L = L(S)$ when the border conditions are superfluous, i.e., the border languages are the corresponding full languages and actually no border condition is imposed.

Example. A concrete example may be useful. Let us look at the finite interactive system Sq in Fig. 1(a) and suppose that the border conditions are irrelevant, i.e., B_n, B_w, B_e and B_s are the full languages $\{1\}^*$, $\{A\}^*$, $\{B\}^*$ and $\{2\}^*$, respectively. A running is:

```
  abb       1 1 1       1 1 1       1 1 1       1 1 1     ...      1 1 1
  cab       Aa b b      AaBb b      AaBbBb      AaBbBb             AaBbBbB
  cca                     2           2 1         2 1              2 1 1

            Ac  a b     Ac  a b     Ac  a b     AcAa  b            AcAaBbB
                                                  2                2 2 1

            Ac  c a     Ac  c a     Ac  c a     Ac  c a            AcAcAaB
                                                                   2 2 2

   w          w₀          w₁          w₂          w₃                w₉
```

The above sequence starts with a planar word w. (1) The word obtaining bordering w with initial states/classes is w_0. (2) At this stage all atoms are unprocessed, but only one is minimal, i.e., the a in the left-upper corner.

In S there is only one transition of the type $\boxed{\begin{array}{c} 1 \\ A\,a \end{array}}$, i.e., $\boxed{\begin{array}{c} 1 \\ A\,a\,B \\ 2 \end{array}}$. After its application one gets w_1. (3) Now there are two minimal unprocessed elements: b (position $(1,2)$ in w) and c (position $(2,1)$ in w). Suppose we choose b. Again, there is only one appropriate transition $\boxed{\begin{array}{c} 1 \\ B\,b\,B \\ 1 \end{array}}$. After its application one gets w_2. (4) Next, the minimal unprocessed elements are b (position $(1,3)$ in w) and c (position $(2,1)$ in w). Suppose we choose c and the corresponding unique transition $\boxed{\begin{array}{c} 2 \\ A\,c\,A \\ 2 \end{array}}$. Then we get w_3. (5) We can continue and finally an accepting running is obtained, leading to w_9.

One may easily see that the recognized language $L(S)$ consists of *square words with a's on the main diagonal, the top-right half filled in with b's and the bottom-left half filled in with c's.*

Finally, one may look at some border conditions. There is only one initial/final state/class, hence the border languages are regular languages over a one letter alphabet. To have an example, if $B_n = (111)^*$, $B_w = (AAAA)^*$, $B_e = B^*$ and $B_s = (22)^*$, then $L(S; B_n, B_w, B_e, B_s)$ is the sub-language of the above $L(S)$ consisting of those squares whose dimension k is a multiple of 12.

State projection and class projection. Some familiar nondeterministic finite automata (nfa's) may be obtained from finite interactive systems neglecting one dimension.

The automaton obtained from a finite interactive system S neglecting the class transforming part is call the *state projection nfa* associated to S and it is denoted by $s(S)$. In UML this is known as the behaviour/statechart diagram.

Similarly, the *class projection nfa*, denoted $c(S)$, is the nfa obtained neglecting the state transforming part of S. UML use a similar class/interaction diagram name for this view of the system.

The class projection nfa $c(S)$ may be used to define *macro-step transitions* in the original finite interactive system S. Dually, the state projection nfa $s(S)$ may be used to define *agent behaviours* (seen as job class transformers) in the original finite interactive system.

Let us return to the finite interactive system of Fig. 1(a). Its class projection nfa $c(S)$ is given by the transitions $A \xrightarrow{a} B$, $A \xrightarrow{c} A$, $B \xrightarrow{b} B$ with A initial and B final. The state projection nfa $s(S)$ is defined by transitions $1 \xrightarrow{a} 2$, $2 \xrightarrow{c} 2$, $1 \xrightarrow{b} 1$ with 1 initial and 2 final.

This particular finite interactive system has an interesting *intersection property*:

$$L(S) = L(s(S)) \cap L(c(S))$$

showing that, in some sense, the interaction between the state and the class transformations of S is not so strong. [6]

---

[6] To avoid some confusions, we emphasize here that the usual languages associated to the projection nfa's are extended to the planar case. For instance, a planar word w is in $L(s(S))$ if every column in w is recognized by $s(S)$ in the usual sense. Similarly for $L(c(S))$, but now the lines are taken into account.

2.3.2. Examples

1. Square words. Let us denote by L_{sq} the language consisting of square words of a's, namely

a	aa	aaa	aaaa	. . .
	aa	aaa	aaaa	
		aaa	aaaa	
			aaaa	

An example of a finite interactive system recognizing L_{sq} is Sq illustrated in Fig. 1(a), where b and c are to be replaced by a.[7] Two examples of running patterns for Sq are shown below (the first one is successful; the second one is unsuccessful, as the southern border contains an occurrence of the non-final state 1):

$$
\begin{array}{ccc}
1 & 1 & 1 \\
A_aB_a&B_a&B \\
2 & 1 & 1 \\
A_aA_a&B_a&B \\
2 & 2 & 1 \\
A_aA_a&A_a&B \\
2 & 2 & 2 \\
\end{array}
\qquad\qquad
\begin{array}{cccc}
1 & 1 & 1 & 1 \\
A_aB_a&B_a&B_a&B \\
2 & 1 & 1 & 1 \\
A_aA_a&B_a&B_a&B \\
2 & 2 & 1 & 1 \\
A_aA_a&A_a&B_a&B \\
2 & 2 & 2 & 1 \\
\end{array}
$$

2. Exponential function.

Another finite interactive system is Pow, drawn in Fig. 1(d). It recognizes the language

1	10	100	. . .
	01	010	
	11	110	
		001	
		101	
		011	
		111	

(a running)

$$
\begin{array}{l}
A\overset{s}{1}B\overset{s}{0}B\overset{s}{0}B \\
A\overset{t}{0}A\overset{s}{1}B\overset{s}{0}B \\
A\overset{s}{1}B\overset{t}{1}B\overset{s}{0}B \\
A\overset{t}{0}A\overset{t}{0}A\overset{s}{1}B \\
A\overset{s}{1}B\overset{s}{0}B\overset{t}{1}B \\
A\overset{t}{0}A\overset{s}{1}B\overset{t}{1}B \\
A\overset{s}{1}B\overset{t}{1}B\overset{t}{1}B \\
\overset{t}{}\overset{t}{}\overset{t}{}
\end{array}
$$

3. Pascal triangle.

(a running)

$$
\begin{array}{l}
A\overset{s}{0}B\overset{s}{1}A\overset{s}{0}B\overset{s}{1}A\overset{s}{0}B\overset{s}{1}A \\
A\overset{t}{1}A\overset{s}{0}B\overset{t}{0}B\overset{s}{1}A\overset{t}{1}A\overset{s}{0}B \\
A\overset{s}{0}B\overset{t}{0}B\overset{s}{0}B\overset{t}{1}A\overset{s}{1}A\overset{t}{0}B \\
A\overset{t}{1}A\overset{t}{1}A\overset{t}{1}A\overset{s}{0}B\overset{t}{0}B\overset{t}{0}B \\
A\overset{s}{0}B\overset{s}{1}A\overset{s}{0}B\overset{t}{0}B\overset{t}{0}B\overset{t}{0}B \\
A\overset{t}{1}A\overset{s}{0}B\overset{t}{0}B\overset{t}{0}B\overset{t}{0}B\overset{t}{0}B \\
\overset{s}{}\overset{t}{}\overset{t}{}\overset{t}{}\overset{t}{}\overset{t}{}
\end{array}
$$

The new finite interactive system Pas in Fig. 1(e) recognizes the set of words over $\{0, 1\}$ whose northern and western borders have alternating sequences of 0 and 1 (starting with 0) and, along the secondary diagonals, it satisfies the recurrence rule of the Pascal triangle, modulo 2 (that is, the value in a cell (i, j) is the sum of the values of the cells $(i-1, j)$ and $(i, j-1)$, modulo 2).

In this case, one can see that the projection automata (on states and classes, respectively) are reset automata having all states/classes final,

---

[7] The same language is recognized by the finite interactive system $Sq2$, drawn in the same figure.

hence they recognizes the corresponding full languages. Their intersection is now the full language of two-dimensional words, very far from the language recognized by *Pas*. This may be interpreted as a strong interaction between the state transformation and the class transformation within *Pas*. It also shows that the 'intersection' property is not generally valid for the full class of finite interactive systems.

4. Spiral words. The finite interactive system drawn in Fig. 1(c) recognizes the language of 'spiral' words

	x	2aa	2aaaa	. . . .
		2x1	22aa1	
		bb1	22x11	
			2bb11	
			bbbb1	

2.3.3. *Finite interaction system languages*

Local Lattice Languages (LLL's, for short), are specified using a finite set of finite (rectangular) words that are allowed as sub-words of the recognized planar words. If one combines this with (letter-to-letter) homomorphisms, than the power of the mechanism does not change by restricting to 2×2 pieces, or even to horizontal and vertical domino-s (i.e., 1×2 and 2×1 pieces). These are also called *tile systems*.

It is easy to show that (1) one may simulate a LLL by a finite interactive system and (2) the class of languages recognized by finite interactive systems is closed under letter-to-letter homomorphisms. From this simple observations it follows that the class of languages recognized by finite interactive systems contains the class of recognizable planar languages. The converse inclusion is also valid: (1) one may see that the runnings of a finite interaction system A, completed with a special letter for the blank space, are locally testable, hence form a LLL; then (2) using an homomorphism one may abstract from the state/class/blank information around the letters, to go to $L(A)$, hence $L(A)$ is a recognizable picture language.

To conclude, finite interactive systems give a different representation for recognizable picture languages, adding more arguments that this class of two-dimensional languages is a robust one.

From this pleasant but unexpected connection one inherits many results known for picture languages. One important fact is the undecidability of the emptiness problem for finite interactive system.[8]

---

[8] This follows from the known fact that if one restrict himself to the top line of the pictures in a recognizable picture language, then one gets precisely context-sensitive (string) languages.

2.4. COOP PSEUDO-CODE PROGRAMS

The finite interactive system in Fig. 1(a) may be used to describe, in pseudo-code, an algorithm for solving the *8-queen problem*. The problem is to place queens on a chess table such that no queen may attack another queen. For convenience, we replace the transition labels from a, b, and c to av (advance and place queen), r (report on a successful attempt) and c (curry on the placement information), respectively.

The interpretation is as follows (from this interpretation one also gets the information regarding the data structures used on the atoms' borders):[9]

r: if $(w = ok$ and $(n = ok$ or $n = new))$,
       then $(s = ok$ and $e = ok)$,
       otherwise $(s = no$ and $e = no)$;
c: if $(n = pos_k$ and $w =< pos_1, \ldots, pos_{k-1} >)$,
       then $(s = pos_k$ and $e =< pos_1, \ldots, pos_{k-1}, pos_k >)$;
av: if $(w =< pos_1, \ldots, pos_k >$ and $(n = ok$ or $n = new)$
           and $free(pos_1, \ldots, pos_k) \neq \emptyset)$,
       then (select $p_{k+1} \in free(pos_1, \ldots, pos_k)$ and make
           $s = pos_{k+1}$ and $e = ok$),
       otherwise $(s = random$ and $e = no)$;

Two running patterns are given below

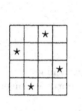

$$
\begin{array}{llllllll}
& new & & new & & new & & new \\
<>\boxed{av}\ OK & \boxed{r} & OK & \boxed{r} & OK & \boxed{r}\ OK \\
\quad 2 & & OK & & OK & OK \\
<>\boxed{c}<2>\boxed{av} & OK & \boxed{r} & OK & \boxed{r}\ OK \\
\quad 2 & 4 & & OK & & OK \\
<>\boxed{c}<2>\boxed{c}<2,4>\boxed{av} & OK & \boxed{r}\ OK \\
\quad 2 & 4 & 1 & & OK \\
<>\boxed{c}<2>\boxed{c}<2,4>\boxed{c}<2,4,1>\boxed{av}\,OK \\
\quad 2 & 4 & 1 & 3
\end{array}
$$

and

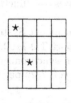

$$
\begin{array}{llllllll}
& new & & new & & new & & new \\
<>\boxed{av}\ OK & \boxed{r} & OK & \boxed{r} & OK & \boxed{r}\ OK \\
\quad 1 & & OK & & OK & OK \\
<>\boxed{c}<1>\boxed{av} & OK & \boxed{r} & OK & \boxed{r}\ OK \\
\quad 1 & 3 & & OK & & OK \\
<>\boxed{c}<1>\boxed{c}<1,3>\boxed{av} & NO & \boxed{r}\ NO \\
\quad 1 & 3 & ? & & NO \\
<>\boxed{c}<1>\boxed{c}<1,3>\boxed{c}<1,3,?>\boxed{av}\,NO \\
\quad 1 & 3 & ? & ??
\end{array}
$$

---

[9] Notice that in order to have defined the *free* predicate, one need to know the chess table dimension.

2.5. CLOSING COMMENTS

The concept of finite interactive systems presented here has evolved from a study of Mixed Network Algebras - see, e.g., [34], [18], [17], [36], or the final part of the recent book [35]. [10]

Our finite interactive systems may be seen as an extension of existing models like finite transition systems [1], Petri nets [14], data-flow networks [5], or asynchronous automata [37], in the sense that a potentially unbound number of processes may interact. In the mentioned models, while the number of processes may be unbounded, their interaction is bounded by the transitions' breadth. This seems to be a key feature for the ability to pass from concurrent to concurrent, object-oriented agents.

Our model has some similarities with the tile model studied by Montanari and his Pisa group (see, e.g., [6], [7], or [13]), but the precise relationship still has to be investigated. We also expect that a logic similar to the spatial logic of [8] may be associated to finite interactive systems and use for the verification of concurrent, object-oriented systems.

Acknowledgment I acknowledge with thanks many useful discussions I had on various mixed network algebra models (including the present finite interaction system model) with various people including M. Broy, C. Dima, R. Grosu, Y. Kawahara, U. Montanari, D. Lucanu, S. Merz, R. Soricut, A. Stefanescu, B. Warinschi, and M. Wirsing.

References

1. Arnold, A. (1994). *Finite transition systems.* Prentice-Hall International.
2. Baeten, J.C.M. and Weijland, W.P. (1990). *Process Algebra.* Cambridge University Press.
3. Bloom, S.L. and Esik, Z. (1993). *Iteration Theories: The Equational Logic of Iterative Processes.* Springer-Verlag.
4. Booch, G., Rumbaugh, J., anf Jacobson, I. (1999) *The Unified Modeling Language User Guide.* Addison Wesley.
5. Broy, M. and Stefanescu, G. (2001). The algebra of stream processing functions. *Theoretical Computer Science*, 258, 95–125.
6. Bruni, R. (1999). Tile Logic for Synchronized Rewriting of Concurrent Systems. PhD thesis, Dipartimento di Informatica, Universita di Pisa. Report TD-1/99.
7. Bruni, R. and Montanari, U. (1997). Zero-safe nets, or transition synchronization made simple. *Electronic Notes in Theoretical Computer Science*, vol. 7(20 pages).

---

[10] In [36] a preliminary more complicated version of the current model was presented. It uses two-dimensional queues associated with states/classes in order to model more complicated running scenarios for concurrent object-oriented systems. The present model of finite interactive systems is simpler, but still raises many difficult mathematical problems.

8. Cardeli, L. and Gordon, A. (2000). Anytime, anywhere: modal logics for mobile ambients. In: *POPL-2000, Symposium on Principles of Programming Languages*, Boston, 2000. ACM Press.

9. Cazanescu, V.E. and Stefanescu, G. (1990). Towards a new algebraic foundation of flowchart scheme theory. *Fundamenta Informaticae*, 13, 171–210.

10. Cazanescu, V.E. and Stefanescu, G. (1994). Feedback, iteration and repetition. In Gh. Păun, editor, *Mathematical aspects of natural and formal languages*, pages 43–62. World Scientific, Singapore.

11. Conway, J.H. (1971). *Regular Algebra and Finite Machines*. Chapman and Hall.

12. Elgot, C.C. (1975). Monadic computation and iterative algebraic theories. In H.E. Rose and J.C. Sheperdson, editors, *Proceedings of Logic Colloquium '73*, volume 80 of *Studies in Logic and the Foundations of Mathematics*, pages 175–230. North-Holland.

13. Gadducci, F. and Montanari, U. (1996). The tile model. In *Papers dedicated to R. Milner festschrift*. The MIT Press, Cambridge, to appear, 1999. Also: Technical Report TR-96-27, Department of Computer Science, University of Pisa.

14. Garg, V. and Ragunath, M.T. (1992). Concurrent regular expressions and their relationship to Petri nets. *Theoretical Computer Science*, 96, 285–304.

15. Giammarresi, D. and Restivo A. (1997). Two-dimensional languages. In G. Rozenberg and A. Salomaa, editors, *Handbook of formal languages. Vol. 3: Beyond words*, pages 215–265. Springer-Verlag.

16. Giammarresi, D., Restivo, A., Seibert, S., and Thomas W. (1996). Monadic second order logic over rectangular pictures and recognizability by tiling systems. *Information and Computation*, 125, 32–45.

17. Grosu, R., Lucanu, D., and Stefanescu G. (2000). Mixed relations as enriched semiringal categories. *Journal of Universal Computer Science*, 6(1), 112–129.

18. Grosu, R., Stefanescu, G., and Broy, M. (1998). Visual formalism revised. In *Proceeding of the CSD'98 (International Conference on Application of Concurrency to System Design, March 23-26, 1998, Fukushima, Japan)*, pages 41–51. IEEE Computer Society Press.

19. Hennessy, M. (1988). *Algebraic theory of processes*. Foundations of Computing. The MIT Press, Cambridge.

20. Inoue, K. and Takanami, I. (1991). A survey of two-dimensional automata theory. *Information Sciences*, 55, 99–121.

21. Joyal, A., Street, R., and Verity, D. (1996). Traced monoidal categories. *Proceedings of the Cambridge Philosophical Society*, 119, 447–468.

22. Kari, J. and Moore, C. (2000). New results on alternating and non-deterministic two-dimensional finite-state automata. In: *Proceedings STACS 2001*, LNCS 2010, 396-406, Springer-Verlag.

23. Kleene, S.C. (1956). Representation of events in nerve nets and finite automata. In: C.E. Shannon and J. McCarthy, eds., *Automata Studies, Annals of Mathematical Studies*, vol. 34, 3–41. Princeton University Press.

24. Lindgren, K., Moore, C., and Nordahl, M. (1998). Complexity of two-dimensional patterns. *Journal of Statistical Physics* 91, 909–951.

25. Latteux, M. and Simplot, D. (1997). Recognizable picture languages and domino tiling. *Theoretical Computer Science* 178, 275–283.

26. Manes, E.G. and Arbib, M.A. (1986). *Algebraic approaches to program semantics*. Springer-Verlag.

78

27. Matz, O. (1997). *Regular expressions and context-free grammars for picture languages.* In: *Proceedings STACS'97*, LNCS 1200, 283–294, Springer-Verlag.
28. Meseguer, J. and Montanari, U. (1990). Petri nets are monoids. *Information and Computation*, 88, 105–155.
29. Milner R. (1989). *Communication and concurrency.* Prentice-Hall International.
30. Petri, C.A. (1962). Kommunikation mit Automaten. PhD thesis, Institute für Instrumentelle Mathematik, Bonn, Germany.
31. Stefanescu, G. (1987). On flowchart theories: Part I. The deterministic case. *Journal of Computer and System Sciences*, 35, 163–191.
32. Stefanescu, G. (1987). On flowchart theories: Part II. The nondeterministic case. *Theoretical Computer Science*, 52, 307–340.
33. Stefanescu, G. (1986). Feedback theories (a calculus for isomorphism classes of flowchart schemes). Preprint Series in Mathematics No. 24, National Institute for Scientific and Technical Creation, Bucharest. Also in: *Revue Roumaine de Mathematiques Pures et Applique*, 35, 73–79, 1990.
34. Stefanescu, G. (1998). Reaction and control I. Mixing additive and multiplicative network algebras. *Logic Journal of the IGPL*, 6(2), 349–368.
35. Stefanescu, G. (2000). *Network algebra.* Springer-Verlag.
36. Stefanescu, G. (2001). Kleene algebras of two dimensional words: A model for interactive systems. *Dagstuhl Seminar on "Applications of Kleene algebras"*, Seminar 01081.
37. Zielonka, W. (1987). Notes on finite asynchronous automata. *Theoretical Informatics and Applications*, 21, 99–135.

COMPUTABILITY AND COMPLEXITY
FROM A PROGRAMMING PERSPECTIVE

NEIL D. JONES*

DIKU, University of Copenhagen, Denmark

neil@diku.dk, http://www.diku.dk/people/NDJ.html

Abstract. A programming approach to computability and complexity theory yields proofs of central results that are sometimes more natural than the classical ones; and some new results as well. These notes contain some high points from the recent book [14], emphasising what is different or novel with respect to more traditional treatments. Topics include:

- Kleene's s-m-n theorem applied to compiling and compiler generation.
- Proof that *constant time factors do matter*: for a natural computation model, problems solvable in linear time have a proper hierarchy, ordered by coefficient values. (In contrast to the "linear speedup" property of Turing machines.)
- Results on which problems possess optimal algorithms, including Levin's Search theorem (for the first time in book form).
- Characterisations in programming terms of a wide range of complexity classes. These are intrinsic: without externally imposed space or time computation bounds.
- Boolean program problems complete for PTIME, NPTIME, PSPACE.

1. Introduction

Book [14] differs didactically and foundationally from traditional treatments of computability and complexity theory. Its didactic aim is to teach the main theorems of these fields to computer scientists. Its approach is to exploit as much as possible the readers' programming intuitions, and

---

* This research was partially supported by the Danish Natural Science Research Council (*DART* project), and the Esprit *Atlantique* project.

H. Schwichtenberg and R. Steinbrüggen (eds.), Proof and System-Reliability, 79–135.
© 2002 *Kluwer Academic Publishers. Printed in the Netherlands.*

to motivate subjects by their relevance to programming concepts. Foundationally it differs by using, instead of the natural numbers $I\!N$, binary trees as data (as in the programming language Lisp). Consequently programs *are* data, avoiding the usual complexities and inefficiencies of encoding program syntax as Gödel numbers.

2. The WHILE language

The simple programming language WHILE is in essence a small subset of LISP or Pascal. Why do we use just this language? Because WHILE seems to have just the right mix of expressive power and simplicity for our purposes. *Expressive power* is important when writing programs that deal with programs as data objects. The data structures of WHILE (binary trees of atomic symbols) are particularly well suited to this. The reason is the WHILE data structures avoid the need for the technically messy tasks of assigning *Gödel numbers* to encode program texts and fragments, and of devising numerical functions to build and decompose Gödel numbers.

Simplicity is also essential to prove theorems about programs and their behaviour. This rules out the use of larger, more powerful languages, since proofs about them would simply be too complex to be easily understood. When proving theorems about programs, we use an equivalent but still simpler language I: identical to WHILE, but limited to programs with one variable and one atom.

2.1. SYNTAX OF WHILE DATA AND PROGRAMS

Values in the set $I\!D_A$ of data values are built up from a fixed finite set A of so-called *atoms* by finitely many applications of the pairing operation. It will not matter too much exactly what A contains, except that we will identify one of its elements, nil for several purposes. To reduce notation we write $I\!D$ instead of $I\!D_{\{nil\}}$ when $A = \{nil\}$.

A value $d \in I\!D_A$ is a binary tree with atoms as leaf labels. An example, written in "fully parenthesised form": ((a.((b.nil).c)).nil).

DEFINITION 2.1.   Let $A = \{a_1, \dots, a_n\}$ be some finite set. Then

1. $I\!D_A$ is the smallest set satisfying $I\!D_A = (I\!D_A \times I\!D_A) \cup A$. The *pairing* or *cons* operation "." yields pair $(d_1.d_2)$ when applied to values d_1, d_2 $\in I\!D$.
2. The *size* $|d|$ of a value $d \in I\!D_A$ is defined as follows: $|d| = 1$ if $a \in A$, and $1 + |d_1| + |d_2|$ if $d = (d_1.d_2)$.

2.1.1. *A compact linear notation for values:*

Unfortunately it is hard to read deeply parenthesised structures (one has to resort to counting), so we will use a more compact "list notation" taken from the Lisp and Scheme languages, in which

$$() \qquad \text{stands for} \qquad \text{nil}$$
$$(d_1\ d_2\ \cdots\ d_n\ ) \qquad \text{stands for} \qquad (d_1.(d_2.\cdots(d_n.\text{nil})\cdots))$$

The syntax of WHILE programs is given by the "informal syntax" part of Figure 1. Programs have only one input/output variable X1, and to manipulate tree structures built by cons from atoms. Informally, "cons" is WHILE code for the pairing operation so $\text{cons}(d, e) = (d.e)$. "hd" and "cotlns" are its left and right inverses, so $\text{hd}(d.e) = d$ and $\text{tl}(d.e) = e$. For completeness, we define $\text{hd}(\text{nil}) = \text{tl}(\text{nil}) = \text{nil}$

Control structures are *sequential composition* C1;C2, the *conditional* if E then C1 else C2, and the *while loop* while E do C. Assignments use := in the usual way. Operations hd and tl (head, tail) decompose such structures, and atom=? tests atoms for equality. In tests, nil serves as "false," and any non-nil value serves as "true." We often write false for nil, and true for (nil.nil).

An example is the following program, reverse:

```
read X;
  Y := nil;
  while X do { Y := cons (hd X) Y;   X := tl X };
write Y
```

This satisfies $[\![\text{reverse}]\!](\text{a}.(\text{b}.(\text{c}.\text{nil}))) = (\text{c}.(\text{b}.(\text{a}.\text{nil})))$, or $[\![\text{reverse}]\!](\text{a b c}) = (\text{c b a})$.

2.1.2. *Concrete syntax for* WHILE-*programs.*

The column "Concrete syntax" in Figure 1 contains a representation of each WHILE-program as an element of ID_A. Notation: symbols

$$:=,\ \text{nil},\ ;,\ \text{if},\ \text{while},\ \text{var},\ \text{quote},\ \text{cons},\ \text{hd},\ \text{tl},\ \text{atom=?}$$

stand for distinct elements of ID (exactly which ones, is not important). Note that with this representation, programs *are* data; no encoding is needed.

82

Syntactic category:		Informal syntax:	Concrete syntax:
P : Program	::=	read X1; C; write X1	C
C : Command	::=	Xi := E	$(\underline{:=}$ nil^i E)
	|	C1; C2	$(\underline{;}$ C1 C2)
	|	if E then C1 else C2	$(\underline{if}$ E C1 C2)
	|	while E do C	$(\underline{while}$ E C)
E : Expression	::=	Xi	$(\underline{var}$ nil^i)
	|	D	$(\underline{quote}$ D)
	|	cons E1 E2	$(\underline{cons}$ E1 E2)
	|	hd E	$(\underline{hd}$ E)
	|	tl E	$(\underline{tl}$ E)
	|	atom=? E1 E2	$(\underline{atom=?}$ E1 E2)
D : Data-value	::=	A | (D.D)	A | (D.D)
A : Atom	::=	nil (| ...)	nil | ...

Figure 1. Program syntax: informal and concrete

2.2. SEMANTICS OF WHILE PROGRAMS

Informally, the net effect of *running a program* p is to compute a partial function $[\![p]\!] : D \to D_\perp$, where $D_\perp$ abbreviates $D \cup \{\perp\}$. We write $[\![p]\!](d) = \perp$ to mean "the computation by p on d does not terminate."

A formal definition of the semantic function $[\![p]\!]$ may be found in [14]. Its approach, here simplified to programs containing only one variable X, is to define two relations, $\mathcal{E} \subseteq$ Expression $\times D \times D$ and $\vdash \subseteq$ Command $\times D \times D$ with these interpretations:

$\mathcal{E}[\![E]\!]d = d'$ If the value of variable X is d, the value of expression E is d'

$C \vdash d \to d'$ If the initial value of X is d, then execution of command

C terminates, with d' as the final value of X

For brevity, running time is only informally defined. This definition is natural, provided the data-sharing implementation techniques used in Lisp and other functional languages is used.

DEFINITION 2.2.   The running time $time_p(d) \in \{0, 1, 2, ...\} \cup \{\perp\}$ is obtained by counting 1 every time any of the following is performed while computing $[\![p]\!](d)$ as defined in the semantics: a variable or constant reference; an operation hd, tl, cons, or :=; or a test in an if or while command. Its value is $\perp$ if the computation does not terminate.

2.3. DECISION PROBLEMS

DEFINITION 2.3.

1. A set $A \subseteq D$ is WHILE decidable (or *recursive*) iff there is a WHILE program p such that for all $d \in D$, $[\![p]\!](d) = $ true if $d \in A$ and $[\![p]\!](d) = $ false if $d \notin A$.
2. A set $A \subseteq D$ is WHILE semi-decidable (or *recursively enumerable*, abbreviated to r.e.) iff for some WHILE-program p

$$A = \text{domain}([\![p]\!]) = \{d \in D \mid [\![p]\!](d) \neq \perp\}$$

Thus a recursive set A's membership question can be answered by a terminating program which, for any input $d \in D$, will answer "true" if d lies in A and "false" otherwise. A recursively enumerable set's membership question is answered by a program p that, for any input $d \in D$, terminates when d lies in A and loops infinitely when it does not.

These concepts lie at the very core of computability theory. They have many alternative chacterisations; and are independent of the programming languages used to define them.

2.4. SOME SIMPLE CONSTRUCTIONS

The following examples illustrate the simplicity of WHILE-program manipulation. All use the concrete syntax of Figure 1.

2.4.1. *The halting problem*
Following is a well-known result: the *undecidability of the halting problem*:

THEOREM 2.4. The set $HALT$ is not recursive, where

$$HALT = \{(p.d) \mid [\![p]\!](d) \neq \perp\}$$

Proof Suppose program q decides $HALT$, so for any WHILE-program p and input d

$$[\![q]\!](p.d) = \begin{cases} \text{true} & \text{if } [\![p]\!](d) \neq \perp \\ \text{false} & \text{if } [\![p]\!](d) = \perp \end{cases}$$

Now q must have form: read X; C; write X. Construct the following program r from q:

```
read X;
X := cons X X;        (*  Does program X stop on input X?  *)
C;                    (*  Apply program q to answer this   *)
if X
    then  while X do X := X  (* Loop if X stops on input X *)
    else  X := nil;   (*  Terminate if it does not         *)
write X
```

Consider the input $X = r$. First, suppose that $[\![r]\!](r) \neq \bot$. Then control in r's computation on input r must reach the **else** branch above (else r would loop on r). But then $X = \texttt{false}$ holds after command C, so $[\![r]\!](r) = \bot$ by the assumption that q decides the halting problem. This is contradictory. Conclusion: $[\![r]\!](r) = \bot$. By similar reasoning, $[\![r]\!](r) = \bot$ is also impossible.

The only unjustified assumption was existence of q, so this must be false.

2.4.2. *Decidable and semi-decidable sets*

THEOREM 2.5.   If A and $\overline{A}$ are both recursively enumerable, then A is recursive.

Proof   idea: Let $A = \text{domain}([\![p]\!])$ and let $\overline{A} = \text{domain}([\![q]\!])$. *Decision procedure* for given input d: run p and q alternately, one step at a time until one stops (and one of them must). If p stops first then $d \in A$, else $d \notin A$. [See Section 4.2.2 for adequate construction details to do this.]

THEOREM 2.6.   The set $HALT$ is recursively enumerable but not recursive. Further, the set $\overline{HALT}$ is not recursively enumerable, where

$$\overline{HALT} = \{(\text{p.d}) \mid [\![p]\!](\text{d}) = \bot\}$$

Proof   First, note that $HALT = \text{domain}([\![u]\!])$, and so is recursively enumerable. Thus $\overline{HALT}$ cannot also be recursively enumerable, since Theorem 2.5 would imply that $HALT$ is also recursive.

2.4.3. *The s-m-n theorem*

A *program specialiser* spec is given a subject program p together with part of its input data, s. Its effect is to construct a new program $p_s = [\![\text{spec}]\!](\text{p.s})$ which, when given p's remaining input d, will yield the same result that p would have produced given both inputs.

DEFINITION 2.7.   Program spec is a *specialiser* for WHILE-programs if $[\![\text{spec}]\!](\_)$ is total, and if for any $p \in$ WHILE-*programs* and s, d $\in$ WHILE-*data*

$$[\![p]\!](\text{s.d}) = [\![[\![\text{spec}]\!](\text{p.s})]\!](\text{d})$$

THEOREM 2.8. (Kleene's *s-m-n* theorem). There exists a program specialiser[1] for WHILE-programs.

Proof Program p has form:

```
read X1; Body; write X1
```

Given known input s, construct program p_s:

```
read X1; X1 := cons s X1; Body; write X1
```

This is obviously (albeit somewhat trivially) correct since p_s, when given input d, will first assign the value (s.d) to X1, and then apply p to the result. It suffices to see how to construct $p_s = [\![spec]\!](p.s)$. A program to transform input s and the concrete syntax of program p into p_s is easily constructed.

3. Robustness of computability

This section briefly describes some computation models used later in these notes: the GOTO language as a variant of WHILE; the TM language for Turing machines; the CM language for counter machines (and the SRAM language for random access machines is briefly mentioned). Further, we show how to measure the running time and space consumption of their programs.

Robustness: all of these models define the same classes of computable functions and decidable sets. Section 3.5 states their equivalence, and proofs may be found in [14]. The reader may wish to skip quickly to Section 3.5 on first reading, returning to the details when or if needed for later sections.

3.1. PROGRAMMING LANGUAGES MORE GENERALLY

DEFINITION 3.1. A *programming language* L consists of

1. Two sets, L-*programs* and L-*data*, and two distinct elements true, false $\in$ L-*data*.
2. L's *semantic function*: for every p $\in$ L-*programs*, a (partial) input-output function $[\![p]\!]^L(\_) : L\text{-}data \to L\text{-}data_\perp$.

---

[1] Kleene's version: $\phi_p^{m+n}(x_1, \ldots, x_m, y_1, \ldots, y_n) = \phi_{s_n^m(p, x_1, \ldots, x_m)}^n(y_1, \ldots, y_n)$. Definition 2.7 is much simpler, partly because the linear notation $[\![p]\!](d)$ instead of $\phi_p^n(d)$ avoids subscripted subscripts. Another reason is the fact that D has a built-in pairing operation. Thus $m = n = 1$ is fully general, so s_n^m for all m, n are not necessary, being replaced by the one function $[\![spec]\!]$.

3. L's *running time function*: for every $p \in$ L-*programs*, a corresponding (partial) time-usage function $time_p^L(\_) :$ L-*data* $\rightarrow I\!N \cup \{\bot\}$ such that for any $p \in$ L-*programs* and $d \in$ L-*data*, $[\![p]\!]^L(d) = \bot$ if and only if $time_p^L(d) = \bot$

The superscript L will be dropped when L is known from context. In the following we alsways have L-*data* $= I\!D$ or L-*data* $= \{0, 1\}^*$.

DEFINITION 3.2. L-program p *decides* a subset $A \subseteq$ L-*data* if for any $d \in$ L-*data*

$$[\![p]\!](d) = \begin{cases} \text{true} & \text{if } d \in A \\ \text{false} & \text{if } d \in \text{L-}data \setminus A \end{cases}$$

3.2. THE GOTO VARIANT OF WHILE

The WHILE language has both "structured" syntax and data. This is convenient for programming, but when constructing one program from another it is often convenient to use a lower-level "flow chart" syntax in which a program is a sequence $p = 1{:}I1 \; 2{:}I2 \; \ldots \; m{:}Im$ of labeled instructions, executed serially except as redirected by control transfers. Instructions are limited to the forms X:=nil, X:=Y, X:=cons Y Z, X:=hd Y, X:=tl Y, or if X goto ℓ else ℓ'.

The semantics is natural and so not presented here. Following is a GOTO program to reverse its input string.[2]

```
0: read X;
1: if X goto 2 else 6
2: Z := hd X;
3: Y := cons Z Y;
4: X := tl X;
5: goto 1;
6: write Y
```

It is easy to see that any WHILE program can be translated into an equivalent GOTO program running at most a constant factor slower (measuring GOTO times by the number of instructions executed). Conversely, any GOTO program can be translated into an equivalent WHILE program running at most a constant factor slower (the factor may depend on the size of the GOTO program).

---

[2] Where goto 1 abbreviates if X goto 1 else 1.

3.3. FUNCTIONAL PROGRAMMING LANGUAGES

3.3.1. FUN1, *a first-order functional language*
This language and its higher-order counterpart FUNHO are functional: there is no assignment operator, and a program is a collection of mutually recursive function definitions.

DEFINITION 3.3. Syntax of FUN1: programs and expressions have forms given by the grammar:

$$
\begin{array}{llll}
\mathtt{p} \in \text{Program} & ::= & \mathtt{def}_1\ \mathtt{def}_2\ \ldots\ \mathtt{def}_m \\
\mathtt{def} \in \text{Definition} & ::= & \mathtt{f}(\mathtt{x}_1, \mathtt{x}_2, \ldots, \mathtt{x}_n) = \mathtt{e}^\mathtt{f} \\
\mathtt{e} \in \text{Expression} & ::= & \mathtt{d} \mid \mathtt{x} \mid \text{if } \mathtt{e}_0\ \mathtt{e}_1\ \mathtt{e}_2 \mid \text{cons } \mathtt{e}_1\mathtt{e}_2 \mid \text{hd } \mathtt{e} \mid \text{tl } \mathtt{e} \\
& \mid & \mathtt{f}(\mathtt{e}_1, \mathtt{e}_2, \ldots, \mathtt{e}_n) \\
\mathtt{d} \in \text{Constant} & ::= & \text{any element of } I\!D \\
\mathtt{x} \in \text{Parameter} & ::= & \text{identifier} \\
\mathtt{f} \in \text{FcnName} & ::= & \text{identifier}
\end{array}
$$

The *main function* $\mathtt{f}_1$ of program p is the one in $\mathtt{def}_1$ above; it must have $n = 1$. Semantics is call-by-value. For readability we often write if $\mathtt{e}_0$ then $\mathtt{e}_1$ else $\mathtt{e}_2$ for if $\mathtt{e}_0\ \mathtt{e}_1\ \mathtt{e}_2$.

The WHILE language corresponds rather precisely to a restriction of FUN1 to *tail recursive.* programs. Briefly: a tail-recursive program is one in which function calls never appear nested inside other calls, or other operations (cons, hd, tl or in if tests). They may however appear in the then or else branches of an if.

THEOREM 3.4. For every WHILE-program p there is a tail-recursive FUN1-program q such that $[\![\mathtt{p}]\!]^{\text{WHILE}} = [\![\mathtt{q}]\!]^{\text{FUN1}}$; and conversely.

Proof. To illustrate the idea, we show a tail recursive functional equivalent of the program reverse seen earlier:

```
reverse(X)  =  step(X, nil)
step(X, Y)  =  if X then  aux(cons (hd X) Y, tl X)
                     else  Y
```

Constructions are very straightforward. Given a WHILE-program p, first construct equivalent GOTO program $\mathtt{p}' = \mathtt{1:I1}\ \mathtt{2:I2}\ \ldots\ \mathtt{m:Im}$. If $\mathtt{p}'$ has variables $\mathtt{X}_1, \ldots, \mathtt{X}_k$, the following is an equivalent FUN1 program:

```
f(x) = step1(X1,nil,...,nil)

step1 (X1,...,Xk)   = <I1>
...
stepm (X1,...,Xk)   = <Im>
stepm+1 (X1,...,Xk) = X1
```

and the expression <I> for GOTO-instruction I is defined as follows:

```
<l:Xj := E>              = stepl+1 (X1,...,E,...,Xk)

<l:goto l'>              = stepl' (X1,...,Xk)

<l:if Xj goto l' else l''> = if Xj then stepl'(X1,...,Xk)
                                    else stepl''(X1,...,Xk)
```

The reverse direction is also easy: each tail call in a FUN1 program is translated to a series of assignments followed by a goto.

3.3.2. FUNHO, *a higher-order functional language*
This includes FUN1, but for convenient extension to higher-order functions uses Haskell's "named combinator" function syntax. The differences from the syntax of FUN1 are few:

DEFINITION 3.5. Syntax: programs and expressions have forms given by the grammar

$$
\begin{aligned}
\text{def} \in \text{Definition} \quad &::= \quad f\, x_1 x_2 \ldots x_n = e^f \\
e \in \text{Expression} \quad &::= \quad d \mid x \mid \text{if } e_0\, e_1\, e_2 \mid \text{cons } e_1 e_2 \mid \text{hd } e \mid \text{tl } e \\
&\quad\; \mid \quad f \mid e_1 e_2 \\
f \in \text{FcnName} \quad &::= \quad \text{identifier, disjoint from Parameter}
\end{aligned}
$$

The *definition of function* f has form $f\, x_1 x_2 \ldots x_n = e^f$, where e^f is called the *body of* f. The number $n > 0$ of parameters in the definition of f is called its *arity*. The main function must have $arity(f_1) = 1$. The formalism is well-known to be equivalent to the lambda calculus with explicit binding and recursion operators λ and μ.

Curried function calls. Multiple-argument functions are handled by "currying." To illustrate informally, consider the program

```
twice f zs   =  f (f zs)
append xs ys =  if xs then cons (hd xs) (append (tl xs) ys)
                else ys
```

Function **append** behaves as expected if applied to all its arguments, for example, call **append xs ys** with **xs** = (1 2) and **ys** = (3 4) returns (1 2 3 4). An *incomplete* call, however, is also legal (in contrast to **FUN1**), and defines a function. For instance call (**append xs**) with **xs** = (1 2) is legal, and defines the function $f(ys)$ = "append 1 2 to the front of ys," i.e., $f(ys) = (1.(2.ys))$. Thus the call (**twice** (**append ys**) **zs**) with **ys** = (1 2) and **zs** = (3 4) returns $f(f((3\ 4)))$, that is, (1 2 1 2 3 4).

A call **f** e_1 e_2 e_3 to a three-argument function can be written: (((**f** e_1) e_2) e_3). In this, **f** by itself is an expression, whose value is the function f that **f** defines. Further, (**f** e_1) is also an expression, whose value is $f(v_1)$ where v_1 is the value of e_1, etc. The idea is that a higher-order value is obtained by an *incomplete function application*. Operationally, the expression body e^f that defines **f** is not activated until all three arguments are present.

DEFINITION 3.6. The **FUNHO** language has a closure-based call-by-value semantics. Expression evaluation is based on inference rules appearing in Figure [omitted in these notes], one for each form of expression in the language. The *principal* judgement form for running program **p** is: $[\![\mathbf{p}]\!](v) = w$, signifying that **p**, when given input v, terminates with output w. The form for evaluating expressions is $\mathbf{p}, env \vdash \mathbf{e} \rightarrow v$, signifying that expression **e** in program **p** evaluates to value v if its free variables have values defined by environment env. Environments, values, etc. are defined by

$$
\begin{aligned}
env \in Env &= \text{Parameter} \rightarrow Value \\
v \in Value &= I\!D \mid Closure \\
cl \in Closure &= \text{FcnName} \times Value^*
\end{aligned}
$$

A "closure" is a device to represent a data value which is a function. In our named combinator syntax, a closure is a function name together with an incomplete parameter list, written $< \mathbf{f}, v_1 \ldots v_i >$.

3.4. TURING MACHINES

By definition these take inputs from the set **TM**-*data* = $\{0,1\}^*$. Since our aims concern only *decision powers* and not computation of functions, a Turing machine output will be a single bit 0 or 1. (Extension to outputs in $\{0,1\}^*$ is routine, by adding a one-way write-only output tape.)

A tape together with its scanning position will be written as L $\underline{S}$ R, where the underline indicates that S is the scanned symbol. As usual the tape is extensible — a new blank appears when a move is made beyond the end of the tape. A *store* σ describes the contents of the machine's one or several tapes, each with a designated scanned symbol. Instruction formats are described below.

A Turing machine program is a sequence p = 1:I1 2:I2 ... m:Im. A *total state* is a pair $s = (\ell, \sigma)$ whose first component ℓ is the number of the instruction about to be executed ($m + 1$ if it has completed execution) and whose second component is a store.

The semantics of each individual instruction is a state transition relation $s \to s'$, as is usual for Turing machines. Together these form define a function. A *computation* on input $d \in \{0, 1\}^*$ is a sequence $s_0 \to s_1 \to \ldots \to s_n$ where $s_i \to s_{i+1}$ for $0 \le i < n$, and s_0 is the initial state for input d, and s_n has instruction $m + 1$ as first component (the "program end").

We define the *time usage* of a Turing machine program p on input d to be

$$time_{\text{p}}^{\text{TM}}(\text{d}) = \begin{cases} n & \text{if } s_0 \to \ldots \to s_n \text{ is p's computation on d} \\ \bot & \text{if p does not terminate on input d} \end{cases}$$

Off-line Turing machines. An off-line Turing machine traditionally has one two-way *read-only input tape*, and one two-way *read-write work tape*. Off-line Turing machine instructions I_ℓ are as follows, where subscript 1 indicates that the input tape 1 is involved; or 2 indicates that work tape 2 is involved. Instruction syntax:

Tape 1:   I.   ::=   right$_1$ | left$_1$ | if$_1$ S goto ℓ else ℓ'

Tape 2:   I   ::=   right$_2$ | left$_2$ | if$_2$ S goto ℓ else ℓ' | write$_2$ S

Symbols: S   ::=   0 | 1 | B

Strings: L,R   ::= ε | L S

A *store* $\sigma = (L_1 \underline{S}_1 R_1, L_2 \underline{S}_2 R_2)$ is a pair whose components describe both of the tapes; underlines mark their scanning positions. The *initial store* for input $d \in I\!\!D$ is $\sigma_0 = (\underline{B} d, \underline{B})$

Two examples of the state transition relation $s \to s'$, instruction 1: right$_2$ causes transition from state $(1, B\underline{1}0, B0\underline{1}11B)$ to $(2, B\underline{1}0, B01\underline{1}1B)$, or from $(1, B\underline{1}0, B0\underline{1})$ to $(2, B\underline{1}0, B01\underline{B})$. We assume the program never moves right or left beyond a blank B, unless a nonblank symbol has first been written.[3]

The output for input d is defined by:

$$[\![p]\!](d) = \begin{cases} \bot & \text{if p has no computation on input d, else} \\ 1 & \text{if p's final state for input d scans worktape symbol 1} \\ 0 & \text{otherwise} \end{cases}$$

We define the *space usage* of a total state $s = (\ell, L_1\underline{S}_1R_1, L_2\underline{S}_2R_2)$ by $|s| = |L_2S_2R_2|$, formally expressing that only the symbols on "work" tape 2 are counted, and not those on tape 1. Finally, we define

$$space_\mathrm{p}^{\mathrm{TM}}(d) = \begin{cases} \max\{|s_i|\} & \text{if } s_0 \to \dots \to s_n \text{ is p's computation on d} \\ \bot & \text{if p does not terminate on input d} \end{cases}$$

One-tape Turing machines. These are as above, except that only one tape is involved. This holds both input and the final output answer. Details are omitted because of brevity and probable familiarity.

Random access machines This machine (RAM or SRAM) more closely resembles current machine languages than counter machines. Further, WHILE programs can be compiled into random access programs in a fairly natural way. The random access machine has a number of storage registers containing natural numbers (zero if uninitialised), and a much richer instruction set than the counter machine. The exact range of instructions allowed differ from one application to another, but nearly always include

1. Copying one register into another.
2. Indirect addressing or indexing, allowing a register whose number has been computed to be fetched from or stored into.
3. Elementary operations on one or more registers, for example adding or subtracting 1, and comparison with zero.
4. Other operations on one or more registers, for example addition, subtraction, multiplication, division, shifting, or bitwise Boolean operations (where register contents are regarded as binary numbers, i.e. bit sequences).

---

[3] This condition simplifies constructions, and causes no loss of generality in computational power, or increase in time beyond a small constant factor.

3.5. THE CHURCH-TURING THESIS: ROBUSTNESS OF THE CONCEPT OF COMPUTABILITY

A wide variety of computing devices and mathematical formalisms (recursive functions, the lambda calculus, and a great number of machine-like computation models including the Turing machine) have been devised to formalise the concept of "solvable problem" or "computable function."

Invariance of computational power. All these computing devices and mathematical formalisms have turned out to be equivalent in their problem-solving power, leading to the *Church-Turing thesis* that they all capture exactly the informal concept of "effective procedure." Conclusion: for many questions it doesn't really matter which computing formalism is used, so the most convenient for the problem at hand may be chosen.

DEFINITION 3.7. Suppose L-*data* = M-*data*. Language L *can simulate* language M, written as: L $\preceq$ M, if for every p $\in$ L-*programs* there is an m-program q such that for all d $\in$ L-*data* we have

$$[\![p]\!]^L(d) = [\![q]\!]^M(d)$$

Language L *is equivalent to* language M, written L $\equiv$ M, if language L and language M can simulate each other, i.e., if L $\preceq$ M and M $\preceq$ L.

THEOREM 3.8. WHILE$\equiv$GOTO

Traditionally Turing and many other machines take strings as inputs, in contrast to the binary trees accepted by WHILE programs. The following resolves the difference and makes it possible to extend the theorem just stated quite broadly.

A bijection between $\{0,1\}^*$ *and a subset of* $I\!\!D$. As defined, Turing machine inputs are in $\{0,1\}^*$ and WHILE inputs are in $I\!\!D$. We resolve this clash by restricting $I\!\!D$ to a subset in bijection with $\{0,1\}^*$. Define the coding $c: \{0,1\}^* \to I\!\!D$ by $c(a_1a_2...a_n) = (a_1' \, a_2' \, ... \, a_n')$ where $0' = $ nil and $1' = $ (nil.nil). An example using Lisp list notation:

$$c(001) = (0'0'1') = (\text{nil nil (nil.nil)}) \in I\!\!D$$

Note that $2n + 1 \leq | c(a_1a_2...a_n) | \leq 4n + 1$, so sizes differ only by a small constant factor.

THEOREM 3.9. ("Robustness of computability") WHILE $\equiv$ GOTO $\equiv$ FUN1 $\equiv$ FUNHO $\equiv$ TM $\equiv$ CM $\equiv$ SRAM, aside from the bijection between $\{0,1\}^*$ and the range of c.

Proof. This is by a series of constructions showing how to simulate a program in one language by one in another language. Details can be found in [14] and [11].

4. Compilation and Interpretation

These concepts provide a natural bridge between the worlds of recursion theory and computer science. Briefly: most of the traditional computability theory constructions are compilations; and the well-known *universal function* is just the function computed by a "self-interpreter."

4.1. COMPILATION

Suppose we are given three programming languages: a *source language* S, a *target language* T, and an *implementation language* L, and that S-*data* = T-*data*, S-*programs* ⊆ L-*data* and T-*programs* ⊆ L-*data*.

DEFINITION 4.1. An L-program comp is a *compiler* from language S to T if [[comp]](p) ∈ T-*programs* for every p ∈ S-*programs*, and for every d ∈ S-*data*,

$$[\![p]\!]^S(d) = [\![[\![comp]\!]^L(p)]\!]^T(d)$$

THEOREM 4.2. i as defined in Figure 2 is an interpreter written in WHILE for the language WHILE^{1var}. (Language WHILE^{1var} is identical to WHILE except that programs are restricted to one variable X.)

```
read PD;                        (* Input is (program . data) *)
  Cd := cons (hd PD) nil;       (* Control stack = (program.nil) *)
  Val := tl PD;                 (* The value of X = data *)
  Stk := nil;                   (* Computation stack is initially empty *)
  while Cd do STEP;             (* Repeat while control stack nonempty *)
write Val
```

Figure 2. Interpreter i for WHILE^{1var}

4.2. INTERPRETATION

Self-interpreters, under the name *universal programs*, play (and have played since the 1930's) a central role in both complexity and computability theory.

Generalising to several languages, an interpreter int ∈ L-programs for a language S has a pair of inputs: a *source program* p ∈ S-programs, and the source program's input data d ∈ S-*data*.

DEFINITION 4.3. Suppose S-*programs* ⊆ L-*data* = S-*data*. Program int is an *interpreter* for S *written in* L if for every p ∈ S-programs and for every d ∈ S-*data*,

$$[\![p]\!]^S(d) = [\![int]\!]^L(p.d)$$

4.2.1. *An interpreter* i *for 1-variable* WHILE *programs*

We now give an example interpreter written in language WHILE. This is nearly a self-interpreter, except that for the sake of simplicity we restrict it to programs containing only one variable X. The interpreter, called i, is in Figure 2, where the STEP macro is in Figure 3. The interpreter will use the concrete syntax of Figure 1, together with unique codes to aid evaluation of cons, hd, tl and execution of assignments and while commands.

These are do_cons, do_hd, do_tl, do_asgn, do_while. They may be arbitrarily chosen from *ID*, except that they must be distinct from each other and from :=,..., atom=? of the "concrete syntax" in Figure 1.

To aid compactness and readability, the STEP part of i is given by a set of transitions in Figure 3, i.e., rewrite rules (Cd, s, v) ⇒ (Cd', s', v') describing transitions from one state of form

(Cd, s, v) = (*Code-stack, Computation-stack, Value*)

to the next state. Expression evaluation is based on the following *net effect property*. Suppose the value of expression E is d, provided that the current value of variable X is v. Then (E.Cd, s, v) ⇒⁺ (Cd, d.s, v) where ⇒⁺ means "can be rewritten in one or more steps to". A similar net effect property characterises execution of commands.

Blank entries in Figure 3 correspond to values that are neither referenced nor changed in a rule. The rules can easily be programmed in the WHILE language. For example, the three while transitions could be programmed as in Figure 4.

4.2.2. *Self-interpretation of* WHILE *and* I

We will call an interpreter for a language L which is written in the same language L a *self-interpreter* or *universal program* for L, and we will often use name u for a universal program. By Definition 4.3, u must satisfy $[\![p]\!]^L(d) = [\![u]\!]^L(p.d)$ for every L-program p and L-data value d.

Cd	Stk	Val	⇒	Cd	Stk	Val
nil			⇒	nil		
(quote D).Cd	S		⇒	Cd	D.S	
(var nil).Cd	S	v	⇒	Cd	v.S	v
(hd E).Cd			⇒	E.do_hd.Cd		
do_hd.Cd	(T.U).S		⇒	Cd	T.S	
do_hd.Cd	nil.S		⇒	Cd	nil.S	
(tl E).Cd			⇒	E.do_tl.Cd		
do_tl.Cd	(T.U).S		⇒	Cd	U.S	
do_tl.Cd	nil.S		⇒	Cd	nil.S	
(cons E1 E2).Cd			⇒	E1.E2.do_cons.Cd		
do_cons.Cd	U.T.S		⇒	Cd	(T.U).S	
(; C1 C2).Cd			⇒	C1.C2.Cd		
(:= X E).Cd			⇒	E.do_asgn.Cd		
do_asgn.Cd	w.S	v	⇒	Cd	S	w
(if E C1 C2).Cd			⇒	E.do_if.C1.C2.Cd		
do_if.C1.C2.Cd	(T.U).S		⇒	C1.Cd	S	
do_if.C1.C2.Cd	nil.S		⇒	C2.Cd	S	
(while E C).Cd			⇒	E.do_while.(while E C).Cd		
do_while.						
(while E C).Cd	(T.U).S		⇒	C.(while E C).Cd	S	
do_while.C1.Cd	nil.S		⇒	Cd	S	

Figure 3. The **STEP** macro, expressed by rewriting rules

The interpreter i just given is *not* a self-interpreter due to the restriction to just one variable in the interpreted programs. Extension to a full self-interpreter for **WHILE** is a straightforward programming exercise.

THEOREM 4.4. There exists a self-interpreter for the **WHILE** language.

96

4.2.3. *The minimal language* I.

DEFINITION 4.5.   The language I is identical to WHILE, with two exceptions. First, $A = \{\texttt{nil}\}$ so data values have only one atom. Second, programs have only one variable X.

The following result will be seen to have interesting complexity consequences.

THEOREM 4.6.   There exists a self-interpreter u for the I language.

Proof   Interpreter i of Theorem 4.2 satisfies $[\![\texttt{i}]\!]^{\texttt{WHILE}}(\texttt{p.d}) = [\![\texttt{p}]\!]^{\texttt{I}}(\texttt{d})$, for any I-program p and input $\texttt{d} \in I\!\!D$, Program i uses only the one atom nil, but has several variables. Now obtain I-program u from i such that $[\![\texttt{u}]\!]^{\texttt{I}}(\texttt{p.d}) = [\![\texttt{p}]\!]^{\texttt{I}}(\texttt{d})$ by packing the several variables of i into one using "cons".

5. Applications of the *s-m-n* theorem

5.1. PARTIAL EVALUATION IS PROGRAM SPECIALISATION

The program specialiser of Theorem 2.8 was very simple, and the programs it ouputs are slightly slower than the ones from which they were derived. (This was also true of Kleene's original construction.) On the other hand, program specialisation can be done so as to yield *efficient* specialised programs. This is known in the programming language community as *partial evaluation*; see [15] for a thorough treatment and a large bibliography.

Applications of program specialisation include *compiling* (done by specialising an interpreter to its source program), and *generating compilers from interpreters,* by using the specialiser to specialise itself to a given interpreter. Surprisingly, this can give quite efficient programs as output.

```
if hd hd Cd = while then              (* Set up iteration *)
    Cd := (cons (hd tl hd Cd) (cons do_while Cd)) else
if hd Cd =do_while then
    if hd Stk then                    (* Do body if test is true *)
    { Cd := cons (hd tl tl tl Cd) (tl Cd);
      Stk := tl Stk } else            (* Else exit while *)
    { Stk := tl Stk; Cd := tl tl Cd } else ...
```

Figure 4.   WHILE code for rewrite rules to implement "while"

A simple but nontrivial example of partial evaluation   Consider Ackermann's function, with program:

```
a(m,n) = if m =? 0 then n+1 else
           if n =? 0 then a(m-1,1)
           else a(m-1,a(m,n-1))
```

Computing $a(2,n)$ involves recursive evaluations of $a(m,n)$ for $m = 0$, 1 and 2, and various values of n. A partial evaluator can evaluate expressions m=?0 and m-1, and function calls of form $a(m-1,...)$ can be unfolded. We can now specialise function a to the values of m, yielding a less general program that is about twice as fast:

```
a2(n) = if n =? 0 then 3 else a1(a2(n-1))
a1(n) = if n =? 0 then 2 else a1(n-1)+1
```

5.2. COMPILING AND COMPILER GENERATION BY THE FUTAMURA PROJECTIONS

This section shows an application of the *s-m-n* theorem in computer science to compiling and, more generally, to generating program generators. For simplicity we elide the name of the language L that is the implementation, input, and output language of the specialiser.

The starting point is an interpreter program **int** for some programming language S. By Definition 4.3 $[\![src]\!]^S(input) = [\![int]\!](src.input)$ for any S-program **src** and data input $\in D$.

First Futamura projection: $target = [\![spec]\!](int.src)$. This shows that *a specialiser can compile.* Correctness is to show that **target** is a program in the specialiser's output language which is equivalent to S-program **src**, i.e., $[\![src]\!]^S = [\![target]\!]$:

$$
\begin{aligned}
[\![src]\!]^S(input) &= [\![int]\!](src.input) && \text{Definition of interpreter} \\
&= [\![[\![spec]\!](int.src)]\!](input) && \text{Definition of specialiser} \\
&= [\![target]\!](input) && \text{Definition of target}
\end{aligned}
$$

Second Futamura projection: $comp = [\![spec]\!](spec.int)$. This shows that *a specialiser can generate a compiler.* Correctness is to show that **comp** is a compiler from S to the specialiser's output language. It constructs the just-mentioned target program from the source program **src**:

$$
\begin{aligned}
\text{target} \;=\;& [\![\text{spec}]\!](\text{int.src}) && \text{Definition of target} \\
=\;& [\![[\![\text{spec}]\!](\text{spec.int})]\!](\text{src}) && \text{Definition of specialiser} \\
=\;& [\![\text{comp}]\!](\text{src}) && \text{Definition of comp}
\end{aligned}
$$

Third Futamura projection: $\text{cogen} = [\![\text{spec}]\!](\text{spec.spec})$. This shows that *a specialiser can generate a compiler generator.*[4] Correctness is to show that cogen constructs the just-mentioned compiler from the interpreter int:

$$
\begin{aligned}
\text{comp} \;=\;& [\![\text{spec}]\!](\text{spec.int}) && \text{Definition of comp} \\
=\;& [\![[\![\text{spec}]\!](\text{spec.spec})]\!](\text{int}) && \text{Definition of specialiser} \\
=\;& [\![\text{cogen}]\!](\text{int}) && \text{Definition of cogen}
\end{aligned}
$$

Perhaps the most surprising thing about this equational reasoning is that it also works well in practice: A variety of partial evaluators (= program specialisers = programs of *s-m-n* functions) have been constructed, and they give good results in practical applications [15].

Speedups from self-application.
Each of program execution, compilation, compiler generation, and compiler generator generation can be done in two different ways:

$$
\begin{aligned}
\text{output} \;&=\; [\![\text{int}]\!](\text{src.input}) \;&=\; [\![\text{target}]\!]^{\text{S}}(\text{input}) \\
\text{target} \;&=\; [\![\text{spec}]\!](\text{int.src}) \;&=\; [\![\text{comp}]\!](\text{src}) \\
\text{comp} \;&=\; [\![\text{spec}]\!](\text{spec.int}) \;&=\; [\![\text{cogen}]\!](\text{int}) \\
\text{cogen} \;&=\; [\![\text{spec}]\!](\text{spec.spec}) \;&=\; [\![\text{cogen}]\!](\text{spec})
\end{aligned}
$$

The exact timings vary according to the design of spec and int, and with the implementation language L. Nonetheless, we have observed in practical computer experiments that *in each case the rightmost run is often about 10 times faster than the leftmost.* Moral: self-application can generate programs that run faster!

6. Constant time factors *do* matter

The *Turing machine constant speedup theorem* is traditionally one of the first learned in complexity courses. In effect it says that from any Turing machine program, one may construct an equivalent one that runs twice as fast (asymptotically, and if its running time is superlinear).

---

[4] A compiler generator transforms an interpreter into a compiler.

Unfortunately, this result gives practically oriented students a bad impression: While the proof is not too complex, what it says is extremely counterintuitive, going against daily programming experience. Further, the construction is useless for speeding up real programs, as it in essence involves changing to a double-density tape. The fact that the theorem is nonetheless taught tends to convey the idea that theory is irrelevant to practice.

The truth of this speedup theorem is an immediate consequence of using an "unfair time measure": one Turing machine state transition is counted as taking one time unit, *regardless of the size of the Turing machine's tape alphabet.*

We show a more satisfying result (at least from a programmer's perspective): a proper hierarchy exists among problems that can be solved by I programs in linear time. In particular there exist constants $0 < c < d$ and set A such that the question $\mathbf{x} \in A$? can be decided in time $c \cdot |\mathbf{x}|$ but not in time $d \cdot |\mathbf{x}|$, *regardless of how clever one is* at programming and/or algorithm design. Such an absolute result is rare in computer science, and attracts the students' attention.

This result is false for Turing machines; and its status for the full WHILE language is an open question.

6.1. TIME-BOUNDED COMPLEXITY CLASSES

DEFINITION 6.1. Given programming language L and a function : $\mathbb{N} \to \mathbb{N}$, we define

$$\text{TIME}^{L}(f) \;=\; \{A \subseteq \mathbb{D} \mid \; A \text{ is decided by some L-program p, such that}$$
$$time_{\mathsf{p}}^{L}(\mathbf{d}) \leq f(|\mathbf{d}|) \text{ for all } \mathbf{d} \in \text{L-}data\}$$

Naming convention: the size of the input, e.g., $|\mathbf{d}|$, is called n. If *exp* is an expression containing n, we write $\text{TIME}^{L}(exp)$ anstead of $\text{TIME}^{L}(\lambda n.exp)$.

THEOREM 6.2. *Linear-time hierarchy for* I: There is a b such that for all $a \geq 1$ there is a set $A \subseteq \mathbb{D}$ which is in $\text{TIME}^{I}(a \cdot b \cdot n)$, but not in $\text{TIME}^{I}(a \cdot n)$.

The key to this result is the existence of an "efficient" interpreter for I, as seen in Definition 6.3. The proof diagonalises (as for undecidability of the halting problem), using a time-bounded extension of the self-interpreter of Theorem 4.6 to stay within the required time bounds. The current framework makes this substantially simpler than traditional time-hierarchy proofs.

6.2. INTERPRETATION OVERHEAD AND "EFFICIENCY"

Let int be an interpreter for S written in L. Assuming both an L-machine and an S-machine are at one's disposal, interpretation is usually rather slower than direct execution of S-programs. In practice, an interpreter's running time on inputs p and d typically satisfies, for all d:

$$time_{int}^L(p.d) \leq a_p \cdot time_p^S(d)$$

Here a_p is a "constant" independent of d, but perhaps depending on source program p. Often $a_p \doteq c + f(p)$, where constant c represents the time taken for "dispatch on syntax" and recursive calls of the evaluation or command execution functions; and $f(p)$ is the time for variable access.

DEFINITION 6.3. An interpreter int (for S written in L) is *efficient* if there is a constant a such that for all $d \in I\!\!D$ and every S-program p

$$time_{int}^L(p.d) \leq a \cdot time_p^S(d)$$

Constant a is here quantified *before* p, so the slowdown caused by an efficient interpreter is independent of p.

THEOREM 6.4. The interpreter u for I written in I from Theorem 4.6 is efficient according to Definition 6.3.

6.3. AN EFFICIENT TIMED UNIVERSAL PROGRAM

DEFINITION 6.5. An I-program tu is an *efficient timed universal program* if there is a constant k such that for all $p \in$ I-*programs*, $d \in I\!\!D$ and $n \geq 1$:

1. If $time_p(d) \leq n$ then $[\![tu]\!](p \ d \ nil^n) = ([\![p]\!](d).nil)$
2. If $time_p(d) > n$ then $[\![tu]\!](p \ d \ nil^n) = nil$
3. $time_{tu}(p \ d \ nil^n) \leq k \cdot \min(n, time_p(d))$.

The effect of $[\![tu]\!](p \ d \ nil^n)$ is to simulate p for $\min(n, time_p(d))$ steps. If $time_p(d) \leq n$, that is, p terminates within n steps, then tu produces a non-nil value containing p's result. If not, the value nil is yielded, indicating "time limit exceeded."

THEOREM 6.6. There exists an efficient timed universal program tu.

Proof. We first construct an efficient timed interpreter tt for I, in the form of a WHILE-program. The idea is to take the interpreter seen before for one-variable WHILE programs, and add some extra code and an extra input, a *time bound* of the form nil^n stored in a variable Cntr, so obtaining a program tt. Every time the simulation of one operation of program input p on data input d is completed, the time bound is decreased by 1. Note the use of the atom encoding function from the proof of Theorem 4.6. Here is tt:

```
read X;                       (* X = (p d niln) *)
Cd := cons (hd X) nil;        (* Code to be executed *)
Val := hd (tl X);             (* Initial value of simulated X *)
Cntr := hd (tl (tl X));       (* Time bound *)
Stk := nil;                   (* Computation stack *)
while Cd do
  if Cntr
  then { if  hd (hd Cd) ∈ {quote, var, do_hd, do_tl, do_cons,
                                      do_asgn, do_while}
        then Cntr := tl Cntr;
        STEP; X := cons Val nil;}
  else { Cd := nil; X := nil};
write X
```

Finally, obtain I-program tu from tt by packing its several variables into one using "cons".

6.4. THE LINEAR-TIME HIERARCHY

This result shows there is a constant b such that for any $a \geq 1$ there is a decision problem which cannot be solved by any I-program that runs in time bounded by $a \cdot n$, *regardless of how clever* one is at programming, or at problem analysis, or both. On the other hand, the problem *can* be solved by an I-program in time $a \cdot b \cdot n$ on inputs of size n. Consequence: given a, there exists an infinite hierarchy

$$\text{TIME}^I(a \cdot n) \subset \text{TIME}^I(a \cdot b \cdot n) \subset \text{TIME}^I(a \cdot b^2 \cdot n) \subset \ldots$$

Proof. of Theorem 6.2. First define program diag by Figure 5.

Claim: the set $A = \{d \mid [\![\text{diag}]\!]^I(d) = \text{true}\}$ is in $\text{TIME}^I(a \cdot b \cdot n)$ for an appropriate b, but is not in $\text{TIME}^I(a \cdot n)$. Further, b is independent of a.

We first analyse the running time of program diag on input p. Since a is fixed, nil$^{a \cdot |d|}$ can be computed in time $c \cdot a \cdot |d|$ for some c and any d. From Definition 6.5, there exists k such that the timed universal program tu of Theorem 6.6 runs in time $time_{tu}(\text{p d nil}^n) \leq k \cdot min(n, time_p(d))$. Thus on input p, the command "X := tu Arg" takes time at most

$$k \cdot min(a \cdot |p|, time_p(p)) \leq k \cdot a \cdot |p|$$

```
read X;
Timebound := nil^{a·|X|};
Arg := cons X (cons X (cons Timebound nil));
X := tu Arg; (* Use tu to run X on X for up to a·|X| steps *)
if hd X then X := false else X := true;
write X
```

Figure 5. Diagonalisation program diag.

so program diag runs in time at most

$$c \cdot a \cdot |p| + k \cdot a \cdot |p| + e$$

where c is the constant factor used to compute $a \cdot |X|$, k is from the timed universal program, and e accounts for the time beyond computing Timebound and running tu. Now $|p| \geq 1$ so

$$c \cdot a \cdot |p| + k \cdot a \cdot |p| + e \leq a \cdot (c + k + e) \cdot |p|$$

which implies that $A \in \text{TIME}^I(a \cdot b \cdot n)$ with $b = c + k + e$.

Now suppose for the sake of contradiction that $A \in \text{TIME}^I(a \cdot n)$. Then there exists a program p which also decides membership in A, and does it quickly, satisfying $time_p(d) \leq a \cdot |d|$ for all $d \in D$. Consider cases of $[\![p]\!](p)$. Then $time_p(p) \leq a \cdot |p|$ implies that tu has sufficient time to simulate p to completion on input p. By Definition 6.5, this implies

$$[\![tu]\!](p \; p \; nil^{a·|p|}) = ([\![p]\!](p) \; nil)$$

If $[\![p]\!](p)$ is false, then $[\![diag]\!](p) = $ true by construction of diag. If $[\![p]\!](p)$ is true, then $[\![diag]\!](p) = $ false. Both cases contradict the assumption that p and diag both decide membership in A. The only unjustified assumption was that $A \in \text{TIME}^I(a \cdot n)$, so this must be false, completing the proof.

This construction has been carried out in detail on the computer by Ben-Amram, who established a stronger result in [?]: that $\text{TIME}^I(232 \cdot a \cdot n)$ properly includes $\text{TIME}^I(a \cdot n)$ for any $a > 1$.

Further, for any non-zero "time constructible" $T(n)$ (using a natural definition), there is a b such that $\text{TIME}^I(b \cdot T(n)) \setminus \text{TIME}^I(T(n))$ is non-empty.

7. Levin's optimal search theorem

A great many familiar problems are searches. Consider the predicate

$$R(\mathcal{F}, \theta) \equiv \text{truth assignment } \theta \text{ makes formula } \mathcal{F} \text{ true}$$

The problem to find θ (if it exists) when given $\mathcal{F}$ is the familar and apparently intractable *satisfiability problem*. As is well known, it is much easier to check truth of $R(\mathcal{F}, \theta)$.

DEFINITION 7.1. A *witness function* for a binary predicate $R \subseteq \mathbb{D} \times \mathbb{D}$ is a partial function $f : \mathbb{D} \to \mathbb{D}_\perp$ such that:

$$\forall x(\exists y . R(x, y)) \Rightarrow R(x, f(x))$$

A *brute-force search* program for finding a witness immediately comes to mind. Given $x \in \mathbb{D}$ we just enumerate elements $y \in \mathbb{D}$, checking one after the other until a witness pair $(x, y) \in R$ has been found.[5] It is quite obvious that this strategy can yield an extremely inefficient program, since it may waste a lot of time on wrong candidates until it finds a witness. Levin's theorem states a surprising fact: for many interesting problems there is another brute-force search strategy that not only is efficient, but *optimal* up to constant factors. The difference is that Levin's strategy generates and tests not *solutions*, but *programs*.

THEOREM 7.2. *Levin's Search Theorem.* Let $R \subseteq \mathbb{D} \times \mathbb{D}$ be a recursively enumerable binary predicate, so $R = \text{dom}(\llbracket r \rrbracket)$ for some program r. Then there is a WHILE program opt such that $f = \llbracket \text{opt} \rrbracket$ is a witness function for R. Further, let q be *any* program that computes a witness function f for R. Then for all x such that $(x, y) \in R$ for some y:

$$time_{\text{opt}}(x) \le a_q(time_q(x) + time_r(x.f(x)))$$

where a_q is a constant that depends on q but not on x. Further, the program opt can be effectively obtained from r.

Sketch of proof of Levin's theorem. Without loss of generality we assume that when program r is run with input $(x.y)$, if $(x, y) \in R$ it gives y as output. Otherwise, it loops forever. Enumerate $\mathbb{D} = \{d_0, d_1, d_2, \ldots\}$ effectively (it can be done in constant time per new element). Build program opt to compute as follows:

---

[5] If R is decidable, this is straightforward. If semi-decidable but not decidable, a "dovetailing" of computations can be used, in effect testing $(x, d_0) \in R$?, $(x, d_1) \in R$?, $\ldots$ in parallel for all values $d_0, d_1, d_2, \ldots \in \mathbb{D}$.

1. A "main loop" to generate all finite trees. At each iteration one new tree is added to list $L = (d_n \ldots d_1 d_0)$. Tree d_n for $n = 0, 1, 2, \ldots$ will be treated as the command part of the n-th I-program p_n.
2. Iteration n will process programs p_k for $k = n, n-1, \ldots, 1, 0$ as follows:

   a) Run p_k on input x for a "time budget" of 2^{n-k} steps.
   b) If p_k stops on x with output y, then run r on input $(x.y)$, so p_k and r together have been executed for at most 2^{n-k} steps.
   c) If p_k or r failed to stop, then replace k by $k - 1$, double the time budget to 2^{n-k+1} steps, and reiterate.

3. If running p_k followed by r terminates within time budget 2^{n-k}, then output $[\![\text{opt}]\!](x) = y$ and stop; else continue with iteration $n + 1$.

Thus the programs are being interpreted concurrently, every one receiving some "interpretation effort." We stop once any one of these programs has both *solved our problem and been checked*, within its given time bounds. Note that opt will loop in case no witness is found. The following table showing the time budgets of the various runs may aid the reader in following the flow of the construction and seeing its correctness.

The keys to "optimality" of opt are the efficiency of the self-interpreter STEP operation, plus a policy of allocating time to the concurrent simulations so that the total time will not exceed, by more than a constant factor, the time of the program that finishes first.

Time budget	p_0	p_1	p_2	p_3	p_4	p_5	$\cdots$
$n = 0$	1	-	-	-	-	-	$\cdots$
$n = 1$	2	1	-	-	-	-	$\cdots$
$n = 2$	4	2	1	-	-	-	$\cdots$
$n = 3$	8	4	2	1	-	-	$\cdots$
$n = 4$	16	8	4	2	1	-	$\cdots$
$n = 5$	32	16	8	4	2	1	$\cdots$
$n = 6$	64	32	16	8	4	2	$\cdots$
$\cdots$	$\cdots$						

Suppose $q = p_k$ computes a witness function f. At iteration n, program p_k and the checker r are run for 2^{n-k} steps. Therefore (assuming $R(x.f(x))$ is true) the process above will not continue beyond iteration n, where

$$2^{n-k} \geq time_{p_k}(x) + time_r(x.f(x))$$

A straightforward time analysis (summing exponentials) yields

$$time_{\text{opt}}(\mathbf{x}) \leq c2^k(time_{\mathbf{q}}(\mathbf{x}) + time_{\mathbf{r}}(\mathbf{x}.f(\mathbf{x})))$$

as required, where c is not excessively large.

We now estimate the value of the constant factor. Imagine (naturally enough) that the programs p_i are enumerated in order of increasing length. Then program $\mathbf{q} = p_k$ would occur at position near[6] $k = 2^{O(|\mathbf{q}|)}$, where the $O()$ represents a small constant. Thus the constant factor can be bounded from above by $c2^k = c2^{2^{O(|\mathbf{q}|)}}$

It must be admitted that this constant factor is enormous. The interesting thing is that it exists at all, i.e., that the construction gives, from an asymptotic viewpoint, the best possible result.

8. Overview of complexity theory

Computability theory concerns only *what* is computable, and pays no attention at all to how much time or space is required to carry out a computation. In the real computing world, however, computational resource usage is of primary importance, as it can determine whether or not a problem is solvable at all in practice.

The central computational concepts of complexity theory all involve resources, chiefly time, space, and nondeterminism. Complexity theory has evolved a substantial understanding of just what the *intrinsic complexity* is of many interesting general and practically motivated problems. This is reflected by a well-developed classification system for "how decidable" a problem is. Theorems 6.2 and 7.2 lie in the realm of complexity theory. Computability theory has similar classification systems, for "how *undecidable*" a problem is, but the subject is beyond the scope of this work.

Complexity theory is mostly about the classification of decision problems (the question $x \in A$? for a set A) by the amount of time, space, etc., required to solve them. Time, space, etc., are usually measured as a function of the size $|x|$ of the question instance. Complexity of function computation is similar but a bit more involved (and not treated here).

As was the case for computability, we will not consider finite problems; instead, we study the *asymptotic complexity* of a program solving a program: a function $f : \mathbb{N} \to \mathbb{N}$ defining how rapidly its resource usage grows, as the size of its input data grows to infinity.

---

[6] If binary strings are enumerated, then number 2^i has length $i + 1$, and it is easy to see $\mathbb{D}$ has a similar property.

The remainder of this work develops a *hierarchy of robust subclasses* within the class of all decidable sets, and investigates its properties. The significance of the hierarchy is greater because of *representation invariants*: the fact that the placement of a problem within it is in general quite independent of the way the problem is described, for example whether graphs are represented by connection matrices or by adjacency lists.

In some cases we prove *proper containments* between adjacent problem classes in the hierarchy: that a sufficient resource increase will *properly increase* the classes of problems that can be solved. In other cases, questions concerning proper containments are still unsolved, and have been for many years.

In lieu of definitive answers, we will characterise certain problems as *complete* for the class of all problems solvable within given resource bounds. A complete problem is both solvable within the given bounds and, in a precise technical sense, "hardest" among all problems so solvable. Many familiar problems will be seen to be complete for various of these complexity classes.

8.1. TIME- AND SPACE-BOUNDED CLASSES OF DECISION PROBLEMS

DEFINITION 8.1. Suppose L is a programming language, and time and space measures $time_p^L(d)$, $space_p^L(d)$ have the property that for any 1-program p and L-data d, $[\![p]\!]^L(d) = \bot$ iff $time_p^L(d) = \bot$, and $[\![p]\!]^L(d) = \bot$ iff $space_p^L(d) = \bot$. Let $f : IN \to IN$ be a nonzero function. Then by definition

$$\text{TIME}^L(f) = \{A \subseteq I\!D \mid \exists \text{ L-program p } (\text{p decides } A, \text{ and} \\ time_p^L(d) \leq f(|d|) \text{ for all L-data d})\}$$

$$\text{SPACE}^L(f) = \{A \subseteq I\!D \mid \exists \text{ L-program p } (\text{p decides } A, \text{ and} \\ space_p^L(d) \leq f(|d|) \text{ for all L-data d})\}$$

Further, define

$$\text{PTIME}^L = \bigcup_{\pi \text{ a polynomial}} \text{TIME}(\pi)$$

$$\text{PSPACE}^L = \bigcup_{\pi \text{ a polynomial}} \text{SPACE}(\pi)$$

$$\text{LOGSPACE}^L = \bigcup_{k \geq 0} \text{SPACE}(\lambda n.k \log n)$$

8.2. INVARIANCE OF RESOURCE-BOUNDED COMPUTATIONAL POWER

Resource-bounded problem-solving power is equivalent over a wide spectrum of computational models. Time and space accounting of simulations as in proof of Theorem 3.9 establish that

1. Computability, *up to linear differences in running time*, is equivalent for WHILE, GOTO, FUN1 and FUNHO.
2. Computability, *up to polynomial differences in running time*, is equivalent for all of: WHILE, GOTO, FUN1, FUNHO SRAM, and TM.
3. Computability, *up to polynomial differences in memory usage*, is equivalent for CM, SRAM, and TM.

This leads to the following "robustness" or invariance result[7], which justifies writing just PTIME:

THEOREM 8.2. $\text{PTIME}^{\text{TM}} = \text{PTIME}^{\text{SRAM}} = \text{PTIME}^{\text{WHILE}} = \text{PTIME}^{\text{FUN1}} = \text{PTIME}^{\text{FUNHO}}$

The class PTIME plays a central role in complexity theory, as it is common to identify "tractable prolem" with "solvable in polynomial time." While this can be argued with it is very natural, and proof that a problem does *not* lie in PTIME is strong evidence that it is indeed intractable.

Similarly, the following result justifies writing just PSPACE:

THEOREM 8.3. $\text{PSPACE}^{\text{TM}} = \text{PSPACE}^{\text{SRAM}} = \text{PSPACE}^{\text{CM}}$

Computation with logarithmic space
The class LOGSPACE^{TM} is the class of problems solvable by offline Turing machines whose space consumption is at most $k \log n$ for inputs of length n. Since the bound makes little sense for other machine types, we henceforth write just LOGSPACE.

LOGSPACE also plays a central role in complexity theory, as it is the beginning of the much-studied hierarchy LOGSPACE $\subseteq$ NLOGSPACE $\subseteq$ PTIME $\subseteq$ NPTIME $\subseteq$ PSPACE $\subseteq$ ... Further, all the well-known *problem reductions* used to show familiar combinatorial problems complete for these classes can be carried out by logspace-bounded Turing machines.

LOGSPACE is in a sense the smallest natural complexity class, because a machine needs storage of at least $O(\log n)$ bits[8] in order to "remember" a position in its input data, or to count a number of input symbols — for instance, to decide whether its input has the form $0^n 1^n 0^n$.

---

[7] Using the bijection to ignore the difference between $\{0, 1\}^*$ and $\mathbb{D}$.

[8] Here $n =$ input size, i.e., string length or tree size.

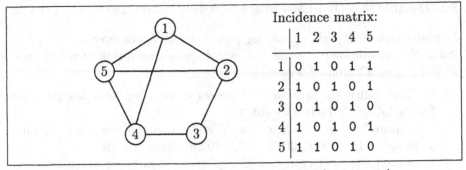

Figure 6. An undirected graph, and its incidence matrix.

8.3. INVARIANCE OF COMPLEXITY WITH RESPECT TO PROBLEM REPRESENTATIONS

An important question: can the way a problem is presented significantly affect the complexity of its solution? The *problem representation invariance property* says no:[9]

> All natural problem representations are inter-translatable by algorithms running in logarithmic space, or polynomials of low degree.

For one example, a number n can be presented either in binary notation, or in the much longer unary notation, such as the list form nil^n. However unary notation for numbers seems unnatural given that we measure complexity as a function of input length, since unary notation is exponentially more space-consuming than binary notation.

Another example is that a directed or undirected graph $G = (V, E)$ with $V = \{v_1, v_2, \ldots, v_n\}$ can be presented in any of several forms:

1. An n by n incidence matrix M with $M_{i,j}$ equal to 1 if $(v_i, v_j) \in E$ and 0 otherwise. Figure 6 contains an example.
2. An adjacency list $(u, u', u'', \ldots)$ for each $v \in V$, containing all vertices u for which $(v, u) \in E$. An example is

   $[1 \mapsto (2, 4, 5), 2 \mapsto (1, 3, 5), 3 \mapsto (2, 4), 4 \mapsto (1, 3, 5), 5 \mapsto (1, 2, 4)]$.
3. A list of all the vertices $v \in V$ and edges $(v_i, v_j) \in E$, in some order. Example: $[1, 2, 3, 4, 5]$, $[(1, 2), (2, 3), (3, 4), (4, 5), (5, 1), (1, 4), (2, 5)]$; or
4. In a compressed format in case the graph is known to be sparse, i.e., to have few edges between vertices.

---

[9] This is not a theorem, but an observation based on experience.

Problem equivalence modulo representation. One example: a k-clique in undirected graph G is a set of k vertices such that G has an edge between every pair in the set. The CLIQUE decision problem is: given (G, k), to decide whether G has at least one k-clique.

It is not known whether the CLIQUE problem is in PTIME or not — and all known algorithms take exponential time in the worst case. However a little thought shows that *the choice of representation will not affect its status*, since one can convert back and forth among the various graph representations in polynomial time; so existence of a polynomial time CLIQUE algorithm for one representation would immediately imply the same for any of the other representations.

From this viewpoint the most "sensible" problem representations are all equivalent, at least up to polynomial-time computable changes in representation. There are a few exceptions to this rule involving degenerate problem instances, for example extremely sparse graphs, but such exceptions only seem to confirm that the rule holds in general.

8.4. NONDETERMINISM

Many practically interesting but apparently intractable problems lie in the class NPTIME, a superset of PTIME including, loosely speaking, programs that can "guess," formally called *nondeterministic*. Such programs can solve many challenging search or optimisation problems by a simple-minded and efficient technique of *guessing* a possible solution and then *verifying*, within polynomial time, whether or not the guessed solution is in fact a correct solution. The ability to guess is formally called "nondeterminism."

Time- and space resource-bounded classes have already been defined, so we describe their nondeterministic counterparts quite briefly. Syntactically, nondeterminism can be expressed by adding one new instruction type:

```
choose ℓ or ℓ'      (* Go to either instruction Iℓ or to Iℓ' *)
```

Semantically, the construct is given an *angelic* interpretation: a nondeterministic program correctly solves a set membership problem "d ∈ A?" if

- d ∈ A implies that p *can possibly take* one or more guess sequences, leading to the answer **true**; and
- d ∉ A implies that p cannot take any *guess sequence at all* that leads to the answer **true**.

Given time resource bound f, we say that p *runs in time f* if for all d, whenever p can yield **true**, then it can do so by a computation with length no longer than $f(|d|)$. An analogous definition applies to space bounds.

DEFINITION 8.4. Resource-bounded problem classes $\text{NTIME}^L(f)$ and $\text{NSPACE}^L(f)$ are exactly as in Definition 8.1, except that the programs p mentioned there are now allowed to be nondeterministic, using the definitions of acceptance and running time just given. Classes NPTIME, NPSPACE and NLOGSPACE are defined analogously.

For practical purposes it is not at all clear *how, or whether*, nondeterministic polynomial-time algorithms can be realised by deterministic polynomial-time computation. This intensely studied problem "PTIME = NPTIME?", often expressed as "P = NP?", has been open for many years. In practice, all solutions to such problems seem to take at least exponential time in worst-case situations. It is particularly frustrating that no one has been able to prove no subexponential worst-case solutions exist.

8.5. A BACKBONE HIERARCHY OF RESOURCE-BOUNDED PROBLEM CLASSES

The combinations of these resources lead to a widely encompassing "backbone" hierarchy (Figure 7):

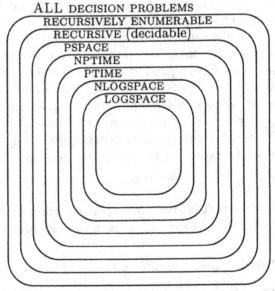

Figure 7. Backbone hierarchy of resource-bounded problem classes.

THEOREM 8.5.

LOGSPACE $\subseteq$ NLOGSPACE $\subseteq$ PTIME $\subseteq$ NPTIME $\subseteq$ PSPACE = NPSPACE $\subset$ REC $\subset$ RE

Here LOGSPACE, PTIME and PSPACE denote those problems solvable by deterministic algorithms in time and space, respectively, bounded by polynomial functions of the problem's input size. Classes NLOGSPACE, NPTIME, NPSPACE denote the problems decidable within the same bounds, but by nondeterministic algorithms that are allowed to or "guess"; and REC, RE are the *recursive* and *recursively enumerable* classes of decision problems already studied in Section 2.4.1.

The significance of the hierarchy is that a great number of practically interesting problems (e.g. maze searching, graph coloring, timetabling, regular expression manipulation, context-free grammar properties) can be precisely located at one or another stage in this progression.

Proof overview: To begin with, LOGSPACE $\subseteq$ NLOGSPACE, and PTIME $\subseteq$ NPTIME, and PSPACE $\subseteq$ NPSPACE. These are trivial by definition, since every ordinary deterministic program is also a nondeterministic program. NPTIME $\subseteq$ NPSPACE is also immediate since a program's space usage cannot exceed its time usage. Further, PSPACE $\subseteq$ REC is immediate by definition, and we have already proven REC $\subset$ RE in Theorem 2.4.

Remaining are: NLOGSPACE $\subseteq$ PTIME and NPSPACE $\subseteq$ PSPACE. These are proven by considering, for a given input d and resource-limited program p, the graph $G_p(d)$ whose vertices are *configurations* encountered in computations of p on d, and whose edges $C \to C'$ correspond to single computation steps.[10]

NLOGSPACE $\subseteq$ PTIME is shown by applying a fast algorithm for graph traversal to $G_p(d)$; and NPSPACE $\subseteq$ PSPACE is shown by applying a memory-economical $O((\log n)^2)$ space algorithm for graph traversal to $G_p(d)$. Details may be found in Chapter 23 of [14].

A collection of open problems

A long-standing open problem is whether these "backbone" inclusions are proper. Many researchers think that every inclusion is proper, but proofs have remained elusive. All that is known for sure (by a theorem analogous to 6.2) is that NLOGSPACE $\subset$ PSPACE, a very weak statement. (However, it implies that at least *one* inclusion among NLOGSPACE $\subseteq$ PTIME $\subseteq$ NPTIME $\subseteq$ PSPACE is proper!)

---

[10] Remark: for a deterministic program p, graph $G_p(d)$ has maximum node out-degree 1. A nondeterministic program p may have many computations on the same input, so nodes in $G_p(d)$ may have greater out-degrees.

9. Expressive power of some programming languages

The *expressivity* of a programming language L can be characterised "extensionally" as the class of all problems that can be solved by L-programs. By this measure, the languages WHILE, GOTO, FUN1, FUNHO, TM, CM and SRAM are all equally expressive: they can accept all the recursively enumerable sets, and can decide all the recursive sets.

In order to obtain nontrivial results on expressiveness we will look at languages L that are less than Turing-complete, and relate the class of L-solvable problems to the complexity classes of Section 8. In this section we overview results linking complexity classes to some sublanguages of the programming languages seen earlier.

Following are some of the more striking linkages, where a *read-only* (or *cons-free*) program is one with no constructor operator "cons". Full proofs are not given but some key ideas are sketched.

THEOREM 9.1. Set $A \subseteq \{0,1\}^*$ is in RE if and only if it is the domain of a read-only FUNHO program (untyped).

THEOREM 9.2. Set $A \subseteq \{0,1\}^*$ is in PTIME if and only if it is decidable by a read-only FUN1 program.

THEOREM 9.3. Set $A \subseteq \{0,1\}^*$ is in LOGSPACE if and only if it is decidable by a tail-recursive read-only FUN1 program.

Applying the construction of Theorem 3.4 to Theorem 9.3, we obtain:

COROLLARY 9.4. Set $A \subseteq \{0,1\}^*$ is in LOGSPACE if and only if it is decidable by a read-only WHILE program.

Remarks on the LOGSPACE characterisation. Despite the centrality of the class LOGSPACE, its Turing machine definition seems unnatural due to hardware restrictions on the number of tapes, their usage, and the external restriction on the run-time length of the work tape. There is, however, a "folklore theorem" that logspace Turing machines have the same decision power as certain *read-only* machines: two-way multihead finite automata. These machines can "see but not touch" their input. The read-only FUN1 and read-only WHILE languages naturally formulate just this idea: Theorems 9.3 and 9.4 are a re-expression of the folk theorem.

Remarks on the PTIME characterisation. Theorem 9.2 asserts that first-order cons-free read-only programs can solve all and only the problems in PTIME. Upon reflection this claim seems quite improbable, since it is easy (without using higher-order functions) to write cons-free read-only programs that run exponentially long before stopping. For example:

```
f x  = if  x = []  then true    else
         if  f(tl x) then f(tl x) else false
```

runs in time $2^{O(n)}$ on an input list of length n (regardless of whether call-by-value or lazy semantics are used), due to computing `f(tl x)` again and again.

What is wrong? The seeming paradox disappears once one understands what it is that the proof accomplishes. It has two parts:

— *Construction 1* shows that any problem in PTIME is computable by some first-order cons-free read-only program. Method: show how to simulate an arbitrary polynomial-time Turing machine by a first-order read-only program.

— *Construction 2* shows that any first-order cons-free read-only program decides a problem in PTIME. Method: show how to simulate an arbitrary first-order cons-free read-only program by a polynomial-time algorithm.

The method of Construction 2 in effect shows how to simulate a cons-free read-only program *faster than it runs*. It is not a step-by-step simulation, but uses a nonstandard "memoizing" semantic interpretation. (For the example above the value of `f(tl x)` would be saved when first computed, and fetched from memory each time the call is subsequently performed.)

The method of Construction 1 yields programs that almost always take exponential time to run; but this is not a contradiction since by Construction 2 the problems they are solving can be decided in polynomial time.

Overview of the following proofs. Each of the theorems is shown by a pair of simulations. Proof for the "only if" part begins with a Turing machine that decides (or accepts) set A. From this, we have to construct a functional program that faithfully simulates the Turing machine.

Proof for the "if" part begins with a functional program that decides (or accepts) set A. From this, we have to construct a Turing machine that faithfully simulates the functional program, and runs within the specified time or space bounds.

9.1. TURING MACHINE SIMULATION BY FUNCTIONAL PROGRAMS

As an intermediate step we show how to simulate a Turing machine by a functional program *equipped with natural numbers*. These numbers will be seen to be have size bound that are functions of tme or space limits on the Turing machine. Finally, it will be shown that the numbers and operations on them can be eliminated, simulating them by read-only functional programs.

Consider a one-tape Turing machine program $tm = 1\!:\!I_1 \; 2\!:\!I_2 \; \ldots m\!:\!I_m$. A "configuration" or state of tm has form $s = (\ell, \sigma)$ with control point $\ell \in \{1, \ldots, m, m+1\}$, and store of form $\sigma = \ldots BL\,\underline{S}\,RB\ldots$ where $S \in \{0, 1, B\}$ is the scanned symbol, and $L, R \in \{0, 1, B\}^*$ are the tape portions lying to its left and right

9.1.1. *Backward Turing machine simulation*
LEMMA 9.5. Let $tm = 1\!:\!I_1 \; 2\!:\!I_2 \; \ldots m\!:\!I_m$ be a one-tape TM-program with input $as = a_1 \ldots a_n \in \{0, 1\}^*$. Let $(\ell_1, \sigma_1) \to \ldots \to (\ell_t, \sigma_t) \to \ldots$ be the (finite or infinite) computation of tm on as, where $\ell_1 = 1$ and the initial tape is $\sigma_1 = \underline{B}\,a_1 \ldots a_n$. Define functions

$$\text{Tape} : \mathbb{Z} \times \mathbb{N} \to \{0, 1, B\}, \text{Label} : \mathbb{N} \to \{1, \ldots, m, m+1\}$$

as follows for any time and tape positions $t \geq 0, i \in \mathbb{Z}$:

$$\text{Tape}(i, t) = b_i \text{ if } s_t = (\ell_t, \sigma_t) \text{ and } \sigma_t = \ldots b_{-2} b_{-1} \underline{b_0}\, b_1\, b_2\, b_3 \ldots$$
$$\text{Label}(t) = \ell_t \text{ if } s_t = (\ell_t, \sigma_t)$$

Then the equations in Figure 8 hold for all $t, i \in \mathbb{Z}$ such that $0 \leq t < time_{tm}(as)$.

Backward simulation by a functional program with numbers
The equations of Figure 8 can be used to construct a functional program simulating Turing program tm. We use a Haskell-like syntax.

```
Run(d) = (Tape(0, Runtime(0)) == 1)  where

    Runtime(t) =  if Label(t) = m+1 then t else Runtime(t+1)

    Label(t)   =  if t=0 then 1 else case Label(t-1) of ...

    Tape(i,t)  =  if t=0 then ... else ...
```

At time $t = 0$:

$$\texttt{Label}(0) \;=\; 1 \qquad \text{At start: about to execute instruction 1}$$

$$\texttt{Tape}(i, 0) \;=\; \begin{cases} \texttt{a}_i & \text{if } 1 \le i \le n \\ \texttt{B} & \text{if } i \le 0 \text{ or } i > n \end{cases}$$

At time $t > 0$:

$$\texttt{Label}(t) \;=\; \begin{cases} \ell' & \text{if } \texttt{I}_{\texttt{Label}(t-1)} = \texttt{if S goto } \ell' \texttt{ else } \ell'' \\ & \text{and } \texttt{Tape}(0, t-1) = \texttt{S} \\ \ell'' & \text{if } \texttt{I}_{\texttt{Label}(t-1)} = \texttt{if S goto } \ell' \texttt{ else } \ell'' \\ & \text{and } \texttt{Tape}(0, t-1) \ne \texttt{S} \\ \texttt{Label}(t-1) + 1 & \text{otherwise} \end{cases}$$

$$\texttt{Tape}(i, t) \;=\; \begin{cases} \texttt{Tape}(i+1, t-1) & \text{if } \texttt{I}_{\texttt{Label}(t-1)} = \texttt{right} \\ \texttt{Tape}(i-1, t-1) & \text{if } \texttt{I}_{\texttt{Label}(t-1)} = \texttt{left} \\ \texttt{S} & \text{if } \texttt{I}_{\texttt{Label}(t-1)} = \texttt{write S and } i = 0 \\ \texttt{Tape}(i, t-1) & \text{otherwise} \end{cases}$$

Figure 8. Relations between the values of **Tape** and **Label**.

The missing parts ... are code for the relations seen in Figure 8. The calls to **Label** and **Tape** terminate since t decreases until it reaches 0. If **tm** terminates on input **as**, the functional program will too, since instruction **m+1** will eventually be reached, so the **Runtime** also terminates.

LEMMA 9.6. A Turing machine program **tm** can be simulated by a read-only number program. Further, if **tm** runs in time $T(n)$ on inputs of length n, then the natural numbers used by $\texttt{Tr}_{\texttt{tm}}$ are at largest $T(n)$.

9.1.2. *Forward Turing machine simulation*

A tape containing $\ldots \texttt{b}_{-2}\texttt{b}_{-1}\,\underline{\texttt{b}_0}\,\texttt{b}_1\,\texttt{b}_2\,\texttt{b}_3 \ldots$ can be represented by a pair of numbers l, r, where

- $l = \sum_{i=0}^{\infty} 2^i \beta_{-i}$ is the value of string $\ldots \texttt{b}_{-2}\texttt{b}_{-1}\texttt{b}_0$ *encoded as a base 3 number*, so $\beta = 0, 1, 2$ for symbols $\texttt{b} = \texttt{B}, 0$ and 1, respectively).
- $r = \sum_{i=1}^{\infty} 2^i \beta_i$ is the value of string $\texttt{b}_1\,\texttt{b}_2\,\texttt{b}_3 \ldots$, also as a base 3 number. (Note the order reversal, and the fact that blanks on left or right end of the tape do not contribute to l, r.)

The simulation is carried out by the program Tr_{tm} of Figure 9, whose variables l, r represent the Turing machine's current tape contents as above. Correctness: It is easy to see that the tape representation invariant is maintained whever a computation step is performed.

```
runTR(as)                  = aux(0,1,as)


aux (s,val,[])             = execute₁(s, val)
aux (s,val,(False:as))     = aux (s+1*val,3*val,as)
aux (s,val,(True:as))      = aux (s+2*val,3*val,as)


execute₁(l,r)              = SIMULATE₁
...                          ...
executeₘ(l,r)              = SIMULATEₘ
executeₘ₊₁(l,r)            = (l'mod'3==2)
```

For $\ell = 1, 2, \ldots, m$, SIMULATE$_\ell$ is defined as follows, where $\bar{S} = 0, 1, 2$, resp., if $S = B$, 0 or 1:

Form of instruction I_ℓ	Expression SIMULATE$_\ell$
right	execute$_{\ell+1}$ (l*3+r'mod'3, r'div'3)
left	execute$_{\ell+1}$ (l'div'3,r*3+l'mod'3)
write S	execute$_{\ell+1}$ (3*(l'mod'3)+$\bar{S}$, r)
if S goto ℓ' else ℓ''	if (l'mod'3 == $\bar{S}$) then execute$_{\ell'}$(l,r)
	else execute$_{\ell''}$(l,r)

Figure 9. Program Tr_{tm}: tail recursive Turing machine simulation

The Turing machine's initial tape contents, given input string $a_1 \ldots a_n \in \{0,1\}^*$, is $\underline{B}a_1 \ldots a_n B \ldots$. Program Tr_{tm} first constructs the pair $(0,r)$ representing the tape, where number r (computed by "aux") encodes input list as = $[a_1, \ldots, a_n]$ as described above. Functions execute$_1$, etc., are executed, each simulating the corresponding Turing instruction.

Tr_{tm} is clearly a first-order read-only number program. If program tm is a space $S(n)$-bounded Turing·machine, then the number of nonblank symbols in any reachable store $\ldots b_{-2} b_{-1} \underline{b_0} b_1 b_2 b_3 \ldots$ is at most $S(n)$. The tape is encoded as a pair(l, r) of ternary numbers, so $l, r \leq 3^{S(n)}$.

LEMMA 9.7. A Turing machine program tm can be simulated by a read-only tail-recursive number program. Further, if tm runs in space $S(n)$ on inputs of length n, and $S(n) \geq n$, then the natural numbers used by Tr_{tm} are at largest $3^{S(n)}$.

An extension of this construction to two-tape offline Turing machines yields the following result. The polynomial comes from the fact that $3^{k \log n} = n^{k \log 3}$.

LEMMA 9.8. Suppose Turing machine program **tm** runs in space $k \log n$ on inputs of length n. Then **tm** can be simulated by a read-only tail-recursive number program Tr_{tm}. Further, the natural numbers used by Tr_{tm} are bounded by a polynomial in n.

9.1.3. *Getting rid of the numbers*

Recall that the n appearing above is the length of the input, which is a list $\text{as} = (a_1 a_2 \ldots a_n)$.

LEMMA 9.9. Suppose read-only number program **p** has numbers bounded by a polynomial in n on inputs of legth n. Then there exists a read-only program **q** without numbers such that $[\![\text{p}]\!] = [\![\text{q}]\!]$. Further, if **q** is tail-recursive if **p** is.

Proof. The idea is to represent a number $i \leq n$ by a suffix of the input **as**. Claim: the following read-only functional program can (in effect) increment and decrement a counter of size up to n, and test for zero.

```
min(as)       = nil
max(as)       = as
decr(i,as)    = tl i
zero?(i,as)   = (i=nil)
incr(i,as)    = aux1(i,as,as)
aux1(i,j,as)  = if i=nil then aux2(tl j,as,as)
                         else aux1(tl i, tl j, as)
aux2(i,j,as)  = if i=nil then j
                         else aux2(tl i, tl j, as)
```

Idea: represent number $i \leq n$ by a length-i suffix i of $\text{as} = (a_1 a_2 \ldots a_n)$. Then **tl** i represents $\max(i-1,0)$, and the zero test is then simply a test on whether i is the empty list. The code for **incr** is trickier but exploits the fact that $i \leq n$. In effect it calls **aux1**$(0,n,n)$, which counts its i, j arguments down until $i = 0, j = n - i$, and then calls **aux2**$(n-i-1,n,n)$. Then **aux2** counts its first and second arguments down together until the first reaches 0, at which time the second is $n - (n - i - 1) = i + 1$.

This "library" allows simulation of any read-only number program **p** whose numbers are bounded by n. If a program's numbers are bounded by, say, n^2, this can be simulated as a pair of digits, each ranging from 0 to n. This generalizes to any polynomial.

Remark: these functions are all tail-recursive.

This plus Lemma 9.6 complete the "only if" part of Theorem 9.2. This plus Lemma 9.8 complete the "only if" part of Theorem 9.3.

9.2. SIMULATING READ-ONLY PROGRAMS BY TURING MACHINES

The "if" parts of Theorems 9.2 and 9.3 require simulating read-only programs by Turing machines.

9.2.1. *Relating* GOTO *and Turing machine states.*
Given a GOTO program, a Turing machine simulating it will store on its work tape the pointer values of all the GOTO variables. Conversely, given a Turing program, a GOTO program simulating it will encode, by means of pointer values, the current contents of the Turing machine's work tape. We first analyse their numbers of states.

Off-line Turing machine: Let **tm** have m instructions, work tape storage bound $k \log n$, and input $a_1 a_1 \ldots a_n \in \{0,1\}^*$. Then **tm** can enter at most

$$ m(n+2)3^{k \log n} = m(n+2)n^{k \log 3} $$

different total states. For fixed **tm** this number is $O(n^{1+k \log 3})$.

COROLLARY 9.10.   LOGSPACE $\subseteq$ PTIME.

Read-only GOTO **program:** Let **q** have m' instructions, k' variables, and input list $(a_1 a_2 \ldots a_n) \in I\!D_{01}$. During the computation every **q**-variable **X** value must be a pointer to one of:

1. The root of a suffix $(a_i a_{i+1} \ldots a_n)$ at a position i along the "spine" of the input list; or
2. The root of (nil.nil), the coding $c(1)$ of some $a_i = 1$; or
3. The atom nil.[11]

---

[11] A value of nil can arise 3 ways, but the effects while executing p are the same: either it is the coding of some $a_i = 0$; or the nil at the end of the input list; or it is the head or tail of $c(1) = $ (nil.nil).

Thus each variable can take on at most $n + 2$ values, so the number of program total states is bounded by: $m'(n + 2)^{k'}$. This is also a polynomial in n and thus a polynomial in $|(a_1 a_2 \ldots a_n)|$, giving hope for Theorem 9.3 since the programs of these two types have comparable numbers of states for a given input.

If part of Corollary 9.4. Let $q = 1{:}I1 \; 2{:}I2 \ldots m{:}Im$ be a cons-free GOTO program. Its instructions are of the form X := nil, X := Y, X := hd Y, X := tl Y, or if X goto ℓ else ℓ'.

The input to q is a list $(a_1 a_2 \ldots a_n) \in I\!D_{01}$ corresponding to string $a_1 a_2 \ldots a_n$ in $\{0, 1\}^*$. Possible variable values in any q state have been analysed in three cases above. Construct an off-line Turing machine p to simulate q, as follows.[12] The idea is to represent each variable X of q by its *position* and its *tag*: numbers (p_X, t_X) with values $(i, 0)$ in case 1, and $(0, 1)$ in case 2, and $(0, 0)$ in case 3. Turing machine program p stores each pair (p_X, t_X) on its work tape in binary, thus using at most $1 + \lceil \log n \rceil$ bits for each of q's variables. GOTO command I is simulated by Turing commands achieving the following effects:

GOTO command I	Effect of corresponding Turing code
X := nil	$(p_X, t_X) := (0, 0)$
X := Y	$(p_X, t_X) := (p_Y, t_Y)$
X := hd Y	$(p_X, t_X) :=$ if $p_Y > 0 \wedge a_{p_Y} = 1$ then$(0, 1)$ else$(0, 0)$
X := tl Y	$(p_X, t_X) :=$ if $p_Y > 0$ then $(p_Y - 1, t_Y)$ else $(0, 0)$
if X goto ℓ else ℓ'	if $t_X = 1 \vee (p_X > 0 \wedge a_{p_X} = 1)$ then goto ℓ else ℓ'

All is straightforward Turing programming; the test "$a_{p_Y} = 1$" is done by scanning the Turing input tape $a_1 a_1 \ldots a_n$ left to right, until p_Y symbols have been seen. It is easy to see that this is a faithful simulation that preserves the representation of the variables in GOTO program q.

If part of Theorem 9.2. Suppose A is decided by a read-only FUN1 program p, and that it is given input as $= a_1 \ldots a_n \in \{0, 1\}^*$, fixed in the rest of this proof. Consider a definition

```
f(x1,...,xk) = exp
```

---

[12] We assume $n > 0$; special case code gives the correct answer for $n = 0$.

Analogous to the preceding proof, let V denote the set of possible values of any parameter of $\mathbf{f}$. A value can only be a pointer to a suffix of $\mathbf{as}$, or to $(\mathbf{nil}.\mathbf{nil})$ or $\mathbf{nil}$, so V has $n + 2$ elementst. Thus there exist at most $(n + 2)^k$ possible argument tuples for $\mathbf{f}$.

We sketch a "memoising" implementation: for each such function $\mathbf{f}$, the simulator maintains a table $mfg(\mathbf{f}) \subseteq V^k \times (V \cup \{\bullet\})$. The program is executed from the start, but the tables $mfg(\mathbf{f})$ for every $\mathbf{f}$ are maintained (and exploited). Initially, all are empty except for the initial function, containing $mfg(\mathbf{f}_1) = ((\mathbf{as}), \bullet)$.

Each time a function call $\mathbf{f}(\mathbf{e}_1 \ldots \mathbf{e}_k)$ occurs, the parameter values $(v_1, \ldots, v_k)$ are obtained. If a pair $((v_1, \ldots, v_k), v) \in mfg(\mathbf{f})$ with $v \neq \bullet$, then $\mathbf{f}$ has already been applied to this argument tuple, and simulation continues with value v. If not, then $\mathbf{f}$ is executed in the normal way, terminating with a value v. Before continuing, tuple $((v_1, \ldots, v_k), v)$ is added to $mfg(\mathbf{f})$.

It is easy to see that no function call is simulated more than once. A straightforward time analysis reveals that the tables contain at most polynomially many elements; and that the total simulation time is also polynomially bounded.

9.3. HIGHER-ORDER FUNCTIONS

First, Theorem 9.1 is just a restatement of the familiar result that expressions in the untyped lambda-calculus can compute arbitrary recursive functions. The situation is quite different, however, when *typed* functional programs are considered. Before stating the results (from the paper [11]) we define a simple monomorphic type system for FUNHO.

9.3.1. *Types*
Each parameter and function in a program is assigned a type (nonpolymorphic). Each type τ denotes a set of values $[\![\tau]\!]$. Type $\bullet$ denotes $[\![\bullet]\!] = \mathbb{D}$, and type $\tau \to \tau'$ denotes $\{f : [\![\tau]\!] \to [\![\tau]\!]'\}$, all functions with one argument of type τ and one result of type τ'. Multiple-argument functions are handled by "currying," for example $f : \tau_1 \to (\tau_2 \to \tau_3)$ is regarded as a function of two arguments, of types τ_1 and τ_2, and result type τ_3.

Judgement $\mathbf{e} :: \tau$, signifying that *expression* $\mathbf{e}$ *has type* τ is defined in Figure 10. A fully type-annotated function definition has form

$$\mathbf{f}^{\tau_1 \to \tau_2 \to \ldots \tau_m \to \tau} \mathbf{x}_1^{\tau_1} \mathbf{x}_2^{\tau_2} \ldots \mathbf{x}_m^{\tau_m} = \mathbf{e}^\tau$$

A program must be well-typed to be syntactically legal. Henceforth, all programs are assumed to be fully annotated. For readability type superscripts are omitted when clear from context.

$$\frac{}{\texttt{d} :: \bullet} \quad \frac{}{\texttt{x}^\tau :: \tau} \quad \frac{}{\texttt{f}^\tau :: \tau} \quad \frac{\texttt{e}_1 :: \bullet \quad \texttt{e}_2 :: \tau \quad \texttt{e}_3 :: \tau}{\texttt{if } \texttt{e}_1\texttt{e}_2\texttt{e}_3 :: \tau}$$

$$\frac{\texttt{e}_1 :: \bullet \quad \texttt{e}_2 :: \bullet}{\texttt{cons } \texttt{e}_1 \texttt{e}_2 :: \bullet} \quad \frac{\texttt{e} :: \bullet}{\texttt{hd } \texttt{e} :: \bullet} \quad \frac{\texttt{e} :: \bullet}{\texttt{tl } \texttt{e} :: \bullet}$$

$$\frac{\texttt{e}_1 :: \tau \to \tau' \quad \texttt{e}_2 :: \tau}{\texttt{e}_1\texttt{e}_2 :: \tau'}$$

Figure 10. Expression types

DEFINITION 9.11. The *order* of a type is defined by $\mathrm{order}(\bullet) = = 0$; and $\mathrm{order}(\tau \to \tau') = \max(1 + \mathrm{order}(\tau), \mathrm{order}(\tau'))$.

DEFINITION 9.12. Program p has *data order k* if every τ, τ_i in any defined function has order k or less. Thus f above has order $k + 1$ if at least one τ_i or τ has order k, justifying the usual term "first-order program" for one that manipulates data of order 0.

Remark: A "first-order program" is one that has data order 0 (it is the functions defined in the program that have order 1).

9.3.2. *The expressive power of higher-order types*

We precisely characterize, in terms of complexity classes, the effects on expressive power of various combinations of three possible limits on programs' *operations on data:* 1) constructors and destructors; 2) the order of their *data values:* 0, 1, or higher; and 3) their *control structures:* general recursion, tail recursion, primitive recursion. The links are summed up in the table of Figure 11, and confirm programmers' intuitions that higher-order types indeed give a greater problem-solving ability.

Many combinations are Turing-complete, so such programs compute all the partial recursive functions. A classic Turing-incomplete language is got by restricting data to order 0 and control to "fold." Such programs compute the *primitive recursive* functions.

Figure 11 shows the effect of higher-order types on the computing power of programs of type $\bullet \to \bullet$. Each entry is a complexity class, i.e., the collection of decision problems solvable by programs restricted by row and column indices. RO stands for "read-only," i.e., programs without constructors, and RW stands for "read-write."

Program class	Data order 0	order 1	Order 2	...	Limit
RO, untyped	–	–	–	...	REC.ENUM
RW, typed	REC.ENUM	REC.ENUM	REC.ENUM	...	REC.ENUM
RW, fold only	PRIM.REC	PRIM^{1}REC	PRIM^{2}REC	...	(System T)
RO, typed	PTIME	EXPTIME	EXP^{2}TIME	...	ELEM'TARY
RO tail recursive	LOGSPACE	PSPACE	EXPSPACE	...	ELEM'TARY
RO, fold only	LOGSPACE	PTIME	PSPACE	...	ELEM'TARY

Figure 11. Expressivity: data order and control. RO = read-only = cons-free.

Explanation of the table. The restrictions RO and RW were explained above. With or without these restrictions, programs may have general recursion, tail recursion, or primitive recursion, yielding 6 combinations. There are only 5 rows, though, since RW=RWTR because an unrestricted program can be converted into a tail recursive equivalent by standard techniques involving a stack of activation records.

The column indices restrict the orders of program data types. An "order $k + 1$" program can have functions of type $\tau \to \tau'$ where data type τ is of order k. Thus, for instance, the first column describes first-order programs, whose parameters are booleans or lists of booleans. Each entry is the collection of decision problems solvable by programs restricted by row and column indices.

Rows 1, 2: These program classes are all Turing complete. Consequently they can accept exactly the recursively enumerable subsets of $\{0, 1\}^*$.

Row 3: These programs have unlimited data operations and types, but control is limited to primitive recursion, familiar to functional programmers under the name "fold right".[13] Such first-order programs accept exactly the sets whose characteristic functions are primitive recursive (true regardless of whether data are strings or natural numbers).

Higher-order primitive recursive functions appeared in Gödel's *System T* many years ago [5], [21]. They are currently of much interest in the field of constructive type theory due to the Curry-Howard isomorphism, which

---

[13] Kleene's definition of primitive recursion is a bit more general than "fold right," but is easily programmed using fold right and composition. See [9] for details.

makes it possible to extract programs from proofs. Primitive recursion comes because of proofs by induction; extraction of programs using general recursion is much less natural.

Row 4: These programs have unlimited control, but allow only *read-only* access to their data. List destructor operations hd and tl are allowed, but not the constructor cons. Even though this may seem a draconian restriction from a programmer's viewpoint, the class of problems that can be solved is respectably large. Order 1 programs can solve any problem that lies in PTIME; order 2 programs, with first-order functions as data values, can solve any problem in the quite large class EXPTIME, etc. In general, any increase in the order of data types leads to a proper increase in the solvable problems, since it is known that PTIME is properly contained in EXPTIME, and so on up the hierarchy.

Row 5 characterizes read-only programs restricted to *tail recursion*, in which no function may call itself in a nested way. Order 1 tail recursive programs accept all and only problems in LOGSPACE, a well-studied subset of PTIME. Higher-order tail recursive programs accept problems in the (properly) larger space-bounded classes PSPACE, EXPSPACE, etc.

(Tail recursion is of operational interest because at run time (assuming eager evaluation, i.e., call-by-value semantics) the call stack depth has a constant depth bound, regardless of input data. Such a program may be converted to nonrecursive imperative form by replacing each function call by a GOTO, and realizing function parameter passing by assignments to global variables.)

Row 6 characterizes read-only programs restricted so all recursion must be expressed using "fold right," i.e., only primitive recursion is allowed. Order 1 read-only primitive recursive programs accept only problems in LOGSPACE and are thus equivalent to tail-recursive programs. At higher orders this equivalence vanishes; the primitive recursive read-only programs' abilities to solve decision problems grow only at "half speed": a data order increase of 2 is needed to achieve the same increase in decision ability that an increase of 1 achieved for general or tail recursive programs.

Limit of rows 3, 4 and 5 It is clear that the union of the classes in row 4 equals the union for row 5 and for row 6. This is the class of problems solvable in time bounded by $2^{2^{\cdots^{2^n}}}$, where the height of the exponent stack is any natural number. This is well-known as the class of *elementary* sets, and was studied by logicians before complexity theory began.

Scope and contribution of these results. The results in Rows 1 and 2 are classical. The results in row 5 appear to be new; and the results in row 4, while in a sense anticipated by [6], are proven in [11] for the first time in a programming language context. The results in Row 6 are obtained from [7] and [8] by re-interpreting results from finite model theory.

On expressivity. It has long been known that order $k+1$ primitive recursive programs are properly more powerful than order k primitive recursive programs, i.e., $\text{PRIM}^k\text{REC.} \subset \text{PRIM}^{k+1}\text{REC.}$ This is of little practical interest, however, since even the order 0 class PRIM.REC. is enormous, properly containing such classes as NPTIME and ELEMENTARY.

Does the use of functions as data values give a greater problem-solving ability? By Figure 11 the answer is "no" for unrestricted programs, and "yes" for all the restricted languages we consider. The only uncertainty is with the read-only primitive recursive programs; for these, an increase in data order of at least 2 is needed in order to guarantee a proper increase in problem-solving power.

Is general recursion more powerful than tail recursion? For first-order read-only programs, this question has classical import since, by the table's first column (rows 4, 5) this is equivalent to the question: Is PTIME a proper superset of LOGSPACE? This is, alas, an unsolved question, open since it was first formulated in the early 1970s. An equivalent question (rows 4, 6): *Is general recursion more powerful than primitive recursion?*.

However the situation is different for second and higher orders. For higher-order read-only programs, the question of whether general recursion is stronger than tail recursion is also open, equivalent to EXPTIME $\supset$ PSPACE? But the answer is "yes" when comparing general recursion to primitive recursion, since it is known that EXPTIME properly includes PTIME.

On strongly normalizing languages. If we assume as usual that programs in a strongly normalizing language have only primitive recursive control, there exist problems solvable by read-only general recursive programs with data order $1, 2, 3, \ldots$, but not solvable by read-only strongly normalizing programs of the same data orders. This suggests an inherent weakness in the extraction of programs from proofs by the Curry-Howard isomorphism.

10. Complete problems

An old slogan: "If you can't solve your problems, then at least you can classify them."

Complete problems for the problem classes in the hierarchy. In spite of the many unresolved questions concerning proper containments in the "backbone," a great many problems have been proven to be *complete* for the various classes. If such a problem X is complete for class $\mathcal{C}$, then X is "hardest" for that class in the sense that if it lay within the next smaller class (call it $\mathcal{B}$ with $\mathcal{B} \subseteq \mathcal{C}$), then *every* problem in class $\mathcal{C}$ would also be in class $\mathcal{B}$, i.e. the hierarchy would "collapse" there, giving $\mathcal{B} = \mathcal{C}$. Complete problems are known to exist and will be constructed for every class in the "backbone" except for LOGSPACE (since no smaller class is present) and REC (for more subtle reasons.)

10.1. REDUCTION OF ONE PROBLEM TO ANOTHER

The questions concerning concerning proper containments within

$$\text{LOGSPACE} \subseteq \text{NLOGSPACE} \subseteq \text{PTIME} \subseteq \text{NPTIME} \subseteq \text{PSPACE} = \text{NPSPACE}$$

are apparently quite difficult, since they have remained open since the 1970s in spite of many reseachers' best efforts to solve them. This has led to an alternative approach: to define *complexity comparison* relations $\leq$, also called *reductions*, between decision problems. Different relations $\leq$ will be appropriate for different complexity classes.

The statement $A \leq B$ can be interpreted as "problem A is no more difficult to solve than problem B," or even better: "given a way to solve B, a way to solve A can be found." One application is to prove a problem undecidable by reducing the halting problem to it. Intuitively: if $HALT$ is thought of as having infinite complexity, proving $HALT \leq B$ shows that B also has infinite complexity.

Further, we can use this idea to break a problem class such as RE or NPTIME into *equivalence subclasses* by defining A and B to be of equivalent complexity if $A \leq B$ and $B \leq A$.

The idea of reduction of one problem to another has been studied for many years, for example quite early in Mathematical Logic as a tool for comparing the complexity of two different unsolvable problems or undecidable sets. Many ways have been devised to reduce one problem to another since Emil Post's pathbreaking work in 1944.

10.2. MANY-ONE REDUCTIONS

Reduction $A \leq B$ where (say) $A, B \subseteq I\!\!D$ can be defined in several ways.

DEFINITION 10.1. Let $A, B \subseteq I\!\!D$. Then A is *many-one* reducible to B by total function $f : I\!\!D \to I\!\!D$ if for any $\mathbf{d} \in I\!\!D$ we have

$\mathbf{d} \in A$ if and only if $f(\mathbf{d}) \in B$

Clearly, if f is *computable* then an algorithm to decide membership in B can be used to decide membership in A. Further, if f is *efficiently* computable, then existence of an efficient B algorithm implies existence of an efficient A algorithm.

For proofs of undecidability, the only essential requirement is that the reduction be computable. Complexity classifications naturally involve bounds on the complexity of the questions that can be asked, for example of the function f used for many-one reducibility. In order to study, say, the class NPTIME, it is natural to limit one's self to reductions that can be computed by deterministic algorithms in polynomial time. Complexity comparison is thus almost always via *resource-bounded reduction* of one problem to another: $A \leq B$ means that one can efficiently transform an algorithm that solves B within given resource bounds into an algorithm that solves A within similar resource bounds. Some instances:

$\leq_{recursive}$ f must be a total computable function

$\leq_{ptime}$ f must be computable in polynomial time

$\leq_{logspace}$ f must be computable in logarithmic space

Some interesting facts, proven in the book, lie at the core of complexity theory:

1. Each of the several complexity classes C in the backbone hierarchy above LOGSPACE possesses *complete problems*. Such a problem (call it H) lies in class C, and is "hardest" for it in the sense that $A \leq H$ for each problem A in C. Class C may have many hard problems.
2. A complete problem H for class D has the property that if $H \in C$ for a lower class C in the hierarchy, then $C = D$: the two classes are identical. Informally said, the hierarchy *collapses* at that point.
3. Even more interesting: Many *natural and practically motivated* problems have been proven to be complete for one or another complexity class C.

10.3. THREE EXAMPLE PROBLEMS

DEFINITION 10.2.

1. A k-clique in undirected graph G is a set of k vertices such that G has an edge between every pair in the set. Figure 6 shows a graph G containing two 3-cliques: one with vertices 1, 2, 5 and another with vertices 1, 4, 5.
2. A boolean expression $\mathcal{F}$ is *closed* if it has no variables. If closed, $\mathcal{F}$ can be *evaluated* by the familiar rules such as $true \wedge false = false$.

3. A *truth assignment* for $\mathcal{F}$ is a function θ mapping variables to truth values such that $\theta(\mathcal{F})$ is a closed boolean expression. $\mathcal{F}$ is *satisfiable* if it evaluates to *true* for some truth assignment θ. $\mathcal{F}$ is in *conjunctive normal form*, abbreviated CNF, if it is a conjunction of disjunctions of variables or their negations.

4. By definition SAT = $\{\mathcal{F} \mid \mathcal{F}$ is a satisfiable boolean CNF expression.$\}$

For an example of a satisfiable formula, the CNF expression

$$(A \vee \neg B) \wedge (B \vee C) \wedge (\neg A \vee \neg C)$$

is satisfied by truth assignment $\theta = [A \mapsto false, B \mapsto false, C \mapsto true]$.

Three combinatorial decision problems. Following are three typical and interesting problems which will serve to illustrate several points. In particular, each will be seen to be complete, i.e. hardest, problems among all those solvable in a nondeterministic time or space class. The problems:

GAP $= \{ G \mid$ directed graph $G = (V, E, v_0, v_{end})$ has a path from vertex v_0 to $v_{end} \}$

CLIQUE $= \{ (G, k) \mid$ undirected graph G has a k-clique $\}$

SAT $= \{ \mathcal{F} \mid \mathcal{F}$ is a satisfiable boolean CNF expression $\}$

THEOREM 10.3. GAP is complete for the class NLOGSPACE with respect to $\leq_{logspace}$ reductions.

THEOREM 10.4. *HALT* is complete for the class RE with respect to $\leq_{recursive}$ reductions, where:

$$HALT = \{(\text{p.d}) \mid \text{p is a GOTO-program and } [\![\text{p}]\!](\text{d}) \neq \bot\}$$

10.4. REDUCING SAT TO CLIQUE IN POLYNOMIAL TIME

Many superficially quite different problems turn out to be "sisters under the skin," in the sense that each can be efficiently reduced to the other. We show an example reduction SAT $\leq_{ptime}$ CLIQUE. This means that there is a polynomial time computable function f which, when given any CNF

128

boolean expression $\mathcal{F}$, will yield a pair $f(\mathcal{F}) = (G, k)$ such that graph G has a k-clique if and only if $\mathcal{F}$ is a satisfiable expression.

This implies that CLIQUE is at least as hard to solve as SAT in polynomial time: given a polynomial time algorithm p to solve CLIQUE, one could answer the question "is $\mathcal{F}$ satisfiable?" by first computing $f(\mathcal{F})$ and then running p on the result.

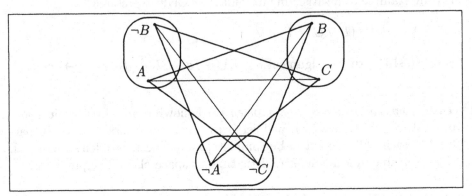

Figure 12.   The graph $f((A \vee \neg B) \wedge (B \vee C) \wedge (\neg A \vee \neg C))$.

Construction.   Given a conjunctive normal form boolean expression $\mathcal{F} = C_1 \wedge \ldots \wedge C_k$, construct a graph $f(\mathcal{F}) = (G, k)$ where graph $G = (V, E)$ and

1. $V = $ the set of occurrences of literals in $\mathcal{F}$
2. $E = \{(a, b) \mid a$ and b are not in the same conjunct of $\mathcal{F}$, and neither negates the other$\}$

For an instance, the expression

$$(A \vee \neg B) \wedge (B \vee C) \wedge (\neg A \vee \neg C)$$

would give graph $f(\mathcal{F})$ as in Figure 12. The expression $\mathcal{F}$ is satisfied by truth assignment $[A \mapsto false, B \mapsto false, C \mapsto true]$, which corresponds to the 3-clique $\{\neg A, \neg B, C\}$. More generally, if $\mathcal{F}$ has n conjuncts, there will be one n-clique in $f(\mathcal{F})$ for every truth assignment that satisfies $\mathcal{F}$, and these will be the only n-cliques in $f(\mathcal{F})$.

It is also possible to show that CLIQUE $\underset{ptime}{\leq}$ SAT, but by a less straightforward construction. Consequence: SAT $\in$ PTIME if and only if CLIQUE $\in$ PTIME.

10.5. COMPLEXITY OF PROBLEMS ABOUT BOOLEAN PROGRAMS

THEOREM 10.5. SAT is complete for the class NPTIME with respect to $\leq_{ptime}$ reductions.

Proof is by reduction. The key idea (due to Cook [4]) is to construct, from a nondeterministic polynomial-time-bounded Turing program p and input d, a CNF formula $\mathcal{F}$ that is satisfiable if and only if p has an accepting computation on input d.

The breakthrough in Cook's proof was to see how to relate Turing machine computations with boolean expressions, so the existence of a choice sequence leading the Turing machine to accept its input corresponds to the existence of variables making the constructed boolean expression true. As a stepping-stone, we show how to reduce Turing machine computations to computations by *boolean programs*.

Program analysis is in general undecidable due the halting problem's recursive unsolvability, and Rice's general result that all nontrivial extensional program properties are undecidable. On the other hand, *finite-memory* programs [17, 18] do have decidable properties since their entire state spaces can be computed. A series of theorems relating well-known complexity classes to finite program analysis appears in [14].

DEFINITION 10.6. (The language BOOLE) A *boolean program* is an input-free program $p = 1:I_1 ... m:I_m$ where each instruction I and expression E is of form given by:

$$I ::= X := E \mid I_1; I_2 \mid goto \ \ell \mid if \ E \ then \ I_1 \ else \ I_2$$
$$E ::= X \mid true \mid false \mid E_1 \lor E_2 \mid E_1 \land E_2 \mid \neg E \mid E_1 \Rightarrow E_2$$
$$X ::= X0 \mid X1 \mid ...$$

Semantics is what one expects, not detailed here.

THEOREM 10.7. The problem of deciding membership in the following subsets of BOOLE-*programs* is characterised as follows:

1. $\{\ p \in \text{BOOLE-}programs \mid [\![p]\!] = true\ \}$: complete for PSPACE (deterministic or nondeterministic polynomial space)
2. $\{\text{goto-free } p \in \text{BOOLE-}programs \mid [\![p]\!] = true\ \}$: complete for PTIME (deterministic polynomial time)
3. $\{\text{goto-free } p \in \text{BOOLE-}programs \mid [\![q;p]\!] = true \text{ for some } q\ \}$: complete for NPTIME (nondeterministic polynomial time)

10.6. REDUCING A PSPACE PROBLEM TO ACCEPTANCE BY A BOOLEAN PROGRAM

We now show Part 1 of Theorem 10.7. Minor variations on the construction suffice for Parts 2 and 3, as well.

LEMMA 10.8. Let Turing machine program p run in polynomial space on any input of length n. Then there is a function

$$f : \{0,1\}^* \rightarrow \{\text{BOOLE programs}\}$$

such that for any input $d = a_1 \dots a_n$ (each $a_i \in \{0,1\}$) we have

$$\text{p accepts d if and only if } [\![f(d)]\!]^{\text{BOOLE}} = \text{true}$$

Further, $f(d)$ is computable in space $O(\log |d|)$.

Proof. Let $p = 1\!:\!I_1 \dots m\!:\!I_m$. Without loss of generality assume that $d \in A$ if and only if p has a computation that terminates at program control point $m+1$. By assumption there is a polynomial $\pi(n)$ bounding from above the run time of p on inputs of length n.

We show how, given p, to construct from d, a Boolean program $q = f(d)$ as desired. The construction is closely analogous to that used in Theorem 9.5, "specialised" to the known data input and space bound. Let $d = a_1 \dots a_n$ with each $a_i \in \{0,1\}$.

Boolean variables. Program $q = f(d)$ has a total of $(m+1)+6\pi(n)$ boolean variables, grouped as follows. Recall that any uninitialized variable has start value false.

Name	Intended interpretation: true iff	Index range
L_ℓ	Instruction ℓ of p is about to be simulated	$1 \leq \ell \leq m+1$
T_i^a	TM square i holds symbol $a \in \{B,0,1\}$	$-\pi(n) \leq i \leq \pi(n)$

Relation to variables of Figure 8. First, the q variables are not indexed by time parameter t: the current time will be implicit in q's current control point. Omitting this gives $\text{Label} : \{1,\dots,m,m+1\}$, which can be represented by $m+1$ boolean variables $L_1,\dots,L_{m+1}$. Similarly, Second, $\text{Tape} : \mathbb{Z} \times \mathbb{N} \rightarrow \{0,1,B\}$ is reduced to a function $T : \mathbb{Z} \rightarrow \{0,1,B\}$. A Boolean program is desired, so this is further split into three Boolean variables $T^0, T^1, T^B : \mathbb{Z} \rightarrow \{\text{true}, \text{false}\}$. Since a computation with time bounded by polynomial $\pi(n)$ will satisfy $-\pi(n) \leq i \leq \pi(n)$, we represent each function $T(i)$ by $6\pi(n)$ Boolean variables T_i^a. The intended relation is thus:

$$L_\ell = \text{true} \quad \text{iff} \quad \text{Label}(t) = \ell$$
$$T_i^a = \text{true} \quad \text{iff} \quad \text{Tape}(i, t) = a$$

Structure of program $q = f(d)$. For brevity we use two abbreviations for multiple assignment ($a \in \{0, 1, B\}$):

$T_i := a;$ stands for $T_i^0:=\text{false};\ T_i^1:=\text{false};\ T_i^B:=\text{false};\ T_i^a:=\text{true};$

$T_i := T_j;$ stands for $T_i^0:=T_j^0;\ T_i^1:=T_j^1;\ T_i^B:=T_j^B;$

In Figure 13, STEP is a sequence of Boole instructions simulating the effect of one Turing machine computational step. It does a "case on instruction" to select the appropriate instruction to be simulated in p. STEP is defined by:

$$\text{STEP} = \text{if } L_1 \text{then } \overline{I}_1;\ \text{if } L_2 \text{then } \overline{I}_2;\dots;\ \text{if } L_m \text{then } \overline{I}_m;$$

Finally, Figure 14 defines the simulation of the individual instructions.

$L_1 := \text{true};$ *Start simulation: Turing instruction 1*

     Assignments for initial tape contents: blanks left and right of input

$T_{-\pi(n)} :\equiv B;\dots;\ T_0 :\equiv B;$
$T_1 :\equiv a_1;\dots;\ T_n :\equiv a_n;\ T_{n+1} :\equiv B;\dots;\ T_{\pi(n)} :\equiv B;$

     Simulate Turing machine instructions

while not L_{m+1} do STEP;   Answer := true

Figure 13. Turing simulation by Boolean program $q = f(d)$.

LEMMA 10.9. Let $1 \leq t \leq \pi(n)$, and consider the state of Boolean program q just before executing the instructions in STEP. Then L_ℓ will be true for exactly one ℓ, and for each i with $-\pi(n) \leq i \leq \pi(n)$, and Boolean variable T_i^a will be true for exactly one a.

Proof. This is immediate for $t = 1$ by the way q was constructed. Further, examination of the cases in Figure 14 shows that these properties are preserved in going from t to $t+1$.

LEMMA 10.10. At the end of execution, q will assign true to Answer if and only if p accepts d.

Proof. This is straightforward. "Only if" is by induction on Turing machine computation length, and "if" is by induction on the length of q.

Turing instruction: I_ℓ	Boolean sequence $\bar{I}_\ell$
right	$T_{-\pi(n)} :\equiv T_{-\pi(n)+1}; \ldots; T_0 :\equiv T_1; \ldots;$ $T_{\pi(n)-1} :\equiv T_{\pi(n)}; T_{\pi(n)} :\equiv B;$ $L_\ell := \text{false}; L_{\ell+1} := \text{true};$
left	$T_{\pi(n)} :\equiv T_{\pi(n)-1}; \ldots; T_1 :\equiv T_0; \ldots;$ $T_{-\pi(n)+1} :\equiv T_{-\pi(n)}; T_{-\pi(n)} :\equiv B;$ $L_\ell := \text{false}; L_{\ell+1} := \text{true};$
write a	$T_0 :\equiv a; L_\ell := \text{false}; L_{\ell+1} := \text{true};$
if a goto ℓ' else ℓ''	$L_\ell := \text{false};$ $L'_\ell := T_0^a; L_{\ell''} := \text{not } T_0^a;$

Figure 14. Turing instructions simulated by Boolean instructions.

LEMMA 10.11. $q = f(d)$ is constructible from d in space $O(\log |d|)$.

Proof. Function f can be computed by one loop to generate the loop initialisations in q, and one loop over $t = 1, \ldots, \pi(n)$. Inside this loop for STEP with $\ell = 1, \ldots, m$, one more loop for $-\pi(n) \le i \le \pi(n)$ generates code for each Turing machine instruction type.
We now show Part 2 of Theorem 10.7.

LEMMA 10.12. Let Turing machine program p run in polynomial time $\pi(n)$ on any input of length n. Then there is a function

$$f : \{0,1\}^* \to \{\text{goto-free BOOLE programs}\}$$

such that for any input $d = a_1 \ldots a_n$ (each $a_i \in \{0,1\}$) we have

$$p \text{ accepts d if and only if } [\![f(d)]\!]^{\text{BOOLE}} = \text{true}$$

Further, $f(d)$ is computable in space $O(\log |d|)$.

Proof. Define q = $f(d)$ as in Figure 13, but with one change: replace the final line

$$\text{while not } L_{m+1} \text{ do STEP; \quad Answer := true}$$

by

$$\text{STEP}^{\pi(n)}; \quad \text{Answer := true}$$

Clearly q is free of goto's, and it is easy to see that it yields **true** if and only if Turing program p accepts input d.

We now show Part 3 of Theorem 10.7.

LEMMA 10.13. Let nondeterministic Turing machine program p run in polynomial time $\pi(n)$ on inputs of length n. Then there is a function

$$g : \{0, 1\}^* \to \{\text{BOOLE programs}\}$$

such that for any input d = $a_1 \ldots a_n$ (each $a_i \in \{0, 1\}$) we have

p accepts d if and only if for some Boolean program b, $[\![b; f(d)]\!]^{\text{BOOLE}}$ = **true**

Further, $g(d)$ is computable in space $O(\log |d|)$.

Proof. The construction of Figure 14 needs extension, because nondeterministic program p may have the choice instruction **choose** ℓ' **or** ℓ''. Extending Figure 14 as follows.

Simulate instruction **choose** ℓ' **or** ℓ'' by Boolean instructions:

$$L_\ell := \text{false; if Oracle}_t \text{ then } L_{\ell'} := \text{true else } L_{\ell''} := \text{true;}$$

Define q = $g(d)$ = Init;$f(d)$ where $f(d)$ is as in Lemma 10.13, and Init; initialises to **false** all variables in $f(d)$ *except* the oracle variables Oracle$_t$.

Now Turing program p accepts input d if and only if there exists a sequence of choices leading to success. Clearly this will be true if and only if for some assignemt b of truth values to the oracle variables, b; Init;$f(d)$ yields **true**.

11. Related Work

The book [19] by Kfoury, Moll and Arbib has similar aims, but [14] goes further in two respects: it covers complexity theory as well as computability;

and it demonstrates the advantages that come from structuring *both programs and data.* [19] deals with structured programs, but uses the natural numbers as data.

Paul Voda is redeveloping recursion theory on the basis of Lisp-like data structures. A book is forthcoming; and [20] is a recent article.

References

1. Andersen, N. and Jones, N. D., Generalizing Cook's construction to imperative stack programs, Lecture Notes in Computer Science 812, pp. 1-18, Springer-Verlag, 1994.
2. Cook, S. A., Linear-time simulation of deterministic two-way pushdown automata, *Information Processing* (IFIP) 71, C.V. Freiman, (ed.), North-Holland, pp. 75-80, 1971.
3. Cook, S. A., Characterizations of pushdown machines in terms of time-bounded computers. Journal of the ACM **18** (1971), 4-18.
4. Cook, S. A., The complexity of theorem-proving procedures, *Proceedings Third Symposium on the Theory of Computing*, pp. 151-158, ACM Press, 1971.
   Ben-Amram, A. and Jones, N. D. A precise version of a time hierarchy theorem. Fundamenta Informaticae, vol. 38, pp. 1-15. 1999.
5. Girard, J.-Y. and Lafont, Y. and Taylor, P. *Proofs and Types*, volume 7 of *Cambridge Tracts in Theoretical Computer Science* Cambridge University Press, 1989.
6. Goerdt, A. Characterizing complexity classes by general recursive definitions in higher types. *Information and Computation* **101** (1992), 201-218.
7. Goerdt, A. Characterizing complexity classes by higher type primitive recursive definitions. *Theoretical Computer Science* **101** (1992), 45-66.
8. Goerdt, A. and Seidl, H. Characterizing complexity classes by higher type primitive recursive definitions, Part II. *Proceedings 6th International Meeting for Young Computer Scientists*, Lecture Notes in Computer Science **464** (1990), 148-158.
9. Hutton, G. A tutorial on the universality and expressiveness of fold. *Journal of Functional Programming* **1** (1):1-17 (1953).
10. Jones, N. D., Space-bounded reducibility among combinatorial problems, *Journal of Computer and System Science*, vol. 11, pp. 68-85, 1975.
11. Jones, N. D., The Expressive Power of Higher-order Types or, Life without CONS, *Journal of Functional Programming* accepted for publication, 2000.
12. Jones, N. D., A note on linear-time simulation of deterministic two-way pushdown automata, *Information Processing Letters* vol. 6, pp. 110-112, 1977.
13. Jones, N. D., Constant time factors *do* matter. In Steven Homer, editor, *STOC '93. Symposium on Theory of Computing*, pages 602-611. ACM Press, 1993.
14. Jones, N. D., *Computability and Complexity from a Programming Perspective.* The MIT Press, 1997.
15. Jones, N. D. and Gomard, C. and Sestoft, P. *Partial Evaluation and Automatic Program Generation.* Prentice-Hall International, 1993.
16. Jones, N. D., LOGSPACE and PTIME characterized by programming languages. *Theoretical Computer Science*, 1998.

17. Jones, N. D., and Muchnick, S. Even simple programs are hard to analyze. *Journal of the Association for Computing Machinery*, 24(2):338–350, 1977.
18. Jones, N. D., and Muchnick, S. Complexity of finite memory programs with recursion. *Journal of the Association for Computing Machinery*, 25(2):312–321, 1978.
19. Kfoury, A. J. and Moll, R. N. and Arbib, M. A. *A Programming Approach to Computability*. Texts and monographs in Computer Science. Springer-Verlag, 1982.
20. Voda, P. Subrecursion as Basis for a Feasible Programming language. In Steven Homer, editor, *Logic in Comp. Science September 94*. Lecture Notes in Computer Science 933, Springer Verlag 1995.
21. Voda, P. A simple ordinal recursive normalization of Gödel's T. *Computer Science Logic*, Lecture Notes in Computer Science **1414** (1997), 491–509.

LOGICAL FRAMEWORKS—A BRIEF INTRODUCTION

FRANK PFENNING (fp+@cs.cmu.edu)
Carnegie Mellon University

Abstract. A logical framework is a meta-language for the formalization of deductive systems. We provide a brief introduction to logical frameworks and their methodology, concentrating on LF. We use first-order logic as the running example to illustrate the representations of syntax, natural deductions, and proof transformations.

We also sketch a recent formulation of LF centered on the notion of canonical form, and show how it affects proofs of adequacy of encodings.

Key words: Logical frameworks, type theory

1. Introduction

Deductive systems, given via axioms and rules of inference, are a common conceptual tool in mathematical logic and computer science. They are used to specify many varieties of logics and logical theories as well as aspects of programming languages such as type systems or operational semantics. A *logical framework* is a meta-language for the specification of deductive systems. A number of different frameworks have been proposed and implemented for a variety of purposes. In this brief introduction we highlight the major themes, concepts, and design choices for logical frameworks and provide pointers to the literature for further reading. We concentrate specifically on the LF type theory and we briefly mention other approaches below and in Section 5.

Logical frameworks are subject to the same general design principles as other specification and programming languages. They should be simple and uniform, providing concise means to express the concepts and methods of the intended application domains. Meaningless expressions should be detected statically and it should be possible to structure large specifications and verify that the components fit together. There are also concerns specific to logical frameworks. Perhaps most importantly, an implementation must

H. Schwichtenberg and R. Steinbrüggen (eds.), Proof and System-Reliability, 137–166.
© 2002 *Kluwer Academic Publishers. Printed in the Netherlands.*

be able to check deductions for validity with respect to the specification of a deductive system. Secondly, it should be feasible to prove (informally) that the representations of deductive systems in the framework are adequate so that we can trust formal derivations. We return to each of these points when we discuss different design choices for logical frameworks.

Historically, the first logical framework was Automath (de Bruijn, 1968) and its various languages, developed during the late sixties and early seventies. The goal of the Automath project was to provide a tool for the formalization of mathematics without foundational prejudice. Therefore, the logic underlying a particular mathematical development was an integral part of its formalization. Many of the ideas from the Automath language family have found their way into modern systems. The main experiment conducted within Automath was the formalization of Landau's *Foundations of Analysis*. In the early eighties the importance of constructive type theories for computer science was recognized through the pioneering work of Martin-Löf (Martin-Löf, 1980). On the one hand, this led to a number of systems for constructive mathematics and the extraction of functional programs from constructive proofs. On the other hand, it strongly influenced the design of LF (Harper et al., 1993), sometimes called the Edinburgh Logical Framework (ELF). Concurrent with the development of LF, frameworks based on higher-order logic and resolution were designed in the form of generic theorem provers (Paulson, 1986) and logic programming languages (Nadathur and Miller, 1988). The type-theoretic and logic programming approaches were later combined in the Elf language (Pfenning, 1991). At this point, there was a pause in the development of new frameworks, while the potential and limitations of existing systems were explored in numerous experiments (see Pfenning (1996)). The mid-nineties saw renewed activity with implementations of frameworks based on inductive definitions such as FS_0 (Feferman, 1988; Basin and Matthews, 1996) and ALF (Altenkirch et al., 1994), partial inductive definitions (Eriksson, 1994), substructural frameworks (Miller, 1994; Cervesato and Pfenning, 1996), rewriting logic (Martì-Oliet and Meseguer, 1993), and labelled deductive systems (Gabbay, 1994). A full discussion of these is beyond the scope of this introduction—the reader can find some remarks in handbook articles on the subject of logical frameworks (Basin and Matthews, 2001; Pfenning, 2001b).

Some researchers distinguish between logical frameworks and *meta-logical frameworks* (Basin and Constable, 1993), where the latter is intended as a meta-language for reasoning *about* deductive systems rather than *within* them. Clearly, any meta-logical framework must also provide means for specifying deductive systems, though with different goals. Space does not permit a discussion of meta-logical frameworks in this survey.

The remainder of this introduction is organized as follows: in Section 2 we discuss the representation of the syntax of a logic and in Section 3 the representation of judgments and deductions. In Section 4 we provide a the details of a formulation of the dependently typed λ-calculus as a point of reference, before concluding in Section 5. As an example throughout we use natural deduction for first-order logic.

2. Abstract syntax

The specification of a deductive system usually proceeds in two stages: first we define the syntax of an object language and then the axioms and rules of inference. In order to concentrate on the meanings of expressions we ignore issues of concrete syntax and parsing and concentrate on specifying abstract syntax. Different framework implementations provide different means for customizing the parser in order to embed the desired object-language syntax.

As an example throughout we consider formulations of intuitionistic and classical first-order logic. In order to keep the length of this survey manageable, we restrict ourselves to the fragment containing implication, negation, and universal quantification. The reader is invited to test his or her understanding by extending the development to include a more complete set of connectives and quantifiers. Representations of first-order intuitionistic and classical logic in various logical frameworks can be found in the literature (see, for example, Harper et al. (1993), Pfenning (2001a)).

Our fragment of first-order logic is constructed from individual variables x, function symbols f, and predicate symbols p in the usual way. We assume each function and predicate symbol has a unique arity, indicated by a superscript, but generally omitted since it will be clear from the context. Individual constants are function symbols of arity 0 and propositional constants are predicate symbols of arity 0.

$$
\begin{array}{lll}
\text{Terms} & t & ::= x \mid f^k(t_1,\ldots,t_k) \\
\text{Atoms} & P & ::= p^k(t_1,\ldots,t_k) \\
\text{Formulas} & A & ::= P \mid A_1 \supset A_2 \mid \neg A \mid \forall x.\, A
\end{array}
$$

We assume that there is an infinite number of variables x. The set of function and predicate symbols is left unspecified in the general development of logic. We therefore view our specification as open-ended. A commitment, say, to arithmetic would fix the available function and predicate symbols. We write x and y for variables, t and s for terms, and

A, B, and C for formulas. There are some important operations on terms and formulas required for the presentation of inference rules. Specifically, we need the notions of free and bound variable, the renaming of bound variables, and the operations of substitution $[t/x]s$ and $[t/x]A$, where the latter may need to rename variables bound in A in order to avoid variable capture. We assume that these operations are understood and do not define them formally. An assumption generally made in connection with variable names is the so-called *variable convention* (Barendregt, 1980) which states that expressions differing only in the names of their bound variables are considered identical. We examine to which extent various frameworks support this convention.

2.1. SIMPLY-TYPED REPRESENTATION

We would like to capture both the variable name convention and the validity of a framework object representing a term or formula *internally*. A standard method to achieve this is to introduce representation types. We begin with *simple types*. The idea is to introduce type constants i and o for object-level terms and formulas, respectively. Implication, for example, is then represented by a constant of type o $\rightarrow$ (o $\rightarrow$ o), that is, a formula constructor taking two formulas as arguments employing the standard technique of Currying. We could now represent variables as strings or integers; instead, we use meta-language variables to model object-language variables. This requires that we enrich the representation language to include higher-order terms, which leads us to the simply-typed λ-calculus, $\lambda^{\rightarrow}$. As we will see from the adequacy theorem for our representation (Theorem 1), the methodology of logical frameworks only requires canonical forms, which are β-normal and η-long. We will capture this in the syntax of our representation language by allowing only β-normal forms; the fact that they are η-long is enforced in the typing rules (see Section 4).

$$
\begin{array}{lll}
\text{Types} & A & ::= a \mid A_1 \rightarrow A_2 \\
\text{Atomic Objects} & R & ::= c \mid x \mid R\, N \\
\text{Normal Objects} & N & ::= \lambda x.\, N \mid R
\end{array}
$$

We use a to range over type constants, c over object constants, and x over object variables. We follow the usual syntactic conventions: $\rightarrow$ associates to the right, and application to the left. Parentheses group subexpressions, and the scope of a λ-abstraction extends to the innermost enclosing parentheses or to the end of the expression. We allow tacit α-conversion (renaming of bound variables) and write $[M/x{:}A]N$ for the β-normal form of the result of capture-avoiding substitution of M for x in N. Constants and variables are declared and assigned types in a signature Σ and context Γ, respectively. Neither is permitted to declare constants or variables more than once. The main judgments of the type theory are

$\Gamma \vdash_\Sigma N \Leftarrow A$ N has type A, and
$\Gamma \vdash_\Sigma R \Rightarrow A$ R has type A.

Here $N \Leftarrow A$ checks N against a given A, while $R \Rightarrow A$ synthesizes A from R or fails. In both cases we assume Σ and Γ are given. We have omitted type labels from λ-abstractions, since they are inherited from the type that a canonical object is checked against.

Returning to the representation of first-order logic, we introduce two declarations

i : type
o : type

for the types of representations of terms and formulas, respectively. For every function symbol f of arity k, we add a corresponding declaration

$$ f : \underbrace{i \rightarrow \cdots \rightarrow i}_{k} \rightarrow i. $$

One of the central ideas in using a λ-calculus for representation is to represent object-language variables by meta-language variables. Through λ-abstraction at the meta-level we can properly delineate the scopes of variables bound in the object language. For simplicity, we give corresponding variables the same name in the two languages.

$$ \ulcorner x \urcorner = x $$
$$ \ulcorner f(t_1, \ldots, t_k) \urcorner = f \ulcorner t_1 \urcorner \ldots \ulcorner t_k \urcorner $$

Predicate symbols are dealt with like function symbols. We add a declaration

$$ p : \underbrace{i \rightarrow \cdots \rightarrow i}_{k} \rightarrow o $$

for every predicate symbol p of arity k. Here are the remaining cases of the representation function.

$$ \ulcorner p(t_1, \ldots, t_k) \urcorner = p \ulcorner t_1 \urcorner \ldots \ulcorner t_k \urcorner $$
$$ \ulcorner A_1 \supset A_2 \urcorner = \text{imp} \ulcorner A_1 \urcorner \ulcorner A_2 \urcorner \qquad \text{imp} : o \rightarrow o \rightarrow o $$
$$ \ulcorner \neg A \urcorner = \text{not} \ulcorner A \urcorner \qquad \text{not} : o \rightarrow o $$
$$ \ulcorner \forall x. A \urcorner = \text{forall} (\lambda x. \ulcorner A \urcorner) \qquad \text{forall} : (i \rightarrow o) \rightarrow o $$

The last case in the definition introduces the concept of *higher-order abstract syntax*. If we represent variables of the object language by variables in the meta-language, then variables bound by a construct in the object language must be bound in the representation as well. The simply-typed λ-calculus has a single binding operator λ, so all variable binding is mapped to binding by λ. This idea goes back to Church's formulation of classical type theory and Martin-Löf's system of arities (Nordström et al., 1990).

This leads to the first important representation principle of logical frameworks employing higher-order abstract syntax: *Bound variable renaming in the object language is modeled by α-conversion in the meta-language*. Since we follow the variable convention in the meta-language, the variable convention in the object language is automatically supported in a framework using the representation technique above. Consequently, it cannot be used directly for binding operators for which renaming is not valid such as occur, for example, in module systems of programming languages.

The variable binding constructor "∀" of the object language is translated into a second-order constructor forall in the meta-language, since delineating the scope of x introduces a function $(\lambda x.\ulcorner A\urcorner)$ of type $i \to o$. What does it mean to apply this function to an argument $\ulcorner t\urcorner$? This question leads to the concept of *compositionality*, a crucial property of higher-order abstract syntax. We can show by a simple induction that

$$[\ulcorner t\urcorner/x{:}i]\ulcorner A\urcorner = \ulcorner [t/x]A\urcorner.$$

Note that the substitution on the left-hand side is in the framework, on the right in first-order logic. Both substitutions are defined to rename bound variables as necessary in order to avoid the capturing of variables free in t. Compositionality also plays a very important role in the representation of deductions in Section 3; we summarize it as: *Substitution in the object language is modeled by substitution in the meta-language*.

The declarations of the basic constants above are *open-ended* in the sense that we can always add further constants without destroying the validity of earlier representations. However, the definition also has an inductive character in the sense that the validity judgment of the meta-language ($\lambda^\to$, in this case) is defined inductively by some axioms and rules of inference. Therefore we can state and prove that there is a *compositional bijection* between well-formed formulas and normal objects of type o. Since a term or formula may have free individual variables, and they are represented by corresponding variables in the meta-language, we must take care to declare them with their proper types in the meta-language context. We refer to the particular signature with the declarations for term and formula constructors as F.

THEOREM 1 (Adequacy).

1. We have

$$x_1{:}\mathsf{i},\dots,x_n{:}\mathsf{i} \vdash_F M \Leftarrow \mathsf{i} \quad \textit{iff} \quad M = \ulcorner t \urcorner \quad \textit{for some } t,$$

where the free variables of term t are among $x_1,\dots,x_n$.
2. We have

$$x_1{:}\mathsf{i},\dots,x_n{:}\mathsf{i} \vdash_F M \Leftarrow \mathsf{o} \quad \textit{iff} \quad M = \ulcorner A \urcorner \quad \textit{for some } A,$$

where the free variables of formula A are among $x_1,\dots,x_n$.
3. The representation function $\ulcorner \cdot \urcorner$ is a compositional bijection *in the sense that*

$$[\ulcorner t \urcorner/x{:}\mathsf{i}]\ulcorner s \urcorner = \ulcorner [t/x]s \urcorner \quad \textit{and} \quad [\ulcorner t \urcorner/x{:}\mathsf{i}]\ulcorner A \urcorner = \ulcorner [t/x]A \urcorner$$

Proof: In one direction we proceed by an easy induction on the structure of terms and formulas. Compositionality can also be established directly by an induction on the structure of s and A, respectively.

In the other direction we carry out an induction over the structure of the derivations of $M \Leftarrow \mathsf{i}$ and $M \Leftarrow \mathsf{o}$. To prove that the representation function is a bijection, we write down its inverse on canonical forms and prove that both compositions are identity functions. □

We summarize the main technique introduced in this section. The technique of *higher-order abstract syntax* represents object language variables by meta-language variables. It requires λ-abstraction in the meta-language in order to properly delineate the scope of bound variables, which suggests the use of the simply-typed λ-calculus as a representation language. In this approach, variable renaming is modeled by α-conversion, and capture-avoiding substitution is modeled by meta-level substitution. Representations in LF are open-ended, rather than inductive.

3. Judgments and deductions

After designing the representation of terms and formulas, the next step is to encode the axioms and inference rules of the logic under consideration. There are several styles of deductive systems which can be found in the literature, such as the axiomatic method, categorical definitions, natural deduction, or sequent calculus.

Logical frameworks are typically designed to deal particularly well with some of these systems, while being less appropriate for others. The Automath languages were designed to reflect and promote good informal mathematical practice. It should thus be no surprise that they were particularly well-suited to systems of natural deduction. The same is true for the LF type theory, so we concentrate on the problem of representing natural deduction first. Other systems, including sequent calculi, can also be directly encoded (Pfenning, 2000).

3.1. PARAMETRIC AND HYPOTHETICAL JUDGMENTS

First, we introduce some terminology used in the presentation of deductive systems introduced with their modern meaning by Martin-Löf (1985). We will generally interpret the notions as proof-theoretic rather than semantic, since we would like to tie them closely to logical frameworks and their implementations. A *judgment* is defined by *inference rules*. An inference rule has zero or more premises and a conclusion; an axiom is an inference rule with no premises. A judgment is *evident* or *derivable* if it can be deduced using the given rules of inference. Most inference rules are *schematic* in that they contain meta-variables. We obtain *instances* of a schematic rule by replacing meta-variables with concrete expressions of the appropriate syntactic category. Each instance of an inference rule may be used in derivations. We write $\mathcal{D} :: J$ or

$$\mathcal{D}$$
$$J$$

when $\mathcal{D}$ is a derivation of judgment J. All derivations we consider must be finite.

Natural deduction further employs *hypothetical judgments*. We write

$$\frac{}{J_1} u$$
$$\vdots$$
$$J_2$$

to express that judgment J_2 is derivable under hypothesis J_1 labelled u, where the vertical dots may be filled by a *hypothetical derivation*. Hypotheses have scope, that is, they may be *discharged* so that they are not available outside a given subderivation. We annotate the discharging inference with the label of the hypothesis. The meaning of a hypothetical judgment can be explained by substitution: We can substitute an arbitrary deduction $\mathcal{E} :: J_1$ for each occurrence of a hypothesis J_1 labelled u in $\mathcal{D} :: J_2$ and obtain a derivation of J_2 that no longer depends on u. We write this substitution as $[\mathcal{E}/u]\mathcal{D} :: J_2$ or two-dimensionally by writing $\mathcal{E}$ above the hypothesis justified by u. For this to be meaningful we assume that multiple occurrences of a label annotate the same hypothesis, and that hypotheses satisfy the structural properties of *exchange* (the order in which hypotheses are made is irrelevant), *weakening* (a hypothesis need not be used) and *contraction* (a hypothesis may be used more than once).

An important related concept is that of a *parametric judgment*. Evidence for a judgment J that is parametric in a variable a is given by a derivation $\mathcal{D} :: J$ that may contain free occurrences of a. We refer to the variable a as a *parameter* and use a and b to range over parameters. We can substitute an arbitrary object O of the appropriate syntactic category for a throughout $\mathcal{D}$ to obtain a deduction $[O/a]\mathcal{D} :: [O/a]J$. Parameters also have scope and their discharge is indicated by a superscript as for hypothesis labels.

3.2. NATURAL DEDUCTION

Natural deduction is defined via a single judgment

$$\vdash^{N} A \qquad \text{formula } A \text{ is true}$$

and the mechanisms of hypothetical and parametric deductions explained in the previous section.

In natural deduction each logical symbol is characterized by its *introduction rule* or *rules* which specify how to infer a conjunction, disjunction, implication, universal quantification, etc. The *elimination rule* or *rules* for the connective then specify how we can use a conjunction, disjunction, etc. Underlying the formulation of the introduction and elimination rules is the principle of *orthogonality*: each connective should be characterized purely by its rules, and the rules should only use judgmental notions and not other logical connectives. Furthermore, the introduction and elimination rules for a logical connective cannot be chosen freely—as explained below, they should match up in order to form a coherent system. We call these conditions *local soundness* and *local completeness*.

Local soundness expresses that we should not be able to gain information by introducing a connective and immediately eliminating it. That is, if we introduce and then eliminate a connective we should be able to reach the same judgment without this detour. We show that this is possible by exhibiting a *local reduction* on derivations. The existence of a local reduction shows that the elimination rules are not too strong—they are locally sound.

Local completeness expresses that we should not lose information by introducing a connective. That is, given a judgment there is some way to eliminate its principal connective and then re-introduce it to arrive at the original judgment. We show that this is possible by exhibiting a *local expansion* on derivations. The existence of a local expansion shows that the elimination rules are not too weak—they are locally complete.

Under the Curry-Howard isomorphism between proofs and programs (Howard, 1980), local reduction corresponds to β-reduction and local expansion corresponds to η-expansion. We express local reductions and expansions via judgments which relate derivations of the same judgment.

$$
\begin{array}{c} \mathcal{D} \\ \vdash^N A \end{array} \quad \Longrightarrow_R \quad \begin{array}{c} \mathcal{D}' \\ \vdash^N A \end{array} \qquad \mathcal{D} \text{ locally reduces to } \mathcal{D}'
$$

$$
\begin{array}{c} \mathcal{D} \\ \vdash^N A \end{array} \quad \Longrightarrow_E \quad \begin{array}{c} \mathcal{D}' \\ \vdash^N A \end{array} \qquad \mathcal{D} \text{ locally expands to } \mathcal{D}'
$$

In the spirit of orthogonal definitions, we proceed connective by connective, discussing introduction and elimination rules and local reductions and expansions.

Implication. To derive $\vdash^N A \supset B$ we assume $\vdash^N A$ to derive $\vdash^N B$. Written as a hypothetical judgment:

$$
\cfrac{\cfrac{\overline{\vdash^N A}^{\,u}}{\vdots} \\ \vdash^N B}{\vdash^N A \supset B} \supset\!I^u
$$

The hypothetical derivation describes a construction by which we can transform a derivation of $\vdash^N A$ into a derivation of $\vdash^N B$. This is accomplished by substituting the derivation of $\vdash^N A$ for every use of the hypothesis $\vdash^N A$ labelled u in the derivation of $\vdash^N B$. The elimination rule expresses just that: if we have a derivation of $\vdash^N A \supset B$ and also a derivation of $\vdash^N A$, then we can obtain a derivation of $\vdash^N B$.

$$
\cfrac{\vdash^N A \supset B \qquad \vdash^N A}{\vdash^N B} \supset\!E
$$

The local reduction carries out the substitution of derivations explained above.

$$
\cfrac{\cfrac{\cfrac{\overline{\vdash^N A}^{\,u}}{\begin{array}{c}\mathcal{D}\\ \vdash^N B\end{array}}}{\vdash^N A \supset B} \supset\!I^u \qquad \begin{array}{c}\mathcal{E}\\ \vdash^N A\end{array}}{\vdash^N B} \supset\!E
\quad \Longrightarrow_R \quad
\cfrac{\cfrac{\mathcal{E}}{\vdash^N A}^{\,u}}{\begin{array}{c}\mathcal{D}\\ \vdash^N B\end{array}}
$$

The derivation on the right depends on all the hypotheses of $\mathcal{E}$ and $\mathcal{D}$ except u, for which we have substituted $\mathcal{E}$. The reduction described above may significantly increase the overall size of the derivation, since the deduction $\mathcal{E}$ is substituted for each occurrence of the assumption labeled u in $\mathcal{D}$ and may therefore be replicated.

Local expansion is specified in a similar manner.

$$
\begin{array}{c} \mathcal{D} \\ \vdash^N A \supset B \end{array} \implies_E \quad
\cfrac{\cfrac{\begin{array}{c}\mathcal{D}\\ \vdash^N A \supset B\end{array} \qquad \cfrac{}{\vdash^N A}\, u}{\vdash^N B}\supset E}{\vdash^N A \supset B}\supset I^u
$$

Here, u must be a new label, that is, it cannot already be used in $\mathcal{D}$.

Negation. In order to derive $\vdash^N \neg A$ we assume $\vdash^N A$ and try to derive a contradiction. This is the usual formulation, but has the disadvantage that it requires falsehood ($\bot$) as a logical symbol, thereby violating the orthogonality principle. Thus, in intuitionistic logic, one ordinarily thinks of $\neg A$ as an abbreviation for $A \supset \bot$. An alternative rule sometimes proposed assumes $\vdash^N A$ and tries to derive $\vdash^N B$ and $\vdash^N \neg B$ for some B. This also breaks the usual pattern by requiring the logical symbol we are trying to define ($\neg$) in a premise of the introduction rule. However, there is another possibility to explain the meaning of negation without recourse to implication or falsehood. We specify that $\vdash^N \neg A$ should be derivable if we can conclude $\vdash^N p$ for any formula p from the assumption $\vdash^N A$. In other words, the deduction of the premise is hypothetical in the assumption $\vdash^N A$ and parametric in the formula p.

$$
\cfrac{\cfrac{}{\vdash^N A}\, u \\ \vdots \\ \vdash^N p}{\vdash^N \neg A}\neg I^{p,u} \qquad\qquad
\cfrac{\vdash^N \neg A \qquad \vdash^N A}{\vdash^N C}\neg E
$$

According to our intuition, the parametric judgment should be derivable if we can substitute an arbitrary concrete formula C for the parameter p and obtain a valid derivation. Thus, p may not already occur in the conclusion $\neg A$, or in any undischarged hypothesis. The reduction rule for negation follows from this interpretation and is analogous to the reduction for implication.

$$\cfrac{\cfrac{\cfrac{\overline{\vdash^N A}\ u}{\begin{array}{c}\mathcal{D}\\ \vdash^N p\end{array}}}{\vdash^N \neg A}\neg\mathrm{I}^{p,u} \qquad \cfrac{\mathcal{E}}{\vdash^N A}}{\vdash^N C}\neg\mathrm{E} \qquad \Longrightarrow_R \qquad \cfrac{\cfrac{\mathcal{E}}{\vdash^N A}\ u}{\begin{array}{c}[C/p]\mathcal{D}\\ \vdash^N C\end{array}}$$

The local expansion is also similar to that for implication.

$$\cfrac{\mathcal{D}}{\vdash^N \neg A} \quad\Longrightarrow_E\quad \cfrac{\cfrac{\cfrac{\mathcal{D}}{\vdash^N \neg A}\qquad \cfrac{\overline{\vdash^N A}\ u}{}}{\vdash^N p}\neg\mathrm{E}}{\vdash^N \neg A}\neg\mathrm{I}^{p,u}$$

Universal quantification. Under which circumstances should we be able to derive $\vdash^N \forall x.\,A$? This clearly depends on the domain of quantification. For example, if we know that x ranges over the natural numbers, then we can conclude $\vdash^N \forall x.\,A$ if we can derive $\vdash^N [0/x]A$, $\vdash^N [1/x]A$, etc. Such a rule is not effective, since it has infinitely many premises. Thus one usually uses induction principles as inference rules. However, in a general treatment of predicate logic we would like to prove statements which are true for *all* domains of quantification. Thus we can only say that $\vdash^N \forall x.\,A$ should be derivable if $\vdash^N [a/x]A$ is derivable for an arbitrary new parameter a. Conversely, if we know $\vdash^N \forall x.\,A$, we know that $\vdash^N [t/x]A$ for any term t.

$$\cfrac{\vdash^N [a/x]A}{\vdash^N \forall x.\,A}\forall\mathrm{I}^a \qquad\qquad \cfrac{\vdash^N \forall x.\,A}{\vdash^N [t/x]A}\forall\mathrm{E}$$

The superscript a is a reminder about the proviso for the introduction rule: the parameter a must be "new", that is, it may not occur in any undischarged hypothesis in the derivation of $[a/x]A$ or in $\forall x.\,A$ itself. In other words, the derivation of the premise is parametric in a. If we know that $\vdash^N [a/x]A$ is derivable for an arbitrary a, we can conclude that $\vdash^N [t/x]A$ should be derivable for any term t. Thus we have the reduction

$$\cfrac{\cfrac{\cfrac{\mathcal{D}}{\vdash^N [a/x]A}}{\vdash^N \forall x.\,A}\forall\mathrm{I}^a}{\vdash^N [t/x]A}\forall\mathrm{E} \qquad\Longrightarrow_R\qquad \cfrac{[t/a]\mathcal{D}}{\vdash^N [t/x]A}$$

Here, $[t/a]\mathcal{D}$ is our notation for the result of substituting t for the parameter a throughout the deduction $\mathcal{D}$. For this to be sensible, we must know that a does not already occur in A, because otherwise the conclusion of $[t/a]\mathcal{D}$ would be $[t/a][t/x]A$. The local expansion just introduces and immediately discharges the parameter.

$$
\begin{array}{ccc}
\mathcal{D} & & \dfrac{\dfrac{\mathcal{D}}{\vdash^N \forall x.\, A}}{\vdash^N [a/x]A} \,\forall E \\[2ex]
\vdash^N \forall x.\, A & \Longrightarrow_E & \rule{0pt}{1pt} \\[1ex]
& & \dfrac{\rule{0pt}{1pt}}{\vdash^N \forall x.\, A} \,\forall I^a
\end{array}
$$

Classical logic. The inference rules so far only model intuitionistic logic, and some classically true formulas such as Peirce's law $((A \supset B) \supset A) \supset A$ (for arbitrary A and B) or double negation $(\neg\neg A) \supset A$ (for arbitrary A) are not derivable. There are a number of equivalent ways to extend the system to full classical logic, typically using negation (for example, the law of excluded middle, proof by contradiction, or double negation elimination). In the fragment without disjunction or falsehood, we might choose either a rule of double negation or proof by contradiction.

$$
\dfrac{\vdash^N \neg\neg A}{\vdash^N A}\;\text{dbneg}
\qquad\qquad
\dfrac{\begin{array}{c}\dfrac{\rule{2em}{0pt}}{\vdash^N \neg A}\,u \\[1ex] \vdots \\[1ex] \vdash^N A\end{array}}{\vdash^N A}\;\text{contr}^u
$$

The rule for classical logic (whichever we choose to adopt) breaks the pattern of introduction and elimination rules. One can still formulate some reductions for classical derivations, but natural deduction is at heart an intuitionistic calculus. The symmetries of classical logic are better exhibited in sequent calculi.

3.3. DEDUCTIONS AS OBJECTS

In the representation of deductions we have a basic choice between simply representing derivable judgments, or giving an explicit representation of deductions as objects. There are many circumstances where we are interested in deductions as explicit objects. For example, we may want to extract functional programs from constructive (or even classical) derivations. Or we may want to implement proof transformation and presentation tools in a theorem proving environment. If we do not trust a complex theorem prover, we may construct it so that it generates proof objects which can be independently verified. In the architecture of proof-carrying code (Necula, 1997), deductions represented in LF are attached to mobile code to certify safety (Necula, 2002). Another class of applications is the implementation of the meta-theory of the deductive systems under consideration. For example, we may want to show that natural deductions and derivations in the sequent calculus define the same theorems and exhibit translations between them. Here, we are interested in formally specifying the local reductions and expansions.

The simply-typed λ-calculus, which we used to represent the terms and formulas of first-order logic, is also a good starting point for the representation of natural deductions. As we will see below we need to refine it further in order to allow an internal validity condition for deductions. This leads us to λ^{Π}, the dependently typed λ-calculus underlying the LF logical framework (Harper et al., 1993).

We begin by introducing a new *type* nd of natural deductions. An inference rule is a constant function from deductions of the premises to a deduction of the conclusion. For example,

impe : nd $\rightarrow$ nd $\rightarrow$ nd

might be used to represent implication elimination. A hypothetical deduction is represented as a function from a derivation of the hypothesis to a derivation of the conclusion.

impi : (nd $\rightarrow$ nd) $\rightarrow$ nd

One can clearly see that this representation requires an *external* validity condition since it does not carry the information about which instance of the judgment is shown by the derivation. For example, we have

$$\vdash \mathsf{impi}\,(\lambda u.\,\mathsf{impe}\,u\,u) \Leftarrow \mathsf{nd}$$

but this term does not represent a valid natural deduction. An external validity predicate can be specified using hereditary Harrop formulas and is executable in λProlog (Felty and Miller, 1988). However, it is not *prima facie* decidable.

Fortunately, it is possible to refine the simply-typed λ-calculus so that validity of the representation of derivations becomes an *internal* property, without destroying the decidability of the type system. This is achieved by introducing *indexed types*. Consider the following encoding of the elimination rule for implication.

$$\text{impe} \; : \; \text{nd} \, (\text{imp} \, A \, B) \to \text{nd} \, A \to \text{nd} \, B$$

In this specification, $\text{nd} \, (\text{imp} \, A \, B)$ is a *type*, the type representing derivations of $\vdash^N A \supset B$. Thus we speak of the *judgments-as-types* principle. The *type family* nd is indexed by objects of type o.

$$\text{nd} \; : \; \text{o} \to \text{type}$$

We call $\text{o} \to \text{type}$ a *kind*. Secondly, we have to consider the status of the free variables A and B in the declaration. Intuitively, impe represents a whole family of constants, one for each choice of A and B. Schematic declarations like the one given above are desirable in practice, but they lead to an undecidable type checking problem (Dowek, 1993). We can recover decidability by viewing A and B as additional arguments in the representation of $\supset$E. Thus impe has four arguments representing A, B, a derivation of $A \supset B$ and a derivation of A. It returns a derivation of B. With the usual function type constructor we could only write

$$\text{impe} \; : \; \text{o} \to \text{o} \to \text{nd} \, (\text{imp} \, A \, B) \to \text{nd} \, A \to \text{nd} \, B.$$

This does not express the dependencies between the first two arguments and the types of the remaining arguments. Thus we name the first two arguments A and B, respectively, and write

$$\text{impe} \; : \; \Pi A{:}\text{o}. \, \Pi B{:}\text{o}. \, \text{nd} \, (\text{imp} \, A \, B) \to \text{nd} \, A \to \text{nd} \, B.$$

This is a closed type, since the *dependent function type* constructor Π binds the following variable. From the consideration above we can see that the typing rule for application of a function with dependent type should be

$$\frac{\Gamma \vdash_\Sigma R \Rightarrow \Pi x{:}A. \, B \qquad \Gamma \vdash_\Sigma N \Leftarrow A}{\Gamma \vdash_\Sigma R \, N : [N/x{:}A^-]B} \; \text{app}$$

Here, A^- is the simple type that arises by erasing all dependencies and indices from A (see Section 4). For example, given a variable $p{:}\text{o}$ we have

$$p{:}\text{o} \vdash_\Sigma \text{impe} \, (\text{not} \, p) \, p \Rightarrow \text{nd} \, (\text{imp} \, (\text{not} \, p) \, p) \to \text{nd} \, (\text{not} \, p) \to \text{nd} \, p$$

where the signature Σ contains the declarations for formulas and inferences rules developed above. The counterexample $\text{impi} \, (\lambda u. \, \text{impe} \, u \, u)$ from above is now no longer well-typed: the instance of A would have to be of the form $A_1 \supset A_2$ (first occurrence of u) and simultaneously be equal to A_1 (second occurrence of u). This is clearly impossible. The rule for λ-abstraction does not change much from the simply-typed calculus.

$$\frac{\Gamma, x{:}A \vdash_\Sigma N \Leftarrow B}{\Gamma \vdash_\Sigma \lambda x. \, N \Leftarrow \Pi x{:}A. \, B} \; \text{lam}$$

The variable x may now appear free in B, whereas without dependencies it could only occur free in N. Note that no type label on λ-abstractions is needed, since the given type $\Pi x{:}A.\,B$ supplies it. From these two rules it can be seen that the rules for $\Pi x{:}A.\,B$ specialize to the rules for $A \to B$ if x does not occur in B. Thus $A \to B$ is generally considered a derived notation that stands for $\Pi x{:}A.\,B$ for a variable x not free in B.

A full complement of rules for the canonical λ^{Π} type theory is given in Section 4. With dependent function types, we can now give a representation for natural deductions with an internal validity condition. This is summarized in Theorem 2 below.

Implication. The introduction rule for implication employs a hypothetical judgment. The derivation of the hypothetical judgment in the premise is represented as a function which, when applied to a derivation of A, yields a derivation of B.

$$
\cfrac{\begin{array}{c} \ulcorner \quad \cfrac{}{\vdash^{N} A}\,u \quad \urcorner \\ \mathcal{D} \\ \vdash^{N} B \end{array}}{\vdash^{N} A \supset B}\supset\!\mathrm{I}^{u} \quad = \mathsf{impi}\ \ulcorner A \urcorner\ \ulcorner B \urcorner\ (\lambda u.\,\ulcorner \mathcal{D} \urcorner)
$$

The assumption A labeled by u which may be used in the derivation $\mathcal{D}$ is represented by the LF variable $u{:}\mathsf{nd}\ \ulcorner A \urcorner$ which ranges over derivations of A.

$$
\ulcorner\ \cfrac{}{\vdash^{N} A}\,u\ \urcorner \ = u
$$

From this we can deduce the type of the impi constant.

$$
\mathsf{impi}\ :\ \Pi A{:}\mathsf{o}.\,\Pi B{:}\mathsf{o}.\,(\mathsf{nd}\ A \to \mathsf{nd}\ B) \to \mathsf{nd}\ (\mathsf{imp}\ A\,B)
$$

The elimination rule is simpler, since it does not involve a hypothetical judgment. The representation of a derivation ending in the elimination rule is defined by

$$
\cfrac{\ulcorner\ \begin{array}{cc} \mathcal{D} & \mathcal{E} \\ \vdash^{N} A \supset B & \vdash^{N} A \end{array}\ \urcorner}{\vdash^{N} B}\supset\!\mathrm{E} \quad = \mathsf{impe}\ \ulcorner A \urcorner\ \ulcorner B \urcorner\ \ulcorner \mathcal{D} \urcorner\ \ulcorner \mathcal{E} \urcorner
$$

where

impe : $\Pi A{:}o.\,\Pi B{:}o.\,\mathsf{nd}\,(\mathsf{imp}\,A\,B) \to \mathsf{nd}\,A \to \mathsf{nd}\,B.$

As an example we consider a derivation of $A \supset (B \supset A)$.

$$\cfrac{\cfrac{\cfrac{}{\vdash^{\!N} A}\,u}{\vdash^{\!N} B \supset A}\,\supset\!I^w}{\vdash^{\!N} A \supset (B \supset A)}\,\supset\!I^u$$

Note that the assumption $\vdash^{\!N} B$ labelled w is not used and therefore does not appear in the derivation. This derivation is represented by the LF object

$$\mathsf{impi}\,\ulcorner A\urcorner\,(\mathsf{imp}\,\ulcorner B\urcorner\,\ulcorner A\urcorner)\,(\lambda u.\,\mathsf{impi}\,\ulcorner B\urcorner\,\ulcorner A\urcorner\,(\lambda w.\,u))$$

which has type

$$\mathsf{nd}\,(\mathsf{imp}\,\ulcorner A\urcorner\,(\mathsf{imp}\,\ulcorner B\urcorner\,\ulcorner A\urcorner)).$$

This example shows clearly some redundancies in the representation of the deduction (there are many occurrences of $\ulcorner A\urcorner$ and $\ulcorner B\urcorner$). Fortunately, it is possible to analyze the types of constructors and eliminate much of this redundancy through term reconstruction (Pfenning, 1991; Necula, 2002).

Negation. The introduction and elimination rules for negation and their representation follow the pattern of the rules for implication.

$$\cfrac{\begin{array}{c}\ulcorner \qquad\qquad \urcorner \\[-2pt] \cfrac{}{\vdash^{\!N} A}\,u \\[2pt] \mathcal{D} \\[2pt] \vdash^{\!N} p\end{array}}{\vdash^{\!N} \neg A}\,\neg I^{p,u} \quad = \mathsf{noti}\,\ulcorner A\urcorner\,(\lambda p.\,\lambda u.\,\ulcorner\mathcal{D}\urcorner)$$

The judgment of the premise is parametric in $p{:}o$ and hypothetical in $u{:}\mathsf{nd}\,\ulcorner A\urcorner$. It is thus represented as a function of two arguments, accepting both a formula p and a deduction of A.

noti : $\Pi A{:}o.\,(\Pi p{:}o.\,\mathsf{nd}\,A \to \mathsf{nd}\,p) \to \mathsf{nd}\,(\mathsf{not}\,A)$

The representation of negation elimination

$$\cfrac{\begin{array}{cc}\ulcorner \qquad\qquad\qquad\qquad \urcorner \\[-2pt] \begin{array}{cc}\mathcal{D} & \mathcal{E} \\ \vdash^{\!N} \neg A & \vdash^{\!N} A\end{array}\end{array}}{\vdash^{\!N} C}\,\neg E \quad = \mathsf{note}\,\ulcorner A\urcorner\,\ulcorner\mathcal{D}\urcorner\,\ulcorner C\urcorner\,\ulcorner\mathcal{E}\urcorner$$

leads to the following declaration

$$\text{note} \; : \; \Pi A{:}\mathsf{o}.\,\mathsf{nd}\;(\text{not}\;A) \to \Pi C{:}\mathsf{o}.\,\mathsf{nd}\;A \to \mathsf{nd}\;C$$

This type just inverts the second argument and result of the noti constant, which is the reason for the chosen argument order. Clearly,

$$\text{note}' \; : \; \Pi A{:}\mathsf{o}.\,\Pi C{:}\mathsf{o}.\,\mathsf{nd}\;(\text{not}\;A) \to \mathsf{nd}\;A \to \mathsf{nd}\;C$$

is an alternative declaration that would work just as well.

Universal quantification. Recall that $\ulcorner \forall x.\,A \urcorner = \text{forall}\;(\lambda x.\ulcorner A \urcorner)$ and that the premise of the introduction rule is parametric in a.

$$
\begin{array}{c}
\begin{array}{c}
\ulcorner \qquad\qquad \urcorner \\
\mathcal{D} \\
\vdash^N [a/x]A \\
\hline
\vdash^N \forall x.\,A
\end{array} \;\forall\mathrm{I}^a \quad = \text{foralli}\;(\lambda x.\ulcorner A \urcorner)\;(\lambda a.\ulcorner \mathcal{D} \urcorner)
\end{array}
$$

Note that $\ulcorner A \urcorner$, the representation of A, has a free variable x which must be bound in the meta-language, so that the representing object does not have a free variable x. Similarly, the parameter a is bound at this inference and must be correspondingly bound in the meta-language. The representation determines the type of the constant foralli.

$$\text{foralli} \; : \; \Pi A{:}\mathsf{i} \to \mathsf{o}.\,(\Pi a{:}\mathsf{i}.\,\mathsf{nd}\;(A\;a)) \to \mathsf{nd}\;(\text{forall}\;(\lambda x.\,A\;x))$$

In an application of this constant, the argument labelled A will be $\lambda x{:}\mathsf{i}.\ulcorner A \urcorner$ and $(A\;a)$ will become $[\ulcorner a \urcorner/x{:}\mathsf{i}]\ulcorner A \urcorner$ which in turn is equal to $\ulcorner [a/x]A \urcorner$ by the compositionality of the representation.

The elimination rule does not employ a hypothetical judgment.

$$
\begin{array}{c}
\begin{array}{c}
\ulcorner \qquad\qquad \urcorner \\
\mathcal{D} \\
\vdash^N \forall x.\,A \\
\hline
\vdash^N [t/x]A
\end{array} \;\forall\mathrm{E} \quad = \text{foralle}\;(\lambda x.\ulcorner A \urcorner)\;\ulcorner \mathcal{D} \urcorner\;\ulcorner t \urcorner
\end{array}
$$

The substitution of t for x in A is representation by the application of the function $(\lambda x.\ulcorner A \urcorner)$ (the first argument to foralle) to $\ulcorner t \urcorner$.

$$\text{foralle} \; : \; \Pi A{:}\mathsf{i} \to \mathsf{o}.\,\mathsf{nd}\;(\text{forall}\;A) \to \Pi t{:}\mathsf{i}.\,\mathsf{nd}\;(A\;t)$$

We now check that

$$
\begin{array}{c}
\begin{array}{c}
\ulcorner \qquad\qquad \urcorner \\
\mathcal{D} \\
\vdash^N \forall x.\,A \\
\hline
\vdash^N [t/x]A
\end{array} \;\forall\mathrm{E} \quad \Leftarrow \mathsf{nd}\;\ulcorner [t/x]A \urcorner,
\end{array}
$$

assuming that $\ulcorner \mathcal{D} \urcorner \Leftarrow$ nd $\ulcorner \forall x.\, A \urcorner$. This is a part in the proof of adequacy of this representation of natural deductions. At each step we verify that the arguments have the expected type and compute the type of the application.

$$\text{foralle} \Rightarrow \Pi A{:}\mathsf{i} \to \mathsf{o}.\, \text{nd (forall } (\lambda x.\, A\ x)) \to \Pi t{:}\mathsf{i}.\, \text{nd } (A\ t)$$
$$\text{foralle } (\lambda x.\ulcorner A \urcorner) \Rightarrow \text{nd (forall } (\lambda x.\ulcorner A \urcorner)) \to \Pi t{:}\mathsf{i}.\, \text{nd } ([t/x{:}\mathsf{i}]\ulcorner A \urcorner)$$
$$\text{foralle } (\lambda x.\ulcorner A \urcorner)\ \ulcorner \mathcal{D} \urcorner \Rightarrow \Pi t{:}\mathsf{i}.\, \text{nd } ([t/x{:}\mathsf{i}]\ulcorner A \urcorner)$$
$$\text{foralle } (\lambda x.\ulcorner A \urcorner)\ \ulcorner \mathcal{D} \urcorner\ \ulcorner t \urcorner \Rightarrow \text{nd } ([\ulcorner t \urcorner/x{:}\mathsf{i}]\ulcorner A \urcorner))$$
$$\text{foralle } (\lambda x.\ulcorner A \urcorner)\ \ulcorner \mathcal{D} \urcorner\ \ulcorner t \urcorner \Leftarrow \text{nd } ([\ulcorner t \urcorner/x{:}\mathsf{i}]\ulcorner A \urcorner)$$

The first step follows by the nature of canonical substitution,

$$[(\lambda x.\ulcorner A \urcorner)/A{:}\mathsf{i} \to \mathsf{o}](A\ t) = [t/x{:}\mathsf{i}]\ulcorner A \urcorner.$$

The last step uses the rule that an atomic object of atomic type P is also canonical at type P. Furthermore, by the compositionality of the representation we have

$$[\ulcorner t \urcorner/x{:}\mathsf{i}]\ulcorner A \urcorner = \ulcorner [t/x]A \urcorner$$

which, together with the last line above, yields the desired

$$\text{foralle } (\lambda x.\ulcorner A \urcorner)\ \ulcorner \mathcal{D} \urcorner\ \ulcorner t \urcorner \Leftarrow \text{nd } (\ulcorner [t/x]A \urcorner).$$

The representation theorem relates canonical objects constructed in certain contexts to natural deductions. The restriction to canonical objects is once again crucial, as are the restrictions on the form of the context. We call the signature consisting of the declarations for first-order terms, formulas, and natural deductions ND.

THEOREM 2 (Adequacy).

1. *If $\mathcal{D}$ is a derivation of A from hypotheses $\vdash^N A_1,\ldots,\ \vdash^N A_n$ labelled $u_1,\ldots,u_n$, respectively, with all free individual parameters among $a_1,\ldots,a_m$ and propositional parameters among $p_1,\ldots,p_k$ then*

$$\Gamma \vdash_{ND} \ulcorner \mathcal{D} \urcorner \Leftarrow \text{nd} \ulcorner A \urcorner$$

for $\Gamma = a_1{:}\mathsf{i},\ldots,a_m{:}\mathsf{i},p_1{:}\mathsf{o},\ldots,p_k{:}\mathsf{o},u_1{:}\text{nd}\ulcorner A_1 \urcorner,\ldots,u_n{:}\text{nd}\ulcorner A_n \urcorner$.

2. *If $\Gamma = a_1{:}\mathsf{i},\ldots,a_m{:}\mathsf{i},p_1{:}\mathsf{o},\ldots,p_k{:}\mathsf{o},u_1{:}\text{nd}\ulcorner A_1 \urcorner,\ldots,u_n{:}\text{nd}\ulcorner A_n \urcorner$ and*

$$\Gamma \vdash_{ND} M \Leftarrow \text{nd} \ulcorner A \urcorner$$

then $M = \ulcorner \mathcal{D} \urcorner$ for a derivation $\mathcal{D}$ as in part 1.

3. *The representation function is a bijection, and is compositional in the sense that the following equalities hold (where* $\mathcal{E} :: \vdash^N A$*):*

$$\ulcorner [t/a]\mathcal{D} \urcorner = [\ulcorner t \urcorner/a{:}i]\ulcorner \mathcal{D} \urcorner$$
$$\ulcorner [C/p]\mathcal{D} \urcorner = [\ulcorner C \urcorner/p{:}o]\ulcorner \mathcal{D} \urcorner$$
$$\ulcorner [\mathcal{E}/u]\mathcal{D} \urcorner = [\ulcorner \mathcal{E} \urcorner/u{:}\mathsf{nd}]\ulcorner \mathcal{D} \urcorner$$

Proof: The proof proceeds by induction on the structure of natural deductions one direction and on the definition of canonical forms in the other direction. $\qquad\square$

Each of the rules that may be added to obtain classical logic can be easily represented with the techniques from above. They are left as an exercise to the reader.

We summarize the LF encoding of natural deductions. First, the syntax.

```
i      : type
o      : type

imp    : o → o → o
not    : o → o
forall : (i → o) → o
```

The second simplification in the concrete presentation is to omit some Π-quantifiers. Free variables in a declaration are then interpreted as a schematic variables whose quantifiers remain implicit. The types of such free variables must be determined from the context in which they appear. In practical implementations such as Twelf (Pfenning and Schürmann, 1999), type reconstruction will issue an error message if the type of free variables is ambiguous.

```
nd      : o → type

impi    : (nd A → nd B) → nd (imp A B).
impe    : nd (imp A B) → nd A → nd B.
noti    : (Πp:o. nd A → nd p) → nd (not A).
note    : nd (not A) → ΠC:o. nd A → nd C.
foralli : (Πa:i. nd (A a)) → nd (forall (λx. A x)).
foralle : nd (forall A) → Πt:i. nd (A t)
```

When constants with implicitly quantified types are used, arguments corresponding to the omitted quantifiers are also left implicit. Again, in practical implementations these arguments are inferred from context. For example, the constant impi now appears to take only two arguments (of type nd A and nd B for some A and B) rather than four, like the fully explicit declaration

impi : ΠA:o. ΠB:o. (nd $A \to$ nd B) $\to$ nd (imp $A\,B$).

The derivation of $A \supset (B \supset A)$ from above has this very concise representation:

$$\text{impi}\,(\lambda u.\,\text{impi}\,(\lambda w.\,u)) \Leftarrow \text{nd(imp}\,A\,(\text{imp}\,B\,A))$$

To recover classical logic, we can add either of the following declarations to the signature, modeling the two rules previously introduced.

dbneg : nd (not (not A)) $\to$ nd A.
contr : (nd (not A) $\to$ nd A) $\to$ nd A.

In summary, the basic representation principle underlying LF is the representation of judgments as types. A deduction of a judgment J is represented as a canonical object N whose type is the representation of J. This basic scheme is extended to represent hypothetical judgments as simple function types and parametric judgments as dependent function types. This encoding reduces the question of validity for a derivation to the question of well-typedness for its representation. Since type-checking in the LF type theory is decidable, the validity of derivations has been internalized as a decidable property in the logical framework.

3.4. HIGHER-LEVEL JUDGMENTS

Next we turn to the local reduction judgment for natural deductions introduced in Section 3.2.

$$\begin{array}{ccc} \mathcal{D} & & \mathcal{D}' \\ \vdash^N A & \Longrightarrow_R & \vdash^N A \end{array}$$

Recall that this judgment witnesses the local soundness of the elimination rules with respect to the introduction rules. We refer to this as a *higher-level judgment* since it relates derivations. The representation techniques underlying LF support this directly, since deductions are represented as objects which can in turn index type families representing higher-level judgments.

In this particular example, reduction is defined only by axioms, one each for implication, negation, and universal quantification. The representing type family in LF must be indexed by the representation of two deductions $\mathcal{D}$ and $\mathcal{D}'$, and consequently also by the representation of A. This shows that there may be dependencies between indices to a type family so that we need a dependent constructor Π for kinds in order to represent judgments relating derivations.

$$\mathsf{redl} \ : \ \Pi A{:}o. \ \mathsf{nd} \ A \to \mathsf{nd} \ A \to \mathsf{type}.$$

As in the representation of inference rules in Section 3.3, we omit the explicit quantifier on A and determine A from context.

$$\mathsf{redl} \ : \ \mathsf{nd} \ A \to \mathsf{nd} \ A \to \mathsf{type}.$$

We show the representation of the reduction rules for each connective in turn.

Implication. This reduction involves a substitution of a derivation for an assumption.

$$
\cfrac{
  \cfrac{
    \cfrac{\dfrac{\overline{\vdash^{\!N} A}\;u}{\begin{array}{c}\mathcal{D}\\ \vdash^{\!N} B\end{array}}}{\vdash^{\!N} A \supset B}\supset\! I^u
    \qquad
    \dfrac{\mathcal{E}}{\vdash^{\!N} A}
  }{\vdash^{\!N} B}\supset\! E
}{}
\qquad\Longrightarrow_R\qquad
\dfrac{\dfrac{\mathcal{E}}{\vdash^{\!N} A}\;u}{\begin{array}{c}\mathcal{D}\\ \vdash^{\!N} B\end{array}}
$$

The representation of the left-hand side is

$$\mathsf{impe} \ (\mathsf{impi} \ (\lambda u. \ D \ u)) \ E$$

where $E = \ulcorner\mathcal{E}\urcorner \Leftarrow \mathsf{nd} \ A$ and $D = (\lambda u.\ulcorner\mathcal{D}\urcorner) \Leftarrow \mathsf{nd} \ A \to \mathsf{nd} \ B$. The derivation on the right-hand side can be written more succinctly as $[\mathcal{E}/u]\mathcal{D}$. Compositionality of the representation (Theorem 2, part 3) yields

$$\ulcorner[\mathcal{E}/u]\mathcal{D}\urcorner = [\ulcorner\mathcal{E}\urcorner/u{:}\mathsf{nd}]\ulcorner\mathcal{D}\urcorner.$$

Thus we can formulate the rule concisely as

$$\mathsf{redl\_imp} \ : \ \mathsf{redl} \ (\mathsf{impe} \ (\mathsf{impi} \ (\lambda u. \ D \ u)) \ E) \ (D \ E)$$

Negation. This is similar to implication. The required substitution of C for p in $\mathcal{D}$ is implemented by application and β-reduction at the meta-level.

$$
\cfrac{
  \cfrac{
    \cfrac{\dfrac{\overline{\vdash^{\!N} A}\;u}{\begin{array}{c}\mathcal{D}\\ \vdash^{\!N} p\end{array}}}{\vdash^{\!N} \neg A}\neg I^{p,u}
    \qquad
    \dfrac{\mathcal{E}}{\vdash^{\!N} A}
  }{\vdash^{\!N} C}\neg E
}{}
\qquad\Longrightarrow_R\qquad
\dfrac{\dfrac{\mathcal{E}}{\vdash^{\!N} A}\;u}{\begin{array}{c}[C/p]\mathcal{D}\\ \vdash^{\!N} C\end{array}}
$$

redl_not : redl (note (noti ($\lambda p. \lambda u. D\ p\ u$)) $C\ E$) ($D\ C\ E$).

Universal quantification. The universal introduction rule involves a para-metric judgment. Consequently, the substitution to be carried out during reduction replaces a parameter by a term.

$$
\cfrac{\cfrac{\begin{array}{c}\mathcal{D}\\ \vdash^{\!\scriptscriptstyle N} [a/x]A\end{array}}{\vdash^{\!\scriptscriptstyle N} \forall x.\,A}\ \forall\mathrm{I}^a}{\vdash^{\!\scriptscriptstyle N} [t/x]A}\ \forall\mathrm{E}
\qquad \Longrightarrow_R \qquad
\begin{array}{c}[t/a]\mathcal{D}\\ \vdash^{\!\scriptscriptstyle N} [t/x]A\end{array}
$$

In the representation we once again exploit the compositionality.

$$\ulcorner[t/a]\mathcal{D}\urcorner = [\ulcorner t\urcorner/a{:}\mathrm{i}]\ulcorner\mathcal{D}\urcorner.$$

This gives rise to the declaration

redl_forall : redl (foralle (foralli ($\lambda a. D\ a$)) T) ($D\ T$).

The adequacy theorem for this encoding states that canonical LF ob-jects of type redl $\ulcorner\mathcal{D}\urcorner\ \ulcorner\mathcal{D}'\urcorner$ constructed over the appropriate signature and in an appropriate parameter context are in bijective correspondence with derivations of $\mathcal{D} \Longrightarrow_R \mathcal{D}'$. We leave the precise formulation and simple proof to the diligent reader.

The encoding of the local expansions employs the same techniques. We summarize it below without going into further detail.

expl   : $\Pi A{:}o.$ nd $A \to$ nd $A \to$ type.
expl_imp   : expl (imp $A\ B$) D (impi ($\lambda u.$ impe $D\ u$)).
expl_not   : expl (not A) D (noti ($\lambda p. \lambda u.$ note $D\ p\ u$)).
expl_forall : expl (forall ($\lambda x.\ A\ x$)) D (foralli ($\lambda a.$ foralle $D\ a$)).

In summary, the representation of higher-level judgments continues to follow the *judgments-as-types* technique. The expressions related by higher-level judgments are now deductions and therefore dependently typed in the representation. Substitution at the level of deductions is implemented by substitution at the meta-level, taking advantage of the compositionality of the representation.

4. A dependently typed λ-calculus

In this section we summarize a recent formulation of the dependently λ-calculus λ^{Π} allowing only canonical forms (Watkins et al., 2002). This avoids an explicit notion of definitional equality (Harper et al., 1993), which is not required for applications of λ^Π as a logical framework. Related systems have been advocated by de Bruijn (1993) and Felty (1991). See Watkins et al. (2002) for further details and properties of this formulation.

λ^Π is predicative calculus with three levels: kinds, families, and objects. We also define signatures and contexts as they are needed for the judgments.

$$\begin{array}{lrcl}
\text{Normal Kinds} & K & ::= & \mathsf{type} \mid \Pi x{:}A.\,K \\
\text{Atomic Types} & P & ::= & a \mid P\,N \\
\text{Normal Types} & A & ::= & P \mid \Pi x{:}A_1.\,A_2 \\
\text{Atomic Objects} & R & ::= & c \mid x \mid R\,N \\
\text{Normal Objects} & N & ::= & \lambda x.\,N \mid R \\
\text{Signatures} & \Sigma & ::= & \cdot \mid \Sigma, a{:}K \mid \Sigma, c{:}A \\
\text{Contexts} & \Gamma & ::= & \cdot \mid \Gamma, x{:}A
\end{array}$$

We write a for type family constants and c for object constants, both declared in signatures Σ with their kind and type, respectively. Variables x are declared in contexts with their type. We make the uniform assumption that no constant or variable may be declared more than once in a signature or context, respectively. We also allow tacit renaming of variables bound by $\Pi x{:}A\ldots$ and $\lambda x.\ldots$. As usual, we avoid an explicit non-dependent function type by thinking of $A \to B$ as an abbreviation for $\Pi x{:}A.\,B$ where x does not occur in B, and similarly for $A \to K$.

From the point of view of natural deduction, atomic objects are composed of destructors corresponding to elimination rules, while normal objects are built from constructors corresponding to introduction rules. The typing rules are *bi-directional* which mirrors the syntactic structure of normal forms: we check a normal object against a type, and we synthesize a type for an atomic object. We write $U \Leftarrow V$ to indicate that U is checked against a given V (which we assume is valid), and $U \Rightarrow V$ to indicate that U synthesizes a V (which we prove is valid).

$\Gamma \vdash_\Sigma K \Leftarrow \mathsf{kind}$ K is a valid kind

$\Gamma \vdash_\Sigma A \Leftarrow \mathsf{type}$ A is a valid type

$\Gamma \vdash_\Sigma P \Rightarrow K$ P is atomic of kind K

$\Gamma \vdash_\Sigma N \Leftarrow A$ N is normal of type A

$\Gamma \vdash_\Sigma R \Rightarrow A$ R is atomic of type A

$\vdash \Sigma\ \mathsf{Sig}$ Σ is a valid signature

$\vdash_\Sigma \Gamma\ \mathsf{Ctx}$ Γ is a valid context

In one rule, we write $A \equiv A'$ for syntactic equality of normal types modulo α-conversion. This is to emphasize the flow of information during type-checking.

$$\frac{}{\vdash \cdot \text{ Sig}} \qquad \frac{\vdash \Sigma \text{ Sig} \qquad \cdot \vdash_{\Sigma} K \Leftarrow \text{kind}}{\vdash \Sigma, a{:}K \text{ Sig}} \qquad \frac{\vdash \Sigma \text{ Sig} \qquad \cdot \vdash_{\Sigma} A \Leftarrow \text{type}}{\vdash \Sigma, c{:}A \text{ Sig}}$$

$$\frac{}{\vdash \cdot \text{ Ctx}} \qquad \frac{\vdash_{\Sigma} \Gamma \text{ Ctx} \qquad \Gamma \vdash_{\Sigma} A \Leftarrow \text{type}}{\vdash_{\Sigma} \Gamma, x{:}A \text{ Ctx}}$$

$$\frac{}{\Gamma \vdash_{\Sigma} \text{type} \Leftarrow \text{kind}} \qquad \frac{\Gamma \vdash_{\Sigma} A \Leftarrow \text{type} \qquad \Gamma, x{:}A \vdash_{\Sigma} K \Leftarrow \text{kind}}{\Gamma \vdash_{\Sigma} \Pi x{:}A.\, K \Leftarrow \text{kind}}$$

$$\frac{}{\Gamma \vdash_{\Sigma} a \Rightarrow \Sigma(a)} \qquad \frac{\Gamma \vdash_{\Sigma} P \Rightarrow \Pi x{:}A.\, K \qquad \Gamma \vdash_{\Sigma} N \Leftarrow A}{\Gamma \vdash_{\Sigma} P\, N \Rightarrow [N/u{:}A^{-}]K}$$

$$\frac{\Gamma \vdash_{\Sigma} P \Rightarrow \text{type}}{\Gamma \vdash_{\Sigma} P \Leftarrow \text{type}} \qquad \frac{\Gamma \vdash_{\Sigma} A \Leftarrow \text{type} \qquad \Gamma, x{:}A \vdash_{\Sigma} B \Leftarrow \text{type}}{\Gamma \vdash_{\Sigma} \Pi x{:}A.\, B \Leftarrow \text{type}}$$

$$\frac{}{\Gamma \vdash_{\Sigma} c \Rightarrow \Sigma(c)} \qquad \frac{}{\Gamma \vdash_{\Sigma} x \Rightarrow \Gamma(x)} \qquad \frac{\Gamma \vdash_{\Sigma} R \Rightarrow \Pi x{:}A.\, B \qquad \Gamma \vdash_{\Sigma} N \Leftarrow A}{\Gamma \vdash_{\Sigma} R\, N \Rightarrow [N/x{:}A^{-}]B}$$

$$\frac{\Gamma \vdash_{\Sigma} N \Rightarrow A' \qquad A' \equiv A}{\Gamma \vdash_{\Sigma} N \Leftarrow A} \qquad \frac{\Gamma, x{:}A \vdash_{\Sigma} N \Leftarrow B}{\Gamma \vdash_{\Sigma} \lambda x.\, N \Leftarrow \Pi x{:}A.\, B}$$

In order to define canonical substitution inductively, we erase all dependences and indices from a type to obtain a simple type τ.

$$\begin{aligned}
(a)^{-} &= a \\
(P\, N)^{-} &= P^{-} \\
(\Pi x{:}A.\, B)^{-} &= A^{-} \to B^{-}
\end{aligned}$$

The canonical substitution $[N/x{:}\tau]B$ and $[N/x{:}\tau]K$ returns a normal type or kind, respectively. It is defined inductively, first on the structure of the simply-typed erasure A^{-} of A and then the structure of B and K, respectively. Modulo the proof of termination, we can also think of it as the

β-normal form of $[N/x]B$ and $[N/x]K$, respectively. For details of the two approaches, the reader may consult Watkins et al. (2002) and Felty (1991).

$$
\begin{aligned}
[N_0/x{:}\tau](\text{type}) &= \text{type} \\
[N_0/x{:}\tau](\Pi y{:}A.\,K) &= \Pi y{:}[N_0/x{:}\tau]A.\,[N_0/x{:}\tau]K \\[4pt]
[N_0/x{:}\tau](a) &= a \\
[N_0/x{:}\tau](P\ N) &= ([N_0/x{:}\tau]P)\,([N_0/x{:}\tau]N) \\
[N_0/x{:}\tau](\Pi y{:}A.\,B) &= \Pi y{:}[N_0/x{:}\tau]A.\,[N_0/x{:}\tau]B \\[4pt]
[N_0/x{:}\tau](\lambda y.\,N) &= \lambda y.\,[N_0/x{:}\tau]N \\
[N_0/x{:}\tau](R) &= [N_0/x{:}\tau]^r(R) \quad \text{or} \quad [N_0/x{:}\tau]^\beta(R) \\[4pt]
[N_0/x{:}\tau]^r(c) &= c \\
[N_0/x{:}\tau]^r(x) &\quad \text{undefined} \\
[N_0/x{:}\tau]^r(y) &= y \qquad \text{provided } x \neq y \\
[N_0/x{:}\tau]^r(R\ N) &= ([N_0/x{:}\tau]^r R)\,([N_0/x{:}\tau]N) \\[4pt]
[N_0/x{:}\tau]^\beta(c) &\quad \text{undefined} \\
[N_0/x{:}\tau]^\beta(x) &= (N_0 : \tau) \\
[N_0/x{:}\tau]^\beta(y) &\quad \text{undefined provided } x \neq y \\
[N_0/x{:}\tau]^\beta(R\ N) &= ([[N_0/x{:}\tau]N/y{:}\tau_1]N' : \tau_2) \\
&\quad \text{where } [N_0/x{:}\tau]^\beta(R) = (\lambda y.\,N' : \tau_1 \to \tau_2)
\end{aligned}
$$

Note that for all atomic terms R, either a case for $[N/x{:}\tau]^r(R)$ or $[N/x{:}\tau]^\beta(R)$ applies, depending on whether the head of R is x or not. Furthermore, if $[N/x{:}\tau]^\beta(R) = (N : \tau')$ then τ' is a subexpression of τ. Hence canonical substitution is a terminating function. Decidability of the LF type theory is then a straightforward consequence. Furthermore, cut is admissible in the sense that if $\Gamma \vdash_\Sigma N_0 \Leftarrow A$ and $\Gamma, x{:}A \vdash_\Sigma N \Leftarrow C$ then $\Gamma \vdash_\Sigma [N_0/x{:}A^-]N \Leftarrow [N_0/x{:}A^-]C$ (Watkins et al., 2002).

5. Conclusion

We have provided an introduction to the techniques of logical frameworks with an emphasis on LF which is based on the dependently typed λ-calculus λ^Π. We now summarize the basic choices that arise in the design of logical frameworks.

Strong vs. weak frameworks. De Bruijn, the founder of the field of logical frameworks, argues (de Bruijn, 1991) that logical frameworks should be foundationally uncommitted and as weak as possible. This allows simple proofs of adequacy for encodings, efficient checking of the correctness of derivations, and allows effective algorithms for unification and proof search in the framework which are otherwise difficult to design (for example, in the presence of iterated inductive definitions). This is also important if we use explicit proofs as a means to increase confidence in the results of a theorem prover: the simpler the logical framework, the more trusted its implementation is likely to be.

Inductive representations vs. higher-order abstract syntax. This is related to the previous question. Inductive representations of logics are supported in various frameworks and type theories not explicitly designed as logical frameworks. They allow a formal development of the meta-theory of the deductive system in question, but the encodings are less direct than for frameworks employing higher-order abstract syntax and functional representations of hypothetical derivations. Present work on combining advantages of both either employ reflection or formal meta-reasoning about the logical framework itself (McDowell and Miller, 1997; Schürmann, 2000).

Logical vs. type-theoretic meta-languages. A logical meta-language such as one based on hereditary Harrop formulas encodes judgments as propositions. Search for a derivation in an object logic is reduced to proof search in the meta-logic. In addition, type-theoretical meta-languages such as LF offer a representation for derivations as objects. Checking the correctness of a derivation is reduced to type-checking in the meta-language. This is a decidable property that enables the use of a logical framework for applications such as proof-carrying code, where an explicit representation for deductions is required (Necula, 2002).

Framework extensions. Logical framework languages can be assessed along many dimensions, as the discussions above indicate. Three of the most important concerns are how directly object languages may be encoded, how easy it is to prove the adequacies of these encodings, and how simple the proof checker for a logical framework can be. A great deal of practical experience has been accumulated, for example, through the use of λProlog, Isabelle, and Elf. These experiments have also identified certain shortcomings in the logical frameworks. Perhaps the most important one is the treatment of substructural logics, or languages with an inherent notion of store or concurrency. Representation of such object languages is possible, but not as direct as one might wish. The logical frameworks Forum (Miller, 1994), linear LF (Cervesato and Pfenning, 1997) and CLF (Watkins et al., 2002) have been designed to overcome these shortcomings by providing linearity intrinsically. Other extensions by subtyping, module constructs, constraints, etc. have also been designed, but their discussion is beyond the scope of this introduction.

Further reading.   There have been numerous case studies and applications carried out with the aid of logical frameworks or generic theorem provers, too many to survey them here. The principal application areas lie in the theory of programming languages and logics, reasoning about specifications, programs, and protocols, and the formalization of mathematics. We refer the interested reader to (Pfenning, 1996) for some further information on applications of logical frameworks. The handbook article (Pfenning, 2001b) provides more detailed development of LF and includes some material on meta-logical frameworks. A survey with deeper coverage of modal logics and inductive definitions can be found in (Basin and Matthews, 2001). The textbook (Pfenning, 2001a) provides a gentler and more thorough introduction to the pragmatics of the LF logical framework and its use for the study of programming languages.

References

Altenkirch, T., V. Gaspes, B. Nordström, and B. von Sydow: 1994, 'A User's Guide to ALF'. Chalmers University of Technology, Sweden.

Barendregt, H. P.: 1980, *The Lambda-Calculus: Its Syntax and Semantics.* North-Holland.

Basin, D. and S. Matthews: 1996, 'Structuring Metatheory on Inductive Definitions'. In: M. McRobbie and J. Slaney (eds.): *Proceedings of the 13th International Conference on Automated Deduction (CADE-13).* New Brunswick, New Jersey, pp. 171–185, Springer-Verlag LNAI 1104.

Basin, D. and S. Matthews: 2001, 'Logical Frameworks'. In: D. Gabbay and F. Guenthner (eds.): *Handbook of Philosophical Logic.* Kluwer Academic Publishers, 2nd edition. In preparation.

Basin, D. A. and R. L. Constable: 1993, 'Metalogical Frameworks'. In: G. Huet and G. Plotkin (eds.): *Logical Environments.* Cambridge University Press, pp. 1–29.

Cervesato, I. and F. Pfenning: 1996, 'A Linear Logical Framework'. In: E. Clarke (ed.): *Proceedings of the Eleventh Annual Symposium on Logic in Computer Science.* New Brunswick, New Jersey, pp. 264–275, IEEE Computer Society Press.

Cervesato, I. and F. Pfenning: 1997, 'Linear Higher-Order Pre-Unification'. In: G. Winskel (ed.): *Proceedings of the Twelfth Annual Sumposium on Logic in Computer Science (LICS'97).* Warsaw, Poland, pp. 422–433, IEEE Computer Society Press.

de Bruijn, N.: 1968, 'The Mathematical Language AUTOMATH, Its Usage, and Some of Its Extensions'. In: M. Laudet (ed.): *Proceedings of the Symposium on Automatic Demonstration.* Versailles, France, pp. 29–61, Springer-Verlag LNM 125.

de Bruijn, N.: 1991, 'A Plea for Weaker Frameworks'. In: G. Huet and G. Plotkin (eds.): *Logical Frameworks.* pp. 40–67, Cambridge University Press.

de Bruijn, N.: 1993, 'Algorithmic Definition of Lambda-Typed Lambda Calculus'. In: G. Huet and G. Plotkin (eds.): *Logical Environment.* Cambridge University Press, pp. 131–145.

Dowek, G.: 1993, 'The Undecidability of Typability in the Lambda-Pi-Calculus'. In: M. Bezem and J. Groote (eds.): *Proceedings of the International Conference on Typed Lambda Calculi and Applications.* Utrecht, The Netherlands, pp. 139–145, Springer-Verlag LNCS 664.

Eriksson, L.-H.: 1994, 'Pi: An Interactive Derivation Editor for the Calculus of Partial Inductive Definitions'. In: A. Bundy (ed.): *Proceedings of the 12th International Conference on Automated Deduction.* Nancy, France, pp. 821–825, Springer Verlag LNAI 814.

Feferman, S.: 1988, 'Finitary Inductive Systems'. In: R. Ferro (ed.): *Proceedings of Logic Colloquium '88.* Padova, Italy, pp. 191–220, North-Holland.

Felty, A.: 1991, 'Encoding Dependent Types in an Intuitionistic Logic'. In: G. Huet and G. D. Plotkin (eds.): *Logical Frameworks.* pp. 214–251, Cambridge University Press.

Felty, A. and D. Miller: 1988, 'Specifying Theorem Provers in a Higher-Order Logic Programming Language'. In: E. Lusk and R. Overbeek (eds.): *Proceedings of the Ninth International Conference on Automated Deduction.* Argonne, Illinois, pp. 61–80, Springer-Verlag LNCS 310.

Gabbay, D. M.: 1994, 'Classical vs Non-Classical Logic'. In: D. Gabbay, C. Hogger, and J. Robinson (eds.): *Handbook of Logic in Artificial Intelligence and Logic Programming,* Vol. 2. Oxford University Press, Chapt. 2.6.

Harper, R., F. Honsell, and G. Plotkin: 1993, 'A Framework for Defining Logics'. *Journal of the Association for Computing Machinery* **40**(1), 143–184.

Howard, W. A.: 1980, 'The formulae-as-types notion of construction'. In: J. P. Seldin and J. R. Hindley (eds.): *To H. B. Curry: Essays on Combinatory Logic, Lambda Calculus and Formalism.* Academic Press, pp. 479–490. Hitherto unpublished note of 1969.

Martì-Oliet, N. and J. Meseguer: 1993, 'Rewriting Logic as a Logical and Semantical Framework'. Technical Report SRI-CSL-93-05, SRI International.

Martin-Löf, P.: 1980, 'Constructive Mathematics and Computer Programming'. In: *Logic, Methodology and Philosophy of Science VI.* pp. 153–175, North-Holland.

Martin-Löf, P.: 1985, 'On the Meanings of the Logical Constants and the Justifications of the Logical Laws'. Technical Report 2, Scuola di Specializzazione in Logica Matematica, Dipartimento di Matematica, Università di Siena. Reprinted in the *Nordic Journal of Philosophical Logic*, **1**(1), 11-60, 1996.

McDowell, R. and D. Miller: 1997, 'A Logic for Reasoning with Higher-Order Abstract Syntax'. In: G. Winskel (ed.): *Proceedings of the Twelfth Annual Symposium on Logic in Computer Science.* Warsaw, Poland, pp. 434–445, IEEE Computer Society Press.

Miller, D.: 1994, 'A Multiple-Conclusion Meta-Logic'. In: S. Abramsky (ed.): *Ninth Annual Symposium on Logic in Computer Science.* Paris, France, pp. 272–281, IEEE Computer Society Press.

Nadathur, G. and D. Miller: 1988, 'An Overview of λProlog'. In: K. A. Bowen and R. A. Kowalski (eds.): *Fifth International Logic Programming Conference.* Seattle, Washington, pp. 810–827, MIT Press.

Necula, G. C.: 1997, 'Proof-Carrying Code'. In: N. D. Jones (ed.): *Conference Record of the 24th Symposium on Principles of Programming Languages (POPL'97).* Paris, France, pp. 106–119, ACM Press.

Necula, G. C.: 2002, 'Proof-Carrying Code: Design and Implementation'. *This volume.* Kluwer Academic Publishers.

Nordström, B., K. Petersson, and J. M. Smith: 1990, *Programming in Martin-Löf's Type Theory: An Introduction.* Oxford University Press.

Paulson, L. C.: 1986, 'Natural Deduction as Higher-order Resolution'. *Journal of Logic Programming* **3**, 237–258.

Pfenning, F.: 1991, 'Logic Programming in the LF Logical Framework'. In: G. Huet and G. Plotkin (eds.): *Logical Frameworks.* pp. 149–181, Cambridge University Press.

Pfenning, F.: 1996, 'The Practice of Logical Frameworks'. In: H. Kirchner (ed.): *Proceed-*

ings of the Colloquium on Trees in Algebra and Programming. Linköping, Sweden, pp. 119–134, Springer-Verlag LNCS 1059. Invited talk.

Pfenning, F.: 2000, 'Structural Cut Elimination I. Intuitionistic and Classical Logic'. *Information and Computation* **157**(1/2), 84–141.

Pfenning, F.: 2001a, *Computation and Deduction.* Cambridge University Press. In preparation. Draft from April 1997 available electronically.

Pfenning, F.: 2001b, 'Logical Frameworks'. In: A. Robinson and A. Voronkov (eds.): *Handbook of Automated Reasoning.* Elsevier Science and MIT Press, Chapt. 16, pp. 977–1061. In press.

Pfenning, F. and C. Schürmann: 1999, 'System Description: Twelf — A Meta-Logical Framework for Deductive Systems'. In: H. Ganzinger (ed.): *Proceedings of the 16th International Conference on Automated Deduction (CADE-16).* Trento, Italy, pp. 202–206, Springer-Verlag LNAI 1632.

Schürmann, C.: 2000, 'Automating the Meta Theory of Deductive Systems'. Ph.D. thesis, Department of Computer Science, Carnegie Mellon University. Available as Technical Report CMU-CS-00-146.

Watkins, K., I. Cervesato, F. Pfenning, and D. Walker: 2002, 'A Concurrent Logical Framework I: Judgments and Properties'. Technical Report CMU-CS-02-101, Department of Computer Science, Carnegie Mellon University. Forthcoming.

LUDICS : AN INTRODUCTION

JEAN-YVES GIRARD
Institut de Mathématiques de Luminy, UPR 9016 - **CNRS**
163, Avenue de Luminy, Case 930,
F-13288 Marseille Cedex 09
girard@iml.univ-mrs.fr

Ludics [1] is a novel approach to logic —especially proof-theory. The present introduction emphasises foundational issues.

1. All quiet on the western front

For ages, not a single *disturbing* idea in the area of « foundations » : the discussion is sort of ossified —as if everything had been said, as if all notions had taken their definite place, in a big cemetery of ideas. One can still refresh the flowers or regild the stone, e.g., prove technicalities, sometimes non-trivial ; but the real debate is still : this paper begins with an *autopsy*, the autopsy of the foundational project.

1.1. REALISM

Up to say 1900, the realist/dualist approach to science was dominant ; during last century some domains like physics evolved so as to become completely anti-realist ; but this evolution hardly concerned logic.

1.1.1. *Hilbert's legacy*
By the turn of the XXth century mathematics was jeopardised by paradoxes, the most famous of them being due to Russell. Hilbert's reaction was to focus on *consistency.* But the reduction of paradoxes —and therefore of foundations— to the sole *antinomies* is highly questionable[1]. For instance the Peano « curve » contradicts our perception of *dimension* ; fortunately, topology has been able to show that dimension m is not the same as dimension n... But just for a second, forget this and imagine *consistent* mathematics in which balls in any dimension are homeomorphic : what a

---

[1] δόξα : dogma, opinion, intuition... : a *paradox* needs not be a formal contradiction.

H. Schwichtenberg and R. Steinbrüggen (eds.), Proof and System-Reliability, 167–211.
© 2002 *Kluwer Academic Publishers. Printed in the Netherlands.*

disaster ! This exclusive focus on consistency —not to speak of the strategic failure of the Programme— should explain why logic, especially *foundations* lost contact with other sciences during last century.

Indeed Hilbert's Programme is not quite realistic, it is *procedural*, see 1.3 below : Hilbert tried to avoid as much as possible the external *reality*. This is wise ; but he made several mistakes, both technical and methodological :

▶ We shall see in 2.1.2 that Hilbert's Programme relies on a duality between proofs of A and proofs of $\neg A$. But the pivot of this duality —the proofs of the absurdity— is empty and this leads to a disaster.

▶ The mere idea of foundations of mathematics *inside* mathematics is questionable, there is an obvious *conflict of interest*, which is not fixed with the controversial idea of *metamathematics*[2]. Indeed these *metamathematics* are just a part of ordinary mathematics —possibly with a hand tied in the back : this remark underlies Gödel's incompleteness. What remains of this is a sort of *diabolus in logica* : the foundational discourse, once embedded in mathematics yields artificial formulas, *secret sharers* —which rather look like sophisms : « I am not provable »[3]. One of the main difficulties of the foundational discourse is that foundations should also take care of these sophisms.

▶ The idea that mathematics is a pure play on symbols... is extremely prejudicial : this is the point on which logic lost contact with the rest of the world. By the way, the statement is presumably correct, only its current reading[4] « meaningless mathematics » is wrong : who told us that a play on symbols is meaningless, has not its own geometry ? By the way, Gentzen —Hilbert's most conspicuous follower— disclosed a structure (sequent calculus) underlying the « play on symbols »... and sequent calculus has a geometrical structure, this is precisely what ludics is about.

The failure of Hilbert's Programme —a reductionist procedural explanation-was felt like the total victory of realism. What remains of Hilbert's spirit in the « official » approach to foundations is a lurking positivism which

---

[2] $\mu\acute{\epsilon}\tau\alpha$ means whatever you like, mainly « besides, after », for instance *metaphysics* means « after-physics ». In logic the expression conveys —together with « intensional »— a magical connotation : this irrationalism is the hidden face of positivism.

[3] Something similar happened in analysis when Bolzano, Weierstraß, ... started to discover secret sharers : the curve with no derivative, ... up to the Peano curve.

[4] And the current formalist ideology, which can be witnessed on —say— questions of notations : if operation $\mathcal{M}$ distributes over operation $\mathcal{A}$, a mathematician will tend to use the notations $\mathcal{M} = \otimes, \mathcal{A} = \oplus$ or something of the like. But many logicians will insist on the opposite choice « you know, notations are arbitrary », implicitly denying any special value to distributivity.

underlies the pregnancy of this strange animal —the meta— which accounts both for the failure of the programme and the cracks in the realistic building.

1.1.2. *Object vs. subject*

The current explanation of logic distinguishes between the world (objective) and its representation (subjective), the *object* and the *subject*. Logical realism relies on an opposition between *semantics* (the world) and *syntax* (its representation) : this opposition is expressed through soundness and (in)completeness. The approach is highly problematic :

▸ It is a bit delicate to say that the integers are « the world ». $\{0, 1, 2\}$ does make a small world, but not $\{0, 1, 2, \ldots\}$: « ... » is a pure fantasy, nobody knowing how to interpret it. What is sure is that we have efficient ways of manipulating « ... » —induction axioms—, but they are on the « subjective » side.

▸ The « subjective » part does not receive any status. This is plain formal bureaucracy, subject to rather arbitrary choices : for instance certain authors will insist on minimising the number of connectives or the number of rules, e.g., do everything with the connective « equivalent ». The fact that predicate calculus is complete becomes a pure miracle : why this bureaucracy should correspond to something natural ?

▸ The same holds when one plugs in incompleteness : not everything is provable, but there are weaker or stronger systems. The image is that of a Big Book in which we can eavesdrop : we can see part of it, but only part of it. Then we can classify the keyholes, think of « reverse mathematics », of the « ordinal strength » of theories. Again there is not the slightest explanation for the coexistence of various systems of arithmetic : simply because the « subjective part » gets no autonomous status.

▸ In fact there is a complete absence of explanation. This is obvious if we look at the Tarskian « definition » of truth « A is true iff A holds ». The question is not to know whether mathematics accepts such a definition but if there is any contents in it... What is disjunction ? Disjunction is disjunction... What is the solution to $x^2 = 3$? It is the set of numbers a such that $a^2 = 3$... The distinction between $\vee$ and a hypothetical meta-$\vee$ is just a way to avoid the problem : you ask for real money but you are paid with meta-money.

▸ Classical logic is the logic of reality, the realistic logic. The realist paradigm, which is criticisable in the classical case, becomes impossible as soon as we step out of classical logic. An interesting exercise is the building of a *Broccoli* logic : it consists in introducing new connectives, new rules, the worse you can imagine, and then to define everything *à la*

Tarski. Miracle of miracles, completeness and soundness still hold : just because the reality has been defined from its representation : typically the semantics is the syntax **in boldface** ; but this nonsense gets a beautiful name, the *free Broccolo.*

▶ Moreover there is a basic mistake in the idea of semantics-as-reality. Everybody has seen a proof, maybe not a formal one, but at least something that can be transformed into a formal proof by a computer. But nobody has ever seen the tail of a model, models are ideal (usually infinite) objects. Moreover, what is the destiny of a proof ? It is to be combined with other proofs : theorem A can be used as a lemma to yield B. This suggests looking for an *internal* explanation.

1.1.3. *The foundational Trinity*

Usual « foundations » therefore involve three layers, just like the Christian Trinity :

Semantics : A first, irreducible, layer is realistic : properties are true or false, they refer to some absolute external reality, the integers —you know $0, 1, 2, \ldots$ Think of the Father.

Syntax : Due to incompleteness, a second partner appears : there are stronger and weaker theories, i.e., ways of accessing this absolute truth. This is the Son (a.k.a. *Verb*) in the Foundational Trinity.

Meta : Truth is « defined » via a pleonasm, Tarskian semantics (a.k.a. *vérité de La Palice*) ; later on, this absolute truth turns into a sort of *Animal Farm* where some are truer than others ; even worse, among the truths some are « predicatively correct », some are not[5] : in the Library of Truth, these incorrect truths are relegated to the *inferno* together with the licentious books of Marquis de Sade. Some sort of glue —or simply smoke[6]— is welcome. This is the Holy Ghost (a.k.a. go-between) of the Trinity.

The meta fixes the Tarskian pleonasm by saying that the Father is only the « meta »[7] of the Son, which in turn can be the meta of somebody else : « Turtles all the way down », each turtle sitting on a bigger metaturtle. The various layers are related through (in)completeness and soundness. It is not impossible that this construction eventually collapses —like the epicycles

---

[5] E.g., assuming that such a system makes sense, the consistency of predicative arithmetic...

[6] « Métaphysicien : sorte d'oiseau qui s'engraisse de brouillard » (Diderot).

[7] The genitor ?

of Ptolemaeus... or simply like the stack of turtles in a famous Japanese riddle. By the way, we shall see in 1.3.1 and 3.4.4 that completeness as an internal property.

1.1.4. *The Thief of Bagdad*

The ludic programme abolishes the distinction syntax/semantics, at least at the deepest level[8]. This of course denies any role to the meta. But this is an obstinate guy : close the door, most likely he will try the window.

One of the most interesting anti-realistic paradigms is game-theoretic :

- ▸ Formulas = Games
- ▸ Proofs = Winning strategies
- ▸ Truth = Existence of a winning strategy

It is designed to exclude the meta... Not quite, the window is open : the rule of the game. Who tells you that this move is legal or not, who determines the winner... For instance Gödel's *Dialectica* interprets A by something like $\exists x^\sigma \forall y^\tau \varphi(x,y) = 0$. x, y can be seen as sort of strategies in a game, whose rule is given by φ. This involves a splitting in two layers : the players vs. the « referee », i.e., the meta. The same can be said of all modern « game semantics » who —although more clever than the antique *Dialectica*— respect this schizophrenia of a rule of the game on top of an interaction[9].

As we shall see in 3.3.3, ludics closes the window too, by allowing the rule to be part of the game, this is the idea of a *game by consensus* : the role of the referee is played by the opponent. This means that the opponent has a lot of losing strategies —called dog's play— whose only effect is to control your moves : if I move in an « incorrect » way, then he is likely to use one of these dog's strategies and produce a *dissensus*, i.e., argue forever ; but if I move « correctly », the same strategy will be consensual and nicely lose. This is symmetrical : I have my own dog's strategies to control his moves...

The door is closed, the window too, but the cellar is open : the thief can still rely on the myth of the « ambient space ». This is simple, I give an explanation, but my explanation is done in a language (e.g., classical set-theory) which is my meta, one cannot escape one's meta... This argument has been heavily used against any kind of non-realistic foundations, e.g., against intuitionistic logic. People will agree on the distinction between $\exists n$ and

---

[8] Practically speaking the distinction may be seen as a polarisation : *my* viewpoint is rather syntax, whereas my *opponent*'s viewpoint is perceived (by me) as semantics.

[9] By the way, the very use of the expression « semantics » is symptomatic of an archaic stage of foundations : do we use « semantics » in a generic acceptation or as part of a dipole (tripole) semantics/syntax (/meta) ?

$\neg\neg\exists n$, one is effective, the other not, but this is a matter of eavesdropping : in the « ambient » space, the distinction vanishes.

1.1.5. *The ambient space*

So everything is embedded in a classical « ambient space » : this is the ultimate weapon of the meta. What to say about this sophism ?

A first remark is that alternative foundations cannot be done in alternative mathematics —think of the unfortunate *ultra-intuitionism*. On the other hand, set-theory is flexible enough to harbour various parts of mathematics, for instance operator algebras : but I would not dare to say that set theory is more basic than operator algebra ! In fact, the use of set-theory is the recognition of a *convenience*, by no way of a foundational status.

Let us take a geometrical analogy : in the XIXth century, non-Euclidian geometries were introduced, the most basic example being that of a sphere embedded in Euclidian space ; later on, mathematicians were able to define *varieties* without embedding them. The « meta » —here the ambient space in which a non-trivial explanation of logic takes place— reminds me of those science-fiction jets travelling faster than light through the ambient Euclidian space. In other terms, « embeddable » is not the same as « embedded ».

Logic is about thought, not about concrete objects. Foundations are impossible *ex nihilo* since they would involve a thought located outside the thought. For similar reasons it is *a priori* impossible to take a photo of the universe since one would need a point of view outside space ; but cosmology made a lot of progress last century, without allowing any room for a « meta-space-time ». This analogy makes us understand that the so-called « meta » could be nothing more than the necessary expression of the foundationalist attitude : the very study of logic generates a strange posture, which perhaps does not make sense.

There is a big difference as to the perception of physics and logic : everybody understands that one cannot tamper with the physical space and embed it into a bigger nonsense. But similar embeddings seem natural in logic. The main thesis of ludics —*locativity*— is that there is a logical space too, with which one cannot tamper either.

1.1.6. *Logic vs. Physics*

We just alluded to general relativity ; the comparison with Physics is always of interest. For instance, up to 1900, the dominant physical paradigm was determinism in the style of Laplace, « everything can be computed from the initial position », which contains in fact two layers, one being about abstract determinism, the other being about our faculty of prediction, semantics and syntax so to speak. The work of Poincaré —later, the theory of *chaos*—

definitely ruined the mere idea of prediction, this must be compared to incompleteness. But the comparison turns short here, for we should imagine a physics surely chaotic, but still deterministic, to match the present state of logic : in this physics, future is written in a Big Book —but this book is out of reach. Fortunately quantum physics left nothing of this frightening fantasy !

1.1.7. *Subjective aspects of realism*

Of course one cannot destroy realism by an act of will, there is a natural tendency to reify... An example : some people introduced long ago the idea of potential (vs. actual), and soon afterwards the set of all potentialities... with the result that 99% of the original idea vanished : the potentialities were *reified*, i.e., actualised. The task in that case would be to define « potential » in such a way that the set of all potentialities would not make sense... moreover the definition should be quite natural, otherwise why shouldn't nature shun such a thing ?

There is tendency to reify, this is why the current interpretation of quantum mechanics is not that convincing. Imagine a sort of *quantum philosopher* arguing with —say— Descartes : the poor guy would not get a chance... but he would be right whereas Descartes and his realism are wrong. The problem is that the global structure of realism is much elaborated and deeply rooted in our minds —especially Western minds.

Coming back to logic, the stronghold of realism is of course natural numbers. It is clear that any foundational reflection should —sooner or later— cope with integers, which are so deeply rooted in our (wrong) intuitions and our (wrong, but efficient) formalisms, that one may question the possibility of achieving anything in that direction. No doubt that this is a problem, but one can imagine indirect approaches, typically through the theory of exponentials, see section 5 below.

1.1.8. *Hands tied in the back*

« *100g pasta, 1 litre water, 10g salt. Add salt to boiling water. Add pasta, stirring occasionally. Cook for 10 minutes. Drain pasta keeping part of water.* » This recipe for *Penne Rigate* is taken from a current brand, *Pasta Barilla*. The obvious question is the meaning of the recommendations (interdictions). First observe that if we follow them, everything works ; but is there a reason to follow them, what is the reference of these rules ? For instance it is legitimate to question the idea of salting the boiling water : could we salt before boiling ? « NO : the salt must be put after »... at least it is what some people will say —in relation to the authority of *Zia Ermenegilda*, a sort of meta-pasta that you cannot taste but to which you must conform, Tarskian *cuisine* so to speak.

This attitude is common in foundations, there are too many artificial restrictions. By the way, this was one of the major objections of Hilbert to Brouwer : intuitionism is a sort of mathematics with « a hand tied in the back »[10]. Nowadays, most attempts at foundations keep a hand in the back. Certain principles —induction, comprehension... — are forbidden under a principle ending with « ism ». There is no doubt that this sort of *bondage* achieves something, but another « ism » may do as well, perhaps better. How can we decide ?

If we exclude divine revelations, the only possibility consists in making things interact with *alter egoes*. In the case of *pasta*, one alters the recipe and see whether it tastes the same. Typically, put the salt before boiling, you will notice no difference[11] ; push the cooking time to 15mn and you get glue.

To sum up, restrictions are not out of a Holy Book, but out of *use*. And use is *internal*, i.e., homogeneous to the object.

1.2. THE PRESENT STATE OF FOUNDATIONS

1.2.1. *The copyright*

Realism says that truth makes sense independently of the way we access it. Accept this and you are bound to compare the « strengths » of various formalisms, i.e., the sizes of the metaturtles. However, commonsense should tell us that nobody has ever seen the class of all integers, not to speak of this « Book of Truth and Falsehood » supposedly kept by Tarski. All these abstractions, truth, standard integers, are handled through proofs —hence the « reality » may involve a re-negotiation of the intertwining between proofs and « truth » —whatever the latter expression means, perhaps nothing at all.

But nothing of the like has been so far done. The compulsory access to « foundations » is through a bleak Trinity *Semantics/Syntax/Meta* : this approach owns the *copyright*, if you don't accept this, you are not interested in foundations, period[12]. The problem with the copyright is that it dispenses one to address the original question : « no new disturbing phenomenon will be disclosed, we concentrate on internal problems ». In an area such as foundations, this attitude should be called *scholastics*, from the medieval

---

[10] Hilbert himself was even more drastic as long as his fantasmatic « metamathematics » was concerned !

[11] But some are likely to taste the difference —if they know—, this is called intensionality.

[12] This sort of situation is not exceptional, remember Communism with its copyright on « Progress ».

philosophers —or rather their followers— who kept on transmitting an ossified approach to logic.

Clearly, if this is the only approach to foundations, one should do something
else... by the way this is what the main stream of mathematics realised long
ago.

1.2.2. *The input of computer science*

Traditional foundations were built as a « logic from mathematics ». By
the end of last century emerged a « logic from computer science » with
its peculiarities : the concept of external —ethereal— Truth (the Father)
was no longer pregnant ; on the other hand —since the computer is a badly
syntactic, bureaucratic, artifact— the Son (Syntax) became omnipresent. Of
course foundations of computer science difficultly departs from the Trinity
—as in the sportive saying : don't tamper with a losing team !—, but
something goes wrong. The mismatch is conspicuous everywhere, think
for instance of « operational semantics » : the semantics of the language
becomes the way you use it, i.e., is syntactical. For instance lambda-terms
are sometimes described as syntax and the normalisation (rewriting) process
as semantics : the idea is far from being stupid ; but try to present it in
Tarskian dressing !

The same can be said of the notorious *Closed World Assumption* « something
is false when not provable », which corresponds to a *procedural* view of logic :
negation is applied to the cognitive process itself and not to some « abstract
contents » : The Wind Blows Where It Wishes[13]...

1.2.3. *The input of proof-theory*

Traditional proof-theory was versatile enough to be transferred *mutatis
mutandis* from the desertic steppes of consistency proofs to theoretical
computer science. Lambda-calculi, denotational semantics, game-theoretic
interpretations... prompted a new universe, reasonably free from foundational anguish, and mainly dedicated to (abstract) programming languages.
It now seems that enough has been gathered to venture a first synthesis —
this is *ludics*— and to transfuse fresh blood into the anaemic foundational
body.

---

[13] But most people don't even hear the sound of It. This correct procedural intuition
was translated into a commutation of negation with provability and stumbled on incompleteness, halting problems : « non-monotonicity », « circumscription »,... tried to
accommodate some procedurality inside classical logic : to figure out the disaster, imagine
a horse-cart powered by a rocket.

1.2.4. *The locative thesis*

When I say « fresh blood », this is not rhetorics, I mean completely new ideas ; really shocking ones, the kind that receives in the best only polite reactions. This is the case for the *locative thesis*, which says that usual logic is wrong, because it is « spiritual », i.e., abstracts from the location. What is location ? Location in logic consists in these apparently irrelevant details known as names of variables, occurrences ; these details are evacuated in the first page of textbooks —they are also the favourite topic for the hangover lecture of Sunday morning. Locativity induces, among others :

▶ A clear approach to subtyping, inheritance, intersection types.

▶ The discovery of *locative* operations. Usual (spiritual) operations are obtained from the locative ones by use of *delocations*, i.e., isomorphic copies which prevent interferences. The typical example is that of a conjunction whose locative form is $A \cap B$ and whose spiritual form is $\varphi(A) \cap \psi(B)$; compare $A \cap A = A$ with $\varphi(A) \cap \psi(A) \simeq A \times A$, see 4.1.4 for a precise statement.

▶ The individuation of second-order quantification as an autonomous operation, not a poor relative of the first order case. In particular second-order logic validates classically wrong formulas.

All this is the evidence that locativity is a positive feature of logic, and not an impossible mess —not to speak of the fact that, maybe I don't know my free-and-bound-variables. The novelty of the idea, i.e., its suspicious flavour, explains why it is not (yet ?) popular. But this is fresh blood.

But we must now unwind the process leading to locativity ; in the beginning stands the uncommon idea of a procedural logic.

1.3. PROCEDURAL LOGIC

1.3.1. *The subformula property*

The domain of validity of usual completeness corresponds to closed Π^1 formulas (roughly : first order formulas universally quantified over their predicate symbols ; this terminology is slightly misleading, since the Σ^0_1 formulas of arithmetic are Π^1). Completeness is the identity between truth and provability for Π^1 formulas, whereas incompleteness is the failure of this property for the dual class Σ^1, which contains the Π^0_1 formulas. Now if we formulate logic in sequent calculus, we discover that the subformula property holds for *the same class* Π^1, and fails outside. What does this mean ? If we consider cut-free proofs, then all possible proofs are already there, there is no way to produce new ones. In other terms, the calculus is complete —nothing is missing. Observe that this completeness does not refer to any sort of model, it is an internal property of syntax. Such a property cannot be an accident, it should be given its real place, the first :

The subformula property is the actual completeness

In order to make sense of this, syntax has to get its autonomy, to become the main object of study, with no relation to anything like a preexisting semantics... and, in the absence of any umbilical link to a semantics, it will eventually lose its character of syntax, it will become a plain mathematical object.

1.3.2. The disjunction property

The idea of a procedural logic must be ascribed to early intuitionism : an existence theorem should *construct* a witness. The most spectacular consequence is the *disjunction property* « A proof of $A \lor B$ induces[14] a proof of A or a proof of B ». Disjunction therefore commutes with provability, i.e., applies to the cognitive process : to prove $A \lor B$ is to prove A *or* to prove B. Intuitionistic logic is *procedural* in the sense that it refers to its own rules, in contrast to classical logic which is realistic, i.e., refers to its own *meta*. Procedurality is considered suspicious, since it opens the door to subjectivism, but this is a superficial impression[15]. For instance realism interprets $\lor$ by meta-$\lor$; depending on the weather, meta-$\lor$ can be classical, intuitionistic, or enjoy intermediate properties like $\neg A \lor \neg\neg A$: there are full handbooks[16] dedicated to such logics, all of them sound and complete w.r.t. their own meta (the same in boldface). On the other hand, if we try to enrich classical logic with a connective enjoying the disjunction property, we badly fail : this connective turns out to be the same as classical disjunction.
There are two positive features of procedurality, first it is absolute (it refers to concrete operations on proofs), whereas realism is relative (it refers to our intuition of the universe, i.e., to what we already have in mind). Second, most procedural interpretations will be inconsistent : it was a true miracle that a connective enjoying the disjunction property could be found, and the price was a drastic modification of logic. Even more difficult was the discovery of an involutive procedural negation —this was the main achievement of linear logic. We eventually discover that procedurality is more demanding than realism.

---

[14] Implicitly : only a moron would state $A \lor B$ if he has obtained A.

[15] Some style this as « intensional », others as rubbish : they basically agree, but miss the point !

[16] Not quite about alternative disjunction : it is too difficult to tamper with this connective ; but this is not the case with modalities if you see what I mean.

1.3.3. *A logic of rules*
Procedurality is not intensional rubbish, it is a change of viewpoint, corresponding to the lineaments of a *logic of rules*. The idea is that disjunction can be applied to the proofs (prove A or prove B) : it is an operation on the representation, not on the « world » —if there is anything like the world.
In ludics, the disjunction property appears as the completeness of disjunction —an internal version of completeness, closely related to the subformula property, see 4.1.6.

1.4. AN EXAMPLE : BARBARA

1.4.1. *Old scholastics*
Barbara is the familiar syllogism :

$$\frac{\forall x(Ax \Rightarrow Bx) \quad \forall x(Bx \Rightarrow Cx)}{\forall x(Ax \Rightarrow Cx)} \; Barbara$$

Scholastics philosophers were basically concerned with the explanation of syllogisms by mutual reduction ; in particular the acronyms *Disamis, Celarent,...* contain information as to these reductions. After centuries of repetitive work, this sort of activity became suspicious —and the expression « scholastics » derogative.

1.4.2. *New scholastics*
By the beginning of last century, Łukasiewicz explained *Barbara* —and all the other figures— by the transivity of inclusion. The usual reaction in front of such an explanation is that of the layman discovering a « sculpture[17] » of the late César : rubbish[18]. With some education, you learn politeness and eventually find some (well-hidden) virtues in the product. But, education or not, this regressive explanation fails at explaining the major point : why is there a rule, why this one precisely, how do we explain the distinctions between the various syllogisms —not to speak of their mutual reductions ?

1.4.3. *Category theory*
A major anti-realistic breakthrough was the introduction of *categories* in logic. The climateric work was the Curry-Howard isomorphism of 1969, but there was something in the air, think of Prawitz's work on natural deduction (1965), Scott domains (1969), my own system $\mathbb{F}$ (1970), Martin-Löf's type

---

[17] A compressed automobile with the signature of the artist.

[18] Myself I remember being taught about inclusion ; three increasing potatoes were drawn on the board, but a comment from the teacher, a sort of Barbara-without-the-name, was still needed.

theory (1974). With the decisive input of computer science, the Boulder meeting (1987) was the apex of this « time of categories ».

Category-theory is often styled as nonsense because of the abuse of diagrams : indeed it is like Japanese *cuisine*, it does not stand mediocrity ; so let us forget the fat and proceed to the meat. Category-theory is —among the established paradigms[19]— the most remarkable attempt at revocating this bleak distinction subject/object. In category, we have objects (objective) and morphisms (subjective), and they live happy together, there is no *secret sharer*.

Now look at the categorical version of the same : A, B, C become objects, proofs become morphisms, i.e., object and subject start to communicate. Eventually *Barbara* is composition of morphisms, something which is definitely not regressive w.r.t. Aristotle.

1.4.4. *Barbara as a proof-net*

Categories induced the first mature re-negotiation of the relation object/subject ; this eventually gave birth to *linear logic*. Linear logic yields in turn a procedural explanation of *Barbara* seen as a *proof-net*[20] :

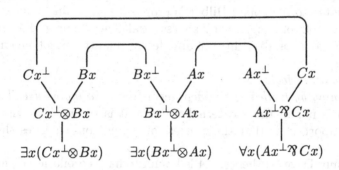

The structure is seen as a graph, in which formulas are read as non-deterministic travel instructions. Whatever choice is made, the resulting trip is connected and acyclic. Such trips correspond to (virtual) « information flows » and connectedness/acyclicity guarantees the possibility of composing syllogisms —i.e., cut-elimination between syllogisms.

This interpretation is really subtler than the Tarskian one, for which all syllogisms are about transitivity of inclusion, period. In particular Tarskism

---

[19] Ludics is even more radical, but still experimental.

[20] Implication is translated as linear implication, not as usual implication... This is not an abuse, one had to wait for Boole to get the idea of idempotency, i.e., the contraction rule.

cannot make sense of Aristotle's original taxonomy of syllogisms (first figure, second figure...). Contrarily to proof-nets : a recent analysis of syllogisms due to Abrusci shows that the figures of Aristotle correspond to crossing numbers : e.g., the syllogisms of the first figure have planar proof-nets.

2. Hilbert revisited

The basic choice is between Hilbert's Programme (updated or not) —a sort of *millennium bug*— and anti-realism : something more experimental, more in the style of... 1900, when people were still trying, had not yet learn to pretend... Hilbert was precisely one of these guys, so let us revisit Hilbert.

2.1. ON ABSURDITY

2.1.1. *Hilbert's program*
Hilbert had to face certain paradoxes like Russell's. Since paradoxes come from our intuition of the real world, the idea was to put reality aside and to emphasise the formal treatment : this leads to another procedural approach to logic : *consistency*. We know since 1931 and the incompleteness theorem that this approach is wrong, but the refutation is rather technical. In what follows, we try to re-explain Hilbert's proposal as a duality between proofs of A and proofs of $\neg A$... with an essential flaw, namely that the pivot (here : the proofs of absurdity) of any decent duality cannot be empty.

2.1.2. *Duality in logic*
Since *meaning by truth* is forbidden, let us try *meaning as use*. The use of a theorem is not in the « external world », it is trough its consequences. What is important is that this process of taking consequences should not be « bugged ».
Play it again, Dave : a theorem A is bound to have corollaries B, hence one should accept as true (i.e., as a theorem) anything having true consequences, whatever this means. In other terms, A is true iff for any B, whenever $A \Rightarrow B$ is true, then B is true :

$$\frac{A \quad A \Rightarrow B}{B} \quad Modus\ Ponens$$

This is a correct characterisation, not an explanation, think of $B = A$: one should eventually get rid of this unknown B, i.e., « close the system ». To make the long story short, various reasons prompt one to simplify the previous *Modus Ponens* into the cut :

$$\frac{\vdash A \quad A \vdash}{\vdash} \quad Cut$$

A proof of A is anything which, combined with a proof of $\neg A$ yields a proof of the absurdity. But what are the proofs of absurdity ?

2.1.3. *Consistency*

If we stick to the literal sense of « proof », there can be no proof of the absurdity. Then A is true (provable) if $\neg A$ is not true, i.e., not provable. In this monist reading, truth becomes the same as consistency. This is the basis of Hilbert's Programme, and the idea —the mistake— was redis-covered in computer science, through non-monotonic logics, closed world assumption... without even the excuse of novelty. Indeed, something in the structure of the duality should warn us in advance : the cut (*Modus Ponens*) establishes a duality between proofs of a formula and proofs of its negation, the output of the duality being in... the empty set. Transpose this in the Euclidian Space and think of the duality defined by $x \perp y := \langle x \mid y \rangle \in \emptyset$, i.e., x never orthogonal to y. Then $X^{\perp}$ is empty when X is non-empty, and $\emptyset^{\perp} = \mathbb{R}^3$, bleak indeed !

The idea of interpreting negation by something like the set-theoretic com-plement is by the way bad taste, for provable formulas should form a closed set, in any reasonable (topological, algebraic...) sense whereas the comple-ment of a closed set can hardly be closed : this is the ultimate reason for incompleteness —and not diagonalisation, which is nothing but a technical tool showing that the complement of a closed set needs not be closed.

2.1.4. *Wrong proofs*

So far so bad... But isn't it because we used a simple-minded notion of proof too much linked with a literal interpretation of reasoning, in particular with the idea that formal reasoning should be correct ? Why not allowing more « proofs » —maybe dubious— so as to replace the traditional duality :

$$\text{\textit{Proofs of} } \vdash A \ / \ \text{\textit{Models of} } A \vdash$$

with something of the form :

$$\text{\textit{Proofs of} } \vdash A \ / \ \text{\textit{Proofs of} } A \vdash$$

In this respect the proofs of the negation could be seen as sorts of « coun-termodels » —but now part of syntax, no longer of « reality ». The only *a priori* objection to such a thing is consistency ; but among proofs some might be more correct than others, so that consistency would be maintained by the exclusion of « wrong » proofs. By the way, if this idea of a wrong proof seems artificial to you, what would you say about transfinite stacks of metaturtles ?

But where to find these additional wrong proofs ? Our only clue is the cut-elimination procedure : a proof of $\vdash A$ and a proof of $A \vdash$ put together should produce, through cut-elimination, a proof of $\vdash$ [21].

2.1.5. *Faith*

Write naive set theory in natural deduction style, what Prawitz did in 1965. Russell's paradox yields a cut between a proof π of $\vdash a \in a$ and λ of $a \in a \vdash$ whose normalisation diverges, what we can note
$\ll \pi \mid \lambda \gg = \Omega$; the symbol Ω —standing for divergence— is reminiscent of the u of recursion theory, or the $\Delta\Delta$ of lambda-calculus. Ω can be seen as a notational convenience, it can also be handled as an object, called *Faith* (i.e., the faith in convergence). But whatever choice we make, this is not in this way that we shall get a proof of the absurdity : our way out of Russell's paradox is not to forbid the cut between π and λ (which can be performed, this is precisely what Prawitz did), it is just to deny their *orthogonality* : if we consider syntax as a way to avoid divergence, π and λ cannot receive simultaneous specifications $\vdash A$ and $A \vdash$ (with the same A).

2.1.6. *Daimon*

We just saw that the basic duality can fail ; this means that it should sometimes succeed, but nothing in the extant proof-theoretic tradition can help us. Coming back to Euclidian space, $x \perp y := \langle x \mid y \rangle = 0$ is a beautiful orthogonality relation (the orthogonal of a line is a plane...) ; absurdity should be given a proof —something like 0—, but can we seriously allow that ?
What seemed absurd at the time of Hilbert is more reasonable at the time of proof-search : *Logic Programming* is organised as the search for cut-free proofs. Starting with the conclusion, the idea is to guess a last rule, then a rule above..., up to completion. Most likely the process eventually aborts, for want of a possible rule to apply, but we anyway did construct a truncated proof. It turns out that these truncated proofs are formal object just as good as usual ones ; in particular one can develop a (straightforward) proof-theory and normalise cuts involving such proofs..., provided abortion is clearly acknowledged as a new rule, the *Daimon* ✠.
The addition of the daimon to syntax by no way produces an inconsistent system —provided we don't play on words. Of course every formula becomes provable with the help of the daimon, but what about daimon-free provable formulas ? They are closed under consequence, i.e., cut-elimination : this

---

[21] It is by the way funny to remark that traditional proof-theory is precisely about this situation —with the frustrating feature that the situation never occurs— : such a work about the empty set !

is not surprising, before ludics, nobody ever heard about a logical daimon, but people knew how to normalise ; in other terms, the daimon cannot be created through normalisation.

Usual proofs therefore appear as daimon-free ones. The new « proofs » using the daimon occupy a space which is usually devoted to *models* : instead of having models of $\neg A$, we shall have proofs of $A^{\perp}$[22]. The difference is as follows :

▶ In the classical paradigm, the notions of proof and models are absolute. A proof of A proves A beyond discussion, a model of $\neg A$ refutes A beyond discussion. To the point that putting together a proof of A and a model of $\neg A$ is absurd. We are back to the empty pivot.

▶ In ludics, proofs —or models— are relative. They try their best to follow the rules, the truth tables, but nobody is perfect. An imperfect *proof* of A can be opposed to an imperfect *proof* of $A^{\perp}$; the interaction eventually yields a daimon —which has been produced by one of the two proofs, which is therefore incorrect. This does not necessarily make the other proof correct, since it may be opposed to another « counterproof » against which it now « fails ».

2.2. FROM PROOFS TO DESIGNS

2.2.1. *Geometry of proofs*

What remains to be found is an object having the structure of a proof, but free from syntactical commitments. This object —a *design*— roughly corresponds to the geometrical structure underlying a sequent calculus proof. We should separate two layers, one being « what is actually performed », the other being useful comments ; useful to us —typically the name of the formula proven— but of no mathematical significance. A toy model of this is typed lambda-calculus : in a typed lambda-term the real guy is the pure lambda-term obtained by erasing the type decorations which can be seen as superfluous comments (they are useful specifications, but they don't participate to normalisation).

The process of finding the object has been very complex ; we already mentioned the invention of the Daimon. But the crucial breakthrough was the discovery of *polarities*.

2.2.2. *Polarities*

Curiously, the notion of polarity existed in logic long before its invention, but with no status. For instance, the « negative fragment » of intuitionistic

---

[22] Notice the use of linear negation ; this conceptual shift is technically impossible with usual negation.

logic regroups $\forall, \Rightarrow, \wedge$, namely the connectives which are well-behaved w.r.t. natural deduction. The notational gimmick of linear logic individualises two classes of connectives, one in « logical style », the negative $\mathscr{R}, \&, \top, \bot, \ldots$ one in « algebraic style », the positive $\otimes, \oplus, 0, 1, \ldots$; in this way it is easy to memorise remarkable isomorphisms, typically distributivity. It is only by 1990 that this convenient distinction turned into a fundamental of logic.

The first remark is that, for any connective, there is always a side of the sequent on which contraction is free. Typically, the contraction rule on $\forall x A$ is redundant on the right of sequents : we can get it from a contraction on A ; the same holds for existence, but on the left. This is a first account of the distinction. The second remark comes from logic programming : negative connectives are *invertible*, i.e., they have a deterministic right rule : typically, proving $\vdash A \Rightarrow B$ is quite the same as proving $A \vdash B$. The fundamental remark of Andreoli —known as *focalisation*— is that the *positive* connectives —the non-invertible ones— enjoy a dual property. This property can be stated as follows : a cluster of operations of the same polarity can be seen as a single connective. Typically we can write complete rules for the double quantifiers $\forall x \forall y$ or $\exists x \exists y$ seen as a single operation, but not for the combinations $\forall x \exists y$ and $\exists x \forall y$. In the first case the two quantifications can be performed as a single step, in the second case, they must be performed sequentially. Polarity is therefore about time in logic, about the intrinsic causality between logical rules.

2.2.3. *Polarised proofs*

To understand how things work, let us take a concrete example, namely the positive formula $A = ((P^{\perp} \oplus Q^{\perp}) \otimes R^{\perp})$, where P, Q, R are positive. A is treated as a single ternary connective $\Phi(P^{\perp}, Q^{\perp}, R^{\perp})$ applying to *negative* subformulas $P^{\perp}, Q^{\perp}, R^{\perp}$, i.e, $A = \Phi(P^{\perp}, Q^{\perp}, R^{\perp})$. We take advantage of the new notion of polarity and only use positive formulas ; this means that a negative formula is handled by means of its negation on the other side of the sequent. Our sequent calculus will therefore consists of sequents $\Gamma \vdash \Delta$ made of positive formulas. Inspection of the rules shows that it is enough to restrict to the case were Γ has at most one formula[23]. The rules for A are

---

[23] This is quite the usual intuitionistic restriction, but left and right have been exchanged... This is because we are using *positive* formulas, whereas intuitionistic logic is merely concerned with negative formulas.

$$\cfrac{P \vdash \Gamma \quad R \vdash \Delta}{\vdash \Gamma, \Delta, A} \; (\vdash A, \{P,R\})$$

$$\cfrac{\vdash \Lambda, P, R \quad \vdash \Lambda, Q, R}{A \vdash \Lambda} \; (A\vdash, \{\{P,R\},\{Q,R\}\})$$

$$\cfrac{Q \vdash \Gamma \quad R \vdash \Delta}{\vdash \Gamma, \Delta, A} \; (\vdash A, \{Q,R\})$$

The right rules are obtained by combining a right Tensor-rule with one of the two possible right Plus-rules and negation, yielding two possibilities distinguished as $(\vdash A, \{P, R\})$ and $(\vdash A, \{Q, R\})$; *the* left rule is obtained by combining *the* Par-rule with *the* With-rule and negation. The rule is written $(A \vdash, \{\{P, R\}, \{Q, R\}\})$ in order to stress the existence of two premises, one involving P, R, the other involving Q, R.

As already explained, besides these « standard » rules, we need another one, the Daimon :

$$\cfrac{}{\vdash \Delta} \; \maltese$$

The daimon is restricted to the case « Γ empty » ; the case $A \vdash \Delta$ is indeed derivable, see the *Negative Daimon* in 3.2.2.

2.2.4. *Normalisation*

Proof-theory is organised along cut-elimination. In the previous case, let us assume that we are given a cut between $\vdash \Gamma, \Delta, A$ and $A \vdash \Lambda$:

1. If $\vdash \Gamma, \Delta, A$ has been obtained through a rule $(\vdash A, \{P, R\})$, replace with two cuts between $\vdash \Lambda, P, R$, $P \vdash \Gamma$ and $R \vdash \Delta$.
2. If $\vdash \Gamma, \Delta, A$ has been obtained through a rule $(\vdash A, \{Q, R\})$, replace with two cuts between $\vdash \Lambda, Q, R$, $Q \vdash \Gamma$ and $R \vdash \Delta$.
3. If $\vdash \Gamma, \Delta, A$ has been obtained by a daimon, replace with the proof of $\vdash \Gamma, \Delta, \Lambda$ consisting of a daimon.
4. Otherwise, apply a commutation of rules.

In the first two cases, the point is that the left rule for A is invertible, hence we do know that $A \vdash \Lambda$ follows from the two premises $\vdash \Lambda, P, R$ and $\vdash \Lambda, Q, R$.

3. Designs and behaviours

What follows is a slightly simplified version of ludics.

3.1. DESIGNS

3.1.1. *Locations*
It remains to remove logical decorations. However, since a proof is basically a sequence of formulas, one should be careful not to remove everything. Indeed we shall keep the location, the *locus* of the formula. This locus is a very concrete notion, it is the place where the name of the formula is written : we shall assume that this space is the usual infinitely branching tree. To come back to our previous example, if A has been assigned the locus σ, then its immediate subformulas P, Q, R will be distinguished by *biases* $3, 4, 7$ —e.g., Q is the subformula of relative location 4 of A— and be respectively located in $\sigma * 3, \sigma * 4, \sigma * 7$. The logical rules just written will be interpreted by the disintegration of σ into its *subloci* $\sigma * 3, \sigma * 7$ and/or $\sigma * 4, \sigma * 7$.

3.1.2. *Occurrences*
It is time to revisit an old nonsense of logic, so-called « occurrences » : a given formula A may « occur » at different places. In theology this familiar property of Saints is known as *bilocation* but we are in mathematics : two objects with distinct locations cannot be quite the same. But they can be isomorphic ; concretely an isomorphism —called a « delocation »— relating the two « occurrences » is provided. This means that :

1. Our formula A can be located everywhere in our tree of loci. But when we change the location we are not quite with the same A, we are with an isomorphic copy.
2. The three subformulas of A have been given the relative locations $3, 4, 7$; we could have chosen as well $9, 6, 22$. The result would have been definitely different since for instance $7 \neq 22$, but of course isomorphic.

This distinction between equality and isomorphism is not a gilding of the lily. It corresponds to the replacement of the dominant *spiritual* treatment of logic with a more refined *locative* approach. It has spectacular positive consequences, such as the expression of the cartesian product as a delocated intersection, see 4.1.4.

3.1.3. *Pitchforks*

A sequent made of loci is called a *pitchfork*[24]. In a *negative* pitchfork $\xi \vdash \Upsilon$, ξ is called the *handle*, in a *positive* pitchfork $\vdash \Upsilon$, there is no handle, only *tines*. What has been so far sketched translates into a sort of « pitchfork » calculus, with the following rules :

Positive rule : I is a ramification, i.e., a non-empty finite set of biases, for $i \in I$ the Λ_i are pairwise disjoint, with $\Lambda = \bigcup \Lambda_i$: one can apply the rule (finite, one premise for each $i \in I$)

$$\frac{\ldots \xi * i \vdash \Lambda_i \ldots}{\vdash \Lambda, \xi} \; (\vdash \xi, I)$$

Negative rule : $\mathcal{N}$ is a set of ramifications, the *directory of the rule* : one can apply the rule (perhaps infinite, one premise for each $I \in \mathcal{N}$)

$$\frac{\ldots \vdash \Lambda, \xi * I \ldots}{\xi \vdash \Lambda} \; (\xi \vdash, \mathcal{N})$$

Daimon :

$$\frac{}{\vdash \Lambda} \maltese$$

ξ is called the *focus* of the rules $(\vdash \xi, I)$ and $(\xi \vdash, \mathcal{N})$. The three rules discovered in 2.2.3, are therefore written $(\xi \vdash, \{\{3, 7\}, \{4, 7\}\})$, $(\vdash \xi, \{3, 7\})$ and $(\vdash \xi, \{4, 7\})$.

3.1.4. *Designs*

A design is anything built in the pitchfork calculus by means of these three rules. No assumption of finiteness, well-foundedness, recursivity, is made ; designs can therefore be badly infinite. However infinite designs can naturally be written as « unions » of finite designs : any design $\mathfrak{D}$ is the directed « union » of the finite designs obtained by restricting all negative rules of $\mathfrak{D}$ to finite directories, all but a finite number of them being empty, see 3.2.4.

Usual sequent calculus contains a distinguished rule, called « identity axiom ». In usual syntax, identity axioms can be replaced with η-expansions, but they are still needed for propositional atoms (variables). In ludics, the η-expansion can proceed « beyond the atoms », yielding the *Fax*, see 4.2.7 ; this is why ludics has nothing like an identity axiom.

---

[24] These loci must be pairwise incomparable w.r.t. the sublocus relation.

3.2. THE ANALYTICAL THEOREMS

3.2.1. *Normalisation revisited*

In 2.2.4 we defined —or rather sketched— normalisation. This translates *mutatis mutandis* to the « pitchfork calculus », provided we avoid some pitfalls :

1. *Cut* is no longer a rule, it is a coincidence handle/tine between two designs. Typically a design of base (conclusion) $\sigma \vdash \tau$ and a design of base $\vdash \sigma$ who share σ form a *net* of base (conclusion) $\vdash \tau$[25].

2. Cases 1,2 of the syntactical process replace a cut on σ by two cuts on $\sigma * 3, \sigma * 7$ (or $\sigma * 4, \sigma * 7$). But this is a mere accident due to logic : we have respected the rules of the formula A and the left rule happens to exactly match the possible right rules. But there is no longer anything like A —for instance, σ does not encode the formula A—, which means that in case of a cut between a positive rule $(\vdash \sigma, I)$ and a negative rule $(\sigma \vdash, \mathcal{N})$ it may happen that $I \notin \mathcal{N}$. In that case the normalisation process diverges, there is no normal form. We can use the symbol Ω to denote a diverging output, but this is a convenience, *Faith* not being a design.

3. We had in head a finite normalisation, but designs need not be finite. In particular what may happen is an infinite sequence of normalisations, a cut is replaced with several cuts, one of these cuts in turn is replaced with other cuts... Such process also diverges, i.e., yields the non-design Ω as an output.

The value of the concept of design lies in a certain number of remarkable properties, which are the respective analogues of Böhm's theorem, Church-Rosser property, and stability.

3.2.2. *Separation*

Meaning is use ; if a design is a really meaningful structure, all of it must be usable, observable. But how do we use a design $\mathfrak{D}$ of base $\vdash \sigma$[26] ? Simply by cutting it with a *counterdesign* $\mathfrak{E}$ of base $\sigma \vdash$ and normalise. The base of the resulting net is the empty pitchfork $\vdash$ and there are very few possibilities :

Consensus : the normalisation converges, and the normal form is the only design of base $\vdash$, namely ✠. In that case, we say that $\mathfrak{D}, \mathfrak{E}$ are *orthogonal*, notation $\mathfrak{D} \perp \mathfrak{E}$.

Dissensus : the normalisation diverges, i.e., yields the non-design Ω.

---

[25] The notion is easily extended to several cuts : the coincidence graph must be connected/acyclic.

[26] The discussion applies to any base.

Given $\mathfrak{D}$ one can consider the set $\mathfrak{D}^\perp$ of those counterdesigns $\mathfrak{E}$ which are consensual with $\mathfrak{D}$. The separation theorem states that designs are determined by their orthogonal, i.e., their use :

Theorem 1 (Separation)
If $\mathfrak{D} \neq \mathfrak{D}'$ then there exists a counterdesign $\mathfrak{E}$ which is orthogonal to one of $\mathfrak{D}, \mathfrak{D}'$ but not to the other.

This analogue of Böhm's theorem has a topological meaning. In the coarsest topology making normalisation continuous, the closure of the singleton $\mathfrak{D}$ is the biorthogonal $\mathfrak{D}^{\perp\perp}$. The separation theorem states that this topology is $\mathcal{T}_0$, i.e., that the preorder $\mathfrak{D} \preceq \mathfrak{D}' \Leftrightarrow \mathfrak{D}'^{\perp\perp} \subset \mathfrak{D}^{\perp\perp}$ is indeed an order. In other terms, we can form a Scott domain with designs : this is our bridge with first generation denotational semantics.

The separation theorem has an effective version, which amounts at giving an explicit description of the relation $\preceq : \mathfrak{D} \preceq \mathfrak{D}'$, i.e., « $\mathfrak{D}'$ is more converging than $\mathfrak{D}$ » if $\mathfrak{D}'$ has been obtained from $\mathfrak{D}$ by means of two operations :

Widen : add more premises to negative rules, i.e., replace $\mathcal{N}$ with $\mathcal{N}' \supset \mathcal{N}$.

Shorten : replace positive rules (ξ, I) with daimons —which has the effect of reducing the depth of branches.

On a positive base, the greatest —most converging element— is the Daimon ; Faith would be the smallest design —but has been excluded from the « official » definition. On a negative base, there is still a greatest design, called the *negative Daimon* :

$$\cfrac{\cdots \quad \cfrac{\maltese}{\vdash \xi * I, \Lambda} \quad \cdots}{\xi \vdash \Lambda} \; (\xi, \wp_f(\mathbb{N})\backslash\{\emptyset\})$$

The smallest design is called the *Skunk* :

$$\frac{}{\xi \vdash} \; (\xi, \emptyset)$$

The name will be explained in 3.3.5.

3.2.3. *Associativity*
Strictly speaking, since normalisation is deterministic, there is no need for a Church-Rosser property. But besides the narrow technical meaning of Church-Rosser, there is a deeper one, namely that in the presence of two cuts, the output of normalisation is the same, whether we normalise them altogether, or one after the other, something like $ABC = (AB)C$:

THEOREM 2 (ASSOCIATIVITY)
Normalisation is associative : let $\{\mathfrak{R}_0, \ldots, \mathfrak{R}_n\}$ *be a net of nets, then*

$$\llbracket \mathfrak{R}_0 \cup \ldots \cup \mathfrak{R}_n \rrbracket = \llbracket \llbracket \mathfrak{R}_0 \rrbracket, \ldots, \llbracket \mathfrak{R}_n \rrbracket \rrbracket \tag{1}$$

Of course $\llbracket \mathfrak{R} \rrbracket$ is short for the normal form of the net $\mathfrak{R}$ and as in recursion theory, the equation also applies in case of divergence.

Associativity is often combined with separation to define adjoints, this is the *closure principle* « everything reduces to closed nets ». Typically, if $\mathfrak{D}, \mathfrak{E}$ are designs of respective bases $\xi \vdash \lambda$ and $\vdash \xi$, the normal form $\llbracket \mathfrak{D}, \mathfrak{E} \rrbracket$ is the unique design $\mathfrak{D}'$ of base $\vdash \lambda$ such that for every $\mathfrak{F}$ of base $\lambda \vdash$:

$$\llbracket \mathfrak{D}', \mathfrak{F} \rrbracket = \llbracket \mathfrak{D}, \mathfrak{E}, \mathfrak{F} \rrbracket \tag{2}$$

The normal form of a net $\mathfrak{S}$ is determined by the normal forms of all completions of $\mathfrak{S}$ into a closed net. The principle is very useful, since closed nets do not need commutative conversions.

3.2.4. *Stability*

Stability was introduced by Berry in the late seventies and is the distinctive feature of second generation denotational semantics —typically coherent spaces— which eventually led to linear negation.

Let us go back to separation and the relation $\mathfrak{D} \preceq \mathfrak{D}'$. The relation means that, if we replace $\mathfrak{D}$ with $\mathfrak{D}'$ in a converging net, then the resulting net still converges. As we saw it, there are two ways to be « more convergent », one is to be shorter, which can be summarised by $(\vdash \xi, I) \preceq \maltese$; the intuition is that if the normalisation process makes use of the positive action $(\vdash \xi, I)$, then the process will converge quicker if we replace it with $\maltese$: it immediately stops. The other way is to be wider, i.e., to replace a non-premise of a negative rule, which can be figured as a premise with nothing —i.e., the non-design Ω— above it, by the same premise with something —typically $(\vdash \xi, I)$— above it. The two ways are summarised by the equation :

$$\Omega \preceq (\vdash \xi, I) \preceq \maltese \tag{3}$$

Widening is of different nature since it does not alter the normalisation process (in contrast to shortening, which makes it... shorter). Stability is about widening, which corresponds to a plain set-theoretic inclusion between designs[27]. Stability says that when $\mathfrak{D} \perp \mathfrak{E}$, then there are well-defined —i.e., minimum w.r.t. inclusion— finite $\mathfrak{D}' \subset \mathfrak{D}, \mathfrak{E}' \subset \mathfrak{E}$ which are

---

[27] Defined as *desseins*, i.e., sets of *chronicles* : this is indeed the official definition, the presentation as a pitchfork calculus being only a (slightly incorrect) convenience.

responsible for this, i.e., such that $\mathfrak{D}' \perp \mathfrak{C}'$. This property is responsible for the major concept of ludics —incarnation, see 3.3.2.

3.3. BEHAVIOURS

3.3.1. *Formulas as specifications*
A formula A can be —up to isomorphism— located everywhere, more precisely, when A is positive (resp. negative) it can receive any location $\vdash \sigma$ (resp. $\sigma \vdash$). Say for instance that A is negative and located in $\sigma \vdash$; then A will be identified with a certain set $\mathbf{G}$ of designs (representing the « proofs » of A) of base $\sigma \vdash$. This set is not arbitrary, it corresponds to a specification, a « how-to », i.e., a certain use of the designs. Remember that we reduced « use » to normalisation with counterdesigns : the use can therefore be represented by a set $\mathbf{G}^u$ of counterdesigns and we therefore obtain that $\mathbf{G} = \mathbf{G}^{u\perp}$. One can get rid of the arbitrary $\mathbf{G}^u$ by rewriting this as $\mathbf{G} = \mathbf{G}^{\perp\perp}$. The idea of formula as specification translates into :

DEFINITION 1
A behaviour is a set $\mathbf{G}$ of designs of a given base equal to its biorthogonal.

Observe that $\mathbf{G}^\perp$, which is one of the possible choices for $\mathbf{G}^u$, is a behaviour too.
A behaviour is never empty : it contains the daimon of the right polarity —this is because $\maltese \perp \mathfrak{C}$ for all $\mathfrak{C}$. Behaviours are closed under $\preceq$: $\mathfrak{D}' \preceq \mathfrak{D}$ and $\mathfrak{D}' \in \mathbf{G}$, then $\mathfrak{D} \in \mathbf{G}$.

3.3.2. *Incarnation*
As a consequence, if $\mathfrak{D}' \subset \mathfrak{D}$ and $\mathfrak{D}' \in \mathbf{G}$, then $\mathfrak{D} \in \mathbf{G}$, but for « bad reasons » : nothing in $\mathfrak{D} - \mathfrak{D}'$ actually matters w.r.t. $\mathbf{G}$. Given $\mathfrak{D} \in \mathbf{G}$ there might be several $\mathfrak{D}' \subset \mathfrak{D}$ which are still in $\mathbf{G}$; but among those $\mathfrak{D}'$, there is a smallest one, the *incarnation* $|\mathfrak{D}_{\mathbf{G}}|$ of $\mathfrak{D}$ w.r.t. $\mathbf{G}$. The existence of the incarnation is just an alternative formulation of stability.
Incarnation is contravariant, i.e.,

$$\mathbf{G} \subset \mathbf{H} \Rightarrow |\mathfrak{D}|_{\mathbf{H}} \subset |\mathfrak{D}|_{\mathbf{G}} \tag{4}$$

Hence the incarnation of $\mathfrak{D}$ is maximum when $\mathbf{G}$ is the smallest (principal) behaviour $\mathfrak{D}^{\perp\perp}$ containing $\mathfrak{D}$; in this case $|\mathfrak{C}| = \mathfrak{C}$ (easy consequence of the separation theorem).

3.3.3. *Formulas as games*

Can we describe ludics as a « game semantics » ? Surely a design is a sort of strategy, which, when put against a counter-design yields a *dispute* — the normalisation process, see 3.4.1— which is a play. Moreover, the use of the daimon in the dispute corresponds to giving up, hence we have a notion of winning. But these notions are defined once for all, there is no way to tamper with them. If we follow the usual game paradigm, little can be added : the first player can act in such a way that the opponent cannot win —what we called somewhere the *atomic weapon*. But wait a minute — the dispute generated by a design and a counterdesign needs not converge, i.e., yield a winner. We shall therefore impose the following : the two players do whatever they want, *provided they stay consensual*, i.e., that all disputes converge. This is the idea of a *game by consensus*.

The rule of the game $\mathbf{G}$ can then be specified by a set of counterdesigns $\mathbf{G}^r$, the only requirement being consensus, i.e., $\mathfrak{D}$ is an admissible strategy for $\mathbf{G}$ iff it is consensual with any $\mathfrak{E} \in \mathbf{G}^r$. If $\mathfrak{D} \perp \mathfrak{E}$ is short for consensus, this rewrites as $\mathbf{G} = \mathbf{G}^{r\perp}$, which implies $\mathbf{G} = \mathbf{G}^{\perp\perp}$; moreover there is a complete symmetry here, since $\mathbf{G}^\perp$ is the most natural $\mathbf{G}^r$. Eventually the game is explained by a consensus between players, as in real life —if real life is a game.

Of course a game by consensus can be seen as a usual game, with a rule, a referee, etc. In that respect, a design becomes a real strategy, but we have to work up to incarnation.

3.3.4. *Typed vs. untyped*

Designs can be handled in two ways :

Untyped : Designs can be considered as pure, i.e., *as themselves.*

Typed : Designs can be considered as part of a behaviour, i.e. with a restriction on their use ; they are no longer considered as themselves, but *as they should be*. In particular, they are only up to incarnation.

Typically, in the case of *subtyping*, i.e., of an inclusion $\mathbf{G} \subset \mathbf{H}$, the untyped viewpoint says that a design in $\mathbf{G}$ is a design in $\mathbf{H}$; the typed viewpoint induces a *coercion map* which replaces a design $\mathfrak{D}$ incarnated in $\mathbf{G}$, by its incarnation $|\mathfrak{D}|_{\mathbf{H}}$, see equation (4) above. This ambiguity is fundamental and has spectacular consequences —typically the *mystery of incarnation* which expresses the cartesian product as a delocated intersection, see 4.1.4.

3.3.5. *The Skunk*

Incarnation is delicate to grasp, so let us give a simple example. The set of all designs of a given base is a behaviour (the orthogonal of the empty set).

When the base is negative, this behaviour is noted $\top$ and it corresponds to the additive neutral « true ». Now what is the *incarnation* of a design w.r.t. $\top$? This incarnation is the smallest design included in $\mathfrak{D}$ —and still in $\top$, but this additional requirement is always satisfied. This design is therefore the *empty design* $\mathfrak{Sk} = \emptyset$, corresponding to a first rule with an empty ramification :

$$\frac{\quad\quad}{\xi \vdash} \, (\xi,\emptyset)$$

This design is called the *Skunk*, and the name is is a comment on his social life :

- ▶ $\top$ is the only behaviour containing $\mathfrak{Sk}$. From this viewpoint, the skunk is highly social, everybody lives in his company...

- ▶ ... But they only pretend ; there is no *incarnated* design in $\top$ other than $\mathfrak{Sk}$.

Here we discover something essential, the biggest behaviour is also the smallest one, depending on the viewpoint —typed or untyped—we adopt. Incidentally, observe that $\top$ is the empty intersection and that $\{\emptyset\}$ is the empty product : this is indeed the 0-ary case of the *mystery of incarnation*. By the way, the orthogonal of $\top$ is the smallest behaviour on a positive base. This behaviour is noted $\mathbf{0}$. As an easy consequence of separation, $\mathbf{0} = \{\maltese\}$.

3.4. TRUTH AND COMPLETENESS

3.4.1. *Winning*
If $\mathfrak{D} \perp \mathfrak{E}$, then the normalisation process —called a *dispute*— is a finite sequence $[\mathfrak{D} \rightleftharpoons \mathfrak{E}]$ consisting of the positive rules performed —up to a final daimon, which corresponds to termination. This is a sort of play, and the daimon corresponds to giving up ; but this daimon comes from one and exactly one of $\mathfrak{D}, \mathfrak{E}$ —this is stability— and this design *loses* the dispute. Let us say $\mathfrak{D}$ is *winning* when it never loses against any $\mathfrak{E}$; by separation, this is the same as saying that it does not use the daimon[28].

3.4.2. *Truth*
Winning induces a notion of *truth* : « a behaviour $\mathbf{G}$ is true when it contains a winning design, false when $\mathbf{G}^\perp$ is true ». Now, observe that, if $\mathfrak{D}, \mathfrak{E}$ are both winning, then they cannot be orthogonal —typically nobody gives up and the dispute becomes infinite. Hence it is impossible for a behaviour $\mathbf{G}$

---

[28] The actual theory of winning involves two other conditions which are beyond the scope of this introduction ; but this one —called *obstination*— is by far the most important.

194

to be both true and false. But it is not the case that a behaviour is either true or false.

3.4.3. *Completeness*

Behaviour correspond to formulas, designs correspond to proofs. We have a rough idea of the translation of a proof into a design, and the next section on connectives will tell us about the translation of formulas into behaviours. Quite naturally, to each proof π of A we shall associate a design in the behaviour associated with A, $\pi \in \boldsymbol{A}$. This is indeed a theorem, called *soundness*. The converse —i.e., whether or not this translation *leaks*— has been called *full completeness* by Abramsky : is every $\mathfrak{D} \in \boldsymbol{A}$, of the form $\mathfrak{D} = \pi$ for some proof π of A ? The idea must be refined so as to avoid pitfalls :

1. Since full completeness admits the forgetful version « true implies provable », A must be restricted to those formulas for which usual completeness works. This class is familiar, it consists of first-order formulas —universally quantified over their predicate symbols to make them closed— the $\boldsymbol{\Pi}^1$ formulas.
2. Due to the daimon, $\mathfrak{D}$ must be winning.
3. Finally, $\mathfrak{D}$ must be incarnated... This last constraint is easy to understand : every design lives in the behaviour T which interprets the logical constant T. But there is only one proof of T, corresponding to the unique incarnated design of T.

Full completeness is the statement :

THEOREM 3
If A is closed and $\boldsymbol{\Pi}^1$, if $\mathfrak{D} \in \boldsymbol{A}$ is a winning incarnated design, then $\mathfrak{D} = \pi$ for some proof π of A.

The theorem has been established in [1] for second-order propositional logic without exponentials, i.e., the contraction-free part of logic.

3.4.4. *Internal completeness*

Full completeness is an important milestone, since it establishes the relevance of our approach —even if some connectives, basically exponentials, are still missing. But it is a step backwards too, since it reintroduces this duality syntax/semantics —this is the story of the Thief of Bagdad, see 1.1.4.
The tradition is to consider completeness as an external thing, syntax is (in)complete w.r.t. a given semantics. We saw in 1.3.1 that the subformula property should be considered as the real completeness « nothing is missing », and by the way what can be the value of a completeness which needs

some external reference ? The task is to justify this mathematically, i.e., to formulate a sort of subformula property independently of any syntactical commitment.

Let us try at giving an internal meaning to full completeness. We start with a formula A. Logic yields rules which govern the syntactical proofs of A ; these proofs can be translated as a set $\mathbf{E}$ of designs ; what can we say about $\mathbf{E}$? Very little indeed, it is a set of designs of the same base ; to remember that it may come from the obedience to some rules, let us call such an arbitrary $\mathbf{E}$ an *ethics*. Now, we can see $\mathbf{E}^{\perp}$ as the (counter-) « semantics » of $\mathbf{E}$; now $\mathbf{E}^{\perp\perp}$ corresponds to what is validated by the « semantics » of $\mathbf{E}$. Completeness is therefore that $\mathbf{E} = \mathbf{E}^{\perp\perp}$, i.e., the fact that we have a direct description of a behaviour, without using the biorthogonal. For technical reasons —especially in the negative case—, the equality is required only up to incarnation : for instance the set $\{\mathfrak{Gl}\}$ is a complete ethics for $\top$.

The typical example of an internal completeness theorem is given by the connective $\oplus$, defined —modulo delocation— as $(\mathbf{G} \cup \mathbf{H})^{\perp\perp}$; one proves that the biorthogonal can be removed, i.e., that $\mathbf{G} \cup \mathbf{H}$ is a complete ethics for $\mathbf{G} \oplus \mathbf{H}$. By the way this internal completeness of $\oplus$ is the familiar disjunction property « a proof of $\mathbf{G} \oplus \mathbf{H}$ is a proof of $\mathbf{G}$ or a proof of $\mathbf{H}$ ».

Internal completeness is the essential ingredient of the proof of external (full) completeness. The task is, given an object of the appropriate type, to build a proof, indeed a cut-free proof. The main difficulty is of course to find the last rule, for we can then iterate the construction for the premises... Typically, in the case of a disjunction, the disjunction property provides one with the last rule, a left or right introduction of the disjunction.

4. The social life of behaviours

4.1. ADDITIVES

4.1.1. *Locative vs. spiritual*
All extant explanations of logic are *spiritual*[29]. This means that the objects are taken up to isomorphism. This is plain in the Tarskian case —a formula refers to an interpretation somewhere on the Moon— this is also the case for more refined paradigms —typically category-theoretic ones. The spiritual treatment of logic can be smart enough to interpret conjunction as a cartesian product in the « constructive » case, and as a lunar intersection in general. But why this duality of interpretations ?... *Circulez, il n'y a rien à voir !* In fact the spiritual straightjacket makes it impossible to imagine

---

[29] An exception : realisability ; but it leaks so badly...

a relation between an intersection and a cartesian product —by cartesian product, I mean a plain set-theoretic product, not a categorical nonsense. What is needed is the possibility of taking intersections of formulas — intersection types so to speak—, but how can we intersect sets defined up to isomorphism ?

Ludics is *locative* : a proof π of $\vdash A$, located in $\vdash \sigma$ interacts with a proof λ of $A \vdash$ located in $\sigma \vdash$, not on the Moon. A behaviour is made of precise objects, which are themselves, not an isomorphism class : in ludics we can take an « intersection of formulas », so let us see what happens. A *detour* through set theory is illuminating.

4.1.2. *Locativity and set theory*

The familiar operation of union admits two versions :

Locative : $X \cup Y$; this operation is commutative, associative, with neutral element $\emptyset$. But it is not spiritual, i.e., compatible with isomorphisms (here : bijections) ; in other terms $\sharp(X \cup Y)$ is not determined by $\sharp(X), \sharp(Y)$, the best we can say is $\sharp(X \cup Y) \leq \sharp(X) + \sharp(Y)$.

Spiritual : a.k.a. disjoint sum, $X + Y := f(X) \cup g(Y)$, where f, g are *ad hoc* injections. This is the total operation satisfying $\sharp(X + Y) = \sharp(X) + \sharp(Y)$. But commutativity, associativity, neutrality fail ; or rather you need a serious training in category-theory to understand in which way something of the like remains.

The same holds for the product, again with two possibilities :

Locative : $X \bowtie Y = \{x \cup y; x \in X, y \in Y\}$; this operation is really commutative, associative, with neutral element $\{\emptyset\}$; it distributes over $\cup$. Unfortunately, we can only state that $\sharp(X \bowtie Y) \leq \sharp(X).\sharp(Y)$.

Spiritual : a.k.a. product[30], $X \times Y := f(X) \bowtie g(Y)$, where f, g are *ad hoc* injections. This is the total operation satisfying $\sharp(X \times Y) = \sharp(X).\sharp(Y)$. Again, commutativity, associativity, neutrality... only survive through (canonical) isomorphisms.

The locative product is —as far as I know— a novel operation ; it is quite good, think of $\wp_f(X \cup Y) = \wp_f(X) \bowtie \wp_f(Y)$, an equality.

---

[30] An alternative definition of the set-theoretic product.

4.1.3. *Delocation*

In the case of set theory, f, g were chosen so as to make $f(X), g(Y)$, « disjoint » in an appropriate sense. Something similar can be done with ludics. The delocations φ, ψ are defined by

$$\varphi(\sigma * i * \tau) = \sigma * 2i * \tau \qquad \psi(\sigma * i * \tau) = \sigma * 2i + 1 * \tau \qquad (5)$$

Since φ respects the tree structure, the image under φ of a design $\mathfrak{D}$ is easily defined and shown to be a design of the same base. We can also define the image under φ of a behaviour $\mathbf{G}$ as $\varphi(\mathbf{G}) = \varphi[\mathbf{G}]^{\perp\perp}$, with $\varphi[\mathbf{G}] = \{\varphi(\mathfrak{D}) ; \mathfrak{D} \in \mathbf{G}\}$. In case the base is positive, this simplifies into $\varphi(\mathbf{G}) = \varphi[\mathbf{G}]$, but what about a negative base ? The answer is negative : if the first (downmost) rule of $\mathfrak{D}$ is $(\sigma \vdash, \mathcal{N})$, then the first rule of $\varphi(\mathfrak{D})$ is $(\sigma \vdash, \varphi(\mathcal{N}))$. But nobody forbids me to « widen » the ramification $\varphi(\mathfrak{D})$ into —say— $\varphi(\mathfrak{D}) \cup \{\{1\}\}$, so as to get $\mathfrak{E} \supset \varphi(\mathfrak{D})$, still in $\varphi(\mathbf{G})$, but clearly not in $\varphi[\mathbf{G}]$. But wait a minute, $\mathfrak{E} \backslash \varphi(\mathfrak{D})$ is useless, i.e., does not contribute to the incarnation. In fact the incarnated designs of $\varphi(\mathbf{G})$ are all in $\varphi[\mathbf{G}]$, i.e., the equality $\varphi(\mathbf{G}) = \varphi[\mathbf{G}]$ holds up to incarnation : this is precisely what we called internal completeness.

In order to understand what φ, ψ actually achieve, observe that $\varphi(I) = \psi(J)$ iff $I = J = \emptyset$, which is impossible, since ramifications are non-empty. This makes the behaviours $\mathbf{G}' = \varphi(\mathbf{G}), \mathbf{H}' = \psi(\mathbf{H})$ *disjoint*, which means :

Positive case : the only design in $\mathbf{G}' \cap \mathbf{H}'$ is the daimon, i.e., $\mathbf{G}' \cap \mathbf{H}' = \mathbf{0}$.

Negative case : if $\mathfrak{D} \in \mathbf{G}' \cap \mathbf{H}'$, then $|\mathfrak{D}|_{\mathbf{G}'} \cap |\mathfrak{D}|_{\mathbf{H}'} = \emptyset$.

This is obvious, for instance, in the negative case, let $(\sigma \vdash, \mathcal{N})$ be the last rule of $\mathfrak{D}$, then $|\mathfrak{D}|_{\mathbf{G}'}$ is obtained by restricting $\mathcal{N}$ to $\{\varphi(I) ; \varphi(I) \in \mathcal{N}\}$, etc.

4.1.4. *The mystery of incarnation*

For any negative $\mathbf{G}', \mathbf{H}'$ of the same base $\sigma \vdash$:

$$|\mathfrak{D}|_{\mathbf{G}' \cap \mathbf{H}'} = |\mathfrak{D}|_{\mathbf{G}'} \cup |\mathfrak{D}|_{\mathbf{H}'} \qquad (6)$$

which means that any incarnated design in $\mathbf{G}' \cap \mathbf{H}'$ is the union of an incarnated design in $\mathbf{G}'$ and an incarnated design in $\mathbf{H}'$. If we introduce the notation $|\mathbf{G}|$ for the set of incarnated designs of $\mathbf{G}$, this rewrites as :

$$|\mathbf{G}' \cap \mathbf{H}'| = |\mathbf{G}'| \boxtimes |\mathbf{H}'| \qquad (7)$$

If $\mathbf{G}', \mathbf{H}'$ are disjoint, then the union (6) is disjoint ; then (7) rewrites as :

$$|\mathbf{G}' \cap \mathbf{H}'| \simeq |\mathbf{G}'| \times |\mathbf{H}'| \qquad (8)$$

This might be the most spectacular result of ludics : the cartesian product is a particular case of the intersection, provided we focus on incarnation.

4.1.5. *An example*

Let us explain this in terms of the syntactical example of 2.2.3, i.e., consider the negative behaviour $\mathbf{K}$ corresponding to the sequent $A \vdash$. It will of course contain designs corresponding to $(A \vdash, \{\{P, R\}, \{Q, R\}\})$, but also designs corresponding to —say— $(A \vdash, \{\{P, R\}, \{Q, R\}, \{P, Q, T\}\})$, where T is another formula not related to A. But the incarnation of such a design will only retain the two premises $\{P, R\}, \{Q, R\}$. This shows something, namely that a negative rule can —in ludics— have useless premises. This is by the way the reason why full completeness is restricted to *incarnated* designs : the additional premises correspond to nothing syntactically visible. Now $\mathbf{K}$ is the orthogonal of the set of proofs starting either with $(\vdash A, \{P, R\})$ or with $(\vdash A, \{Q, R\})$, i.e., it is the orthogonal of a union —which is the same as an intersection of orthogonals. We can therefore write $\mathbf{K} = \mathbf{G} \cap \mathbf{H}$. To take the incarnation in $\mathbf{G}$ (resp. $\mathbf{H}$) consists in retaining only the premise corresponding to $\{P, R\}$ (resp. $\{Q, R\}$). In that case the behaviours $\mathbf{G}, \mathbf{H}$ are disjoint, hence, up to incarnation, $\mathbf{K}$ appears as the cartesian product of $\mathbf{G}$ and $\mathbf{H}$.

4.1.6. *The disjunction property*

If $\mathbf{G}', \mathbf{H}'$ are positive behaviours of the same base $\vdash \sigma$, then the set (ethics) $\mathbf{G}' \cup \mathbf{H}'$ is not a behaviour. However, when $\mathbf{G}', \mathbf{H}'$ are disjoint

$$(\mathbf{G}' \cup \mathbf{H}')^{\perp\perp} = \mathbf{G}' \cup \mathbf{H}' \qquad (9)$$

4.1.7. *Additive connectives*

Coming to the point of connectives —i.e., the social life of behaviours—, we discover that each connective can be presented —like the union and the product— in two different ways, a basic locative connective, and a spiritual one obtained by delocation from the basic case. Let us treat an example ; if $\mathbf{G}, \mathbf{H}$ are behaviours of base $\sigma \vdash$, then we can define the (negative) conjunction —i.e., a behaviour $\mathbf{K}$ of the same base—, in two different ways :

Locative : $\mathbf{K} = \mathbf{G} \cap \mathbf{H}$. This operation enjoys very good properties, typically (real) commutativity, associativity ; it has a real neutral element, namely $\top$. But completeness —internal, hence external— fails ; this is the well-known problem of the syntax of « intersection types ».

Spiritual : $\mathbf{G} \& \mathbf{H} = \varphi(\mathbf{G}) \cap \psi(\mathbf{H})$. The delocations φ, ψ are used to force the behaviours to be disjoint. As a result, we get a total and complete connective. But equalities are weakened into canonical isomorphisms.

The completeness of & is the fact that there is a simple description of $\mathbf{G} \& \mathbf{H}$, *in terms of incarnation*, obtained from (8) :

$$|\mathbf{G} \& \mathbf{H}| \simeq |\mathbf{G}| \times |\mathbf{H}| \tag{10}$$

Usual (negative) conjunction appears as a particular case of intersection ; the « intersection type » is more primitive.

If we turn our attention towards disjunction, then there are again two ways to form the disjunction of positive behaviours $\mathbf{G}, \mathbf{H}$, one being the locative sum $(\mathbf{G} \cup \mathbf{H})^{\perp\perp}$, the other being the spiritual sum
$\mathbf{G} \oplus \mathbf{H} = (\varphi(\mathbf{G}) \cup \psi(\mathbf{H}))^{\perp\perp}$. In the spiritual case, the disjunction property enables us to remove the biorthogonal : this is the completeness of the sum, which can be written as :

$$\mathbf{G} \oplus \mathbf{H} \backslash \mathbf{0} \simeq (\mathbf{G} \backslash \mathbf{0}) + (\mathbf{H} \backslash \mathbf{0}) \tag{11}$$

4.1.8. *Incarnation and records*
This explanation of conjunction as a delocated intersection is indeed the final answer to a small mystery. In the eighties, computer scientists developed various theories of objects ; but they insisted on the point that records were not quite products. In a record, we have fields with *labels* ; in a cartesian product, we have two projections. In a record, we can decide to ignore part of the data, we still get a record of the same type : « a coloured point is still a point » ; the same is logically impossible, a pair (point,colour) is not a point.
Ludics is very close to the « record spirit ». For instance, its locative features are the exact analogues of the field labels ; similarly, inheritance is a natural property of designs. It is projection, coercion (the fact for a point to lose its colour) that become more complex : coercion corresponds to incarnation « we have no use for the colour, let's erase it », and projection involves a delocation.

4.2. MULTIPLICATIVES

4.2.1. *Shifts*
As we said, one of the novelties of ludics is the use of polarity. This means that behaviours are divided into two classes, the positive and the negative ones. We just defined & as a connective sending negative behaviours

to negative behaviours ; similarly $\oplus$ sends positive behaviours to positive behaviours. But the tradition is that connectives apply, independently of polarity. In order to allow this, it is enough to define connectives allowing a change of polarity : if **G** is of base $\sigma * 0$ (positive or negative), then $\updownarrow$**G** is of base σ and of opposite polarity. More precisely, if **G** is of base $\sigma * 0 \vdash$, then $\downarrow$**G** is of base $\vdash \sigma$ (resp. if **G** is of base $\vdash \sigma * 0$, then $\uparrow$**G** is of base $\sigma \vdash$). The construction amounts at adding to the bottom of each design in **G** a rule $(\vdash \sigma, \{0\})$ (resp. a rule $(\sigma \vdash, \{\{0\}\})$).

In terms of a naive game-theoretic intuition, the shift corresponds to a sort of dummy move. But this move is not quite dummy, because $\downarrow\uparrow$**G** is not isomorphic to **G** : « operationally » speaking, two dummy moves have been added, with the possibility for each « player » to « give up » (i.e., use Daimon) at these early stages.

4.2.2. *The locative tensors*

The tensor is the positive product, the adjoint of (linear) implication. The general form of the definition is :

$$\mathbf{G} \circledast \mathbf{H} = \{\mathfrak{D} \circledast \mathfrak{D}'; \mathfrak{D} \in \mathbf{G}, \mathfrak{D}' \in \mathbf{H}\}^{\perp\perp} \qquad (12)$$

i.e., the tensor of behaviours is defined from a tensor of designs. The basic question is therefore : given any two designs $\mathfrak{D}, \mathfrak{D}'$ of base $\vdash \sigma$, how do we form their tensor product ? The answer depends on the respective first rules of $\mathfrak{D}, \mathfrak{D}'$:

1. If one of $\mathfrak{D}, \mathfrak{D}'$ is a daimon, then $\mathfrak{D} \circledast \mathfrak{D}' = \maltese$.
2. If the first rules of $\mathfrak{D}, \mathfrak{D}'$ are $(\vdash \sigma, I), (\vdash \sigma, I')$ and $I \cap I' = \emptyset$, then one can define $\mathfrak{D} \circledast \mathfrak{D}'$ as follows : the first rule is $(\vdash \sigma, I \cup I')$, with premises $\sigma * i$, for $i \in I$, on which we proceed like in $\mathfrak{D}$ and with premises $\sigma * i'$, for $i' \in I'$, on which we proceed like in $\mathfrak{D}'$.
3. But in case $I \cap I' \neq \emptyset$, there is no obvious answer. There is of course a spiritual answer : $\mathfrak{D} \otimes \mathfrak{D}' = \varphi(\mathfrak{D}) \circledast \psi(\mathfrak{D}')$. Since $\varphi(I) \cap \psi(I) = \emptyset$, we are back, up to a delocation, to the previous case.

But imagine that we want a *locative* tensor, something which should be to $\otimes$ what $\cap$ is to &. We have to solve a locative conflict —to take an exact image, imagine a flight, $\mathfrak{D}$ has booked rows $\{1, 12, 13\}$, $\mathfrak{D}'$ has booked rows $\{7, 13, 21\}$, with a conflict on the coveted row 13. The spiritual solution just mentioned amounts at delocating $\mathfrak{D}$ on rows $\{2, 24, 26\}$ and $\mathfrak{D}'$ on rows $\{15, 27, 43\}$, but this supposes an airplane big enough. If we turn our attention towards locative solutions, there are four possibilities —indeed four solutions admitting adjoints.

$\mathfrak{D} \otimes \mathfrak{D}'$: $\mathfrak{D}$ gets seats $1, 12$, $\mathfrak{D}'$ gets $7, 13, 21$.

$\mathfrak{D} \ominus \mathfrak{D}'$: $\mathfrak{D}$ gets seats $1, 12, 13$, $\mathfrak{D}'$ gets $7, 21$.

$\mathfrak{D} \odot \mathfrak{D}'$: the flight is cancelled : $\mathfrak{D} \odot \mathfrak{D}' = \maltese$.

$\mathfrak{D} \oslash \mathfrak{D}'$: $\mathfrak{D}$ gets seats $1, 12$, $\mathfrak{D}'$ gets $7, 21$; 13 given to $\mathfrak{Sk}$. Better to travel with a skunk than not travelling at all !

The first two solutions are the symmetric expressions of the same non-commutative idea : one of the the two designs has priority, e.g., when $i \in I \cap I'$, proceed as in $\mathfrak{D}'$ (protocol $\ominus$). The last protocol ($\oslash$) gives the disputed row to anybody : when $i \in I \cap I'$, proceed as you want. For reasons of incarnation, « as you want » may be replaced with the « worse » case, the empty design $\mathfrak{Sk}$. But the most natural protocol remains $\odot$ which solves the conflict in a drastic way. These four protocols are associative, but this does not imply that the resulting tensor of behaviours is associative : if $\mathbf{G} \copyright \mathbf{H}$ is short for $\{\mathfrak{D} \circledast \mathfrak{D}'; \mathfrak{D} \in \mathbf{G}, \mathfrak{D}' \in \mathbf{H}\}$, there is no reason why $((\mathbf{G} \copyright \mathbf{H})^{\perp\perp} \copyright \mathbf{K})^{\perp\perp}$ should equal $((\mathbf{G} \copyright (\mathbf{H} \copyright \mathbf{K})^{\perp\perp})^{\perp\perp}$. Associativity comes from the existence of adjoints.

4.2.3. *The adjoint implications*
Each of the locative tensors has adjoints, for instance

$$\ll \mathfrak{F} \mid \mathfrak{D} \ominus \mathfrak{D}' \gg = \ll \mathfrak{F}[\mathfrak{D}] \mid \mathfrak{D}' \gg = \ll [\mathfrak{D}']\mathfrak{F} \mid \mathfrak{D} \gg \tag{13}$$

$\mathfrak{F}, \mathfrak{F}[\mathfrak{D}], [\mathfrak{D}']\mathfrak{F}$ are of base $\sigma \vdash$ and $\mathfrak{D}, \mathfrak{D}'$ are of base $\vdash \sigma$. By the separation theorem, equation (13) *defines* —say— $\mathfrak{F}[\mathfrak{D}]$ in terms of $\mathfrak{F}, \mathfrak{D}$.
The adjoints enable one to give a characterisation of the orthogonals of the various tensors. Typically :

$$\mathfrak{F} \in (\mathbf{G} \ominus \mathbf{H})^{\perp} \Leftrightarrow \forall \mathfrak{D}(\mathfrak{D} \in \mathbf{G} \Rightarrow \mathfrak{F}[\mathfrak{D}] \in \mathbf{H}^{\perp}) \tag{14}$$

This equation is used, together with the dual form :

$$\mathfrak{F} \in (\mathbf{G} \ominus \mathbf{H})^{\perp} \Leftrightarrow \forall \mathfrak{D}'(\mathfrak{D}' \in \mathbf{H} \Rightarrow [\mathfrak{D}']\mathfrak{F} \in \mathbf{G}^{\perp}) \tag{15}$$

to prove the associativity of $\ominus$ as well as its distributivity over the locative sum $(\mathbf{G} \cup \mathbf{H})^{\perp\perp}$. The connectives $\oslash, \odot$ enjoy similar properties, and, of course, commutativity.

4.2.4. *The spiritual tensor*
$\mathbf{G} \otimes \mathbf{H}$ is defined as $\varphi(\mathbf{G}) \circledast \psi(\mathbf{H})$, where $\circledast$ is any of the locative tensors $\ominus, \ominus, \oslash, \odot$ (the choice is irrelevant). The spiritual tensor inherits all associativity, commutativity, distributivity from the locative case —but only up to isomorphism.

4.2.5. *The meaning of commutativity*

The fact that the spiritual tensor is not commutative is nothing but the fact that $f(x, y) \neq f(y, x)$. *A contrario*, the commutativity of the locative tensors $①, ⊙$ mean something like $f(x, y) = f(y, x)$. Can we clarify this nonsense ? First, let us introduce the notation for the adjoint of $⊙$:

$$\ll \mathfrak{F} \mid \mathfrak{D} \odot \mathfrak{D}' \gg \; = \; \ll (\mathfrak{F})\mathfrak{D} \mid \mathfrak{D}' \gg \; = \; \ll (\mathfrak{F})\mathfrak{D}' \mid \mathfrak{D} \gg \qquad (16)$$

from which we actually get $((\mathfrak{F})\mathfrak{D})\mathfrak{D}' = ((\mathfrak{F})\mathfrak{D}')\mathfrak{D}$: the locative application is quite commutative ! In the case of usual (spiritual) application, we get $((\mathfrak{F})\varphi(\mathfrak{D}))\psi(\mathfrak{D}') \neq ((\mathfrak{F})\varphi(\mathfrak{D}'))\psi(\mathfrak{D})$, which explains the mystery. To sum up, locative application is commutative because the arguments are given *together with their locations* : there is no need for an order of application. You may think of the function as a module with several plugs : if application is plugging, then application is commutative. Delocation is indeed the possibility of shuffling plugs, no wonder that this induces a non-commutativity —but to some extent this non-commutativity is external to application.

4.2.6. *Completeness properties*

The completeness of the tensor consists in finding a complete ethics for $\mathbf{G} ⊛ \mathbf{H}$; the obvious candidate is $\mathbf{G} ⓒ \mathbf{H}$. To make the long story short, completeness holds in the case of $\varphi(\mathbf{G}) ⓒ \psi(\mathbf{H})$, which means that the spiritual conjunction $\mathbf{G} \otimes \mathbf{H}$ is complete. Contrarily to the additive case, this completeness result is highly non-trivial.

4.2.7. *Sequents of behaviours*

The behaviours so far considered have atomic bases $\vdash \sigma$ or $\sigma \vdash$. One can define behaviours on any base ; furthermore, one can form, on the model of sequents of formulas, sequents of behaviours. For instance, if $\mathbf{G}, \mathbf{H}$ are behaviours of respective bases $\vdash \sigma$ and $\vdash \tau$, one can define the behaviour $\mathbf{G} \vdash \mathbf{H}$ of base $\sigma \vdash \tau$ by :

$$\mathfrak{F} \in \mathbf{G} \vdash \mathbf{H} \Leftrightarrow \forall \mathfrak{D} \in \mathbf{G} \quad [\![\mathfrak{F}, \mathfrak{D}]\!] \in \mathbf{H} \qquad (17)$$

equivalently :

$$\mathfrak{F} \in \mathbf{G} \vdash \mathbf{H} \Leftrightarrow \forall \mathfrak{E} \in \mathbf{H}^{\perp} \quad [\![\mathfrak{F}, \mathfrak{E}]\!] \in \mathbf{G}^{\perp} \qquad (18)$$

The obvious question is « what about $\mathbf{G} \vdash \mathbf{G}$ and the *identity axiom* ? » First notice that the answer is —strictly speaking— negative : $\mathbf{G} \vdash \mathbf{G}$ has the base $\sigma \vdash \sigma$, which is not a pitchfork, remember that the loci must be pairwise incomparable. But we can reformulate the question with $\mathbf{G} \vdash \theta(\mathbf{G})$, where θ is the delocation $\theta(\sigma * \tau) = \sigma' * \tau$, and σ, σ' are incomparable. Game-theoretically, the answer is well-known, it is the « copycat » strategy, which consists in recopying the last move of the opponent. Here we shall call it *Fax*, to stress the fact that it implements a delocation, here θ ; by the way, real copycats are at work everyday on the Web, and it is essential that they do their cheating at distance. But there is an even older intuition, namely the $\eta - expansion$ of the identity axiom, which reduces an identity on A to identities on P, Q, R, which in turn are reduced to identities... The process needs not stop : just understand that a propositional atom is indeed a variable, quantified universally or existentially ; once it has been replaced with a witness (an actual behaviour) the η-expansion can be resumed. The fax $\mathfrak{Fax}_{\sigma,\sigma'}$ is the following design :

$$
\cfrac{\cfrac{\cfrac{\vdots \ \mathfrak{Fax}_{\sigma'*i,\sigma*i}}{\dots \sigma'*i \vdash \sigma*i \dots}}{\dots \quad \vdash \sigma', \sigma * I} (\sigma',I)}{\sigma \vdash \sigma'} (\sigma,\wp_f(\mathbb{N})\setminus\{\emptyset\})
$$

The operationality of the fax is expressed by the following :

$$[\![\mathfrak{Fax}, \mathfrak{D}]\!] = \theta(\mathfrak{D}) \tag{19}$$

$$[\![\mathfrak{Fax}, \mathfrak{E}]\!] = \theta^{-1}(\mathfrak{E}) \tag{20}$$

$$[\![\mathfrak{Fax}, \mathfrak{D}, \mathfrak{E}]\!] = [\![\mathfrak{D}, \theta^{-1}(\mathfrak{E})]\!] = [\![\theta(\mathfrak{D}), \mathfrak{E}]\!] \tag{21}$$

The three equations (19) (20) (21) are equivalent definitions of the fax : this is plain from separation and associativity.

4.3. QUANTIFIERS

4.3.1. *Locative quantifiers*

Locative quantifiers are quantifiers which does not follow the truth tables, which shun category theory. They are defined from the idea of intersection, not the idea of infinite product, and since no delocation can work, they are definitely different. Now the question is : does this become a mess, or do we get something nice out of it ? In fact something wonderful arises from the unexpected shock between locative quantification and spiritual connectives : these operations commute —sometimes beyond what seems reasonable, i.e., up to the violation of certain classical principles.
Let $\mathbf{G}_d, d \in \mathbb{D}$ be a family of behaviours of the same base ; then one defines the behaviours

$$\forall d \in \mathbb{D} \ \mathbf{G}_d := \bigcap_{d \in \mathbb{D}} \mathbf{G}_d \tag{22}$$

and

$$\exists d \in \mathbb{D} \; \mathbf{G}_d := (\bigcup_{d \in \mathbb{D}} \mathbf{G}_d)^{\perp\perp} \tag{23}$$

The typical example is second-order quantification : $\mathbb{D}$ consists in all the behaviours of a given positive base (typically $\vdash \langle\rangle$) ; if a formula $A[X]$ contains several « occurrences » of $X, X^\perp$, we interpret these occurrences as the images under delocations $\theta_1, \ldots, \theta_n$ of an unknown behaviour $\mathbf{G}$ (or its negation $\mathbf{G}^\perp$) based on $\vdash \langle\rangle$. The typical example $\forall X(X^\perp \,\mathfrak{N}\, \uparrow X)$, makes uses of the « occurrences » : $X^\perp = \theta(\mathbf{G}^\perp)$, with $\theta(\langle\rangle) = \langle\rangle$, $\theta(n * \sigma) = 2n * \sigma$, and $X = \rho(\mathbf{G})$, with $\rho(\sigma) = 1 * \sigma$.

Another example is the plain intersection type, $\mathbf{G}_0 \cap \mathbf{G}_1$; in that case, $\mathbb{D} = \{0, 1\}$. But first-order quantification is not locative, see below.

4.3.2. *Prenex forms*

The basic —and completely unexpected result— is that quantifiers commute with all *spiritual connectives* —but exponentials, not yet treated in ludics ; these commutations enable one to write *prenex forms*, a facility usually restricted to classical logic. Some of these commutations are obvious, there are just the result of polarity : $\forall$ commutes to negative, $\exists$ commute to positive, typically $(\forall X A X) \,\mathfrak{N}\, B$ is the same as $\forall X(AX \,\mathfrak{N}\, B)$. Other commutations are a real surprise ; let us give two examples :

$$\forall d(\mathbf{G}_d \otimes \mathbf{H}_d) = (\forall d \mathbf{G}_d) \otimes (\forall d \mathbf{H}_d) \tag{24}$$

This equation implies the second order formula $\exists X \forall Y (AX \Rightarrow AY)$.

$$\forall d(\mathbf{G}_d \oplus \mathbf{H}_d) = (\forall d \mathbf{G}_d) \oplus (\forall d \mathbf{H}_d) \tag{25}$$

This equation is even more violent, since it contradicts classical logic : typically it implies $\neg \forall X(X \vee \neg X)$. By the way this shows that a « constructive » —i.e., procedural— interpretation needs not be weaker than a classical one : no hand tied in the back !

4.3.3. *Locative phenomenons*

It is with second order quantification that locative phenomenons become the most prominent —shocking or promising, depending on one's attitude towards « foundations ».

Equation (25) can be understood from second-order realisability :

$$c \, \textcircled{R} \, A \vee B \;\Leftrightarrow\; \exists d \, ((c = 1 * d \wedge d \, \textcircled{R} \, A) \vee (c = 2 * d \wedge d \, \textcircled{R} \, B)) \tag{26}$$

$$c \, \textcircled{R} \, \forall X A \;\Leftrightarrow\; \forall C \, c \, \textcircled{R} \, A[C/X] \tag{27}$$

From this :

$$c \,\circledR\, \forall X(A \vee B) \;\Leftrightarrow\; (c \,\circledR\, \forall X A) \vee (c \,\circledR\, \forall X B) \tag{28}$$

The reason is simple, if $c \,\circledR\, \forall X(A \vee B)$, then $c \,\circledR\, A[C_0/X] \vee B[C_0/X]$, hence is of the form $1 * d$, in which case $d \,\circledR\, A[C/X]$ for *all* C, or $2 * d$, in which case $d \,\circledR\, B[C/X]$ for *all* C. Observe that (27) refers to a quantification on something whose generic member has been called C, but we didn't need any information about those C but perhaps the fact that we can name one of them, C_0. In fact we are using the locative aspects of realisability : c is « located » among the $1 * d$ or among the $2 * d$, and this location is a feature of c not of the parameter C. The crucial point is therefore that the realisers c_C are *equal* (to c) ; should they be isomorphic, the property would fail. What we just explained is an exact rephrasing of the real argument in the archaic language of realisability : if we replace the maps $d \rightsquigarrow 1 * d$, $d \rightsquigarrow 2 * d$ with φ, ψ, c with a design, C with a positive behaviour, we get the correct proof of (25).

Of course, if somebody had stumbled 30 years ago on something like (28) no conclusion would have been drawn —for realisability was leaking a lot, especially as soon as implication —or simply negation— was concerned : just one more leakage ! But in ludics, the situation is quite different, there is no leakage up to $\mathbf{\Pi}^1$; a sequent like $\forall X(A \oplus B) \vdash (\forall X A) \oplus (\forall X B)$ is not $\mathbf{\Pi}^1$: it belongs to the incomplete realm where everything is possible, including a depart from classical logic. The usual rules for second order logic are correct, but incomplete ; the usual prejudice is to say « they are incomplete for want of enough comprehension axioms », and by the way it is true that, if we enlarge the syntax so as to get more comprehension, we get more $\mathbf{\Sigma}^1$ theorems. But here we got something which cannot be fixed in this way —remember that (25) implies the classically false $\neg\forall X(X \vee \neg X)$, that no comprehension axiom will entail.

The classical explanation of quantification is that of a big conjunction — maybe uniform in some sense. Classically speaking, $\forall X(A \oplus B)$ refers to each X separately. In ludics, this is not the case ; not because we absolutely want to produce warped effects, but because « there is not enough space ». If we want to interpret $\forall X(A \oplus B)$ as a sort of Bbbig conjunction, then we need disjoint delocations, one for each behaviour $\mathbf{G}$ for which the variable X stands. But there can be at most $\aleph_0$ such delocations[31], whereas the number of behaviours is $2^{2^{\aleph_0}}$.

The formula $\exists X \forall Y(AX \Rightarrow AY)$ is obtained from $\exists X \forall Y(AX \multimap AY)$ which in turn comes from $\forall X AX \multimap \forall Y AY$ by prenex operations, justified by (24).

---

[31] The same argument applies to old style realisability.

The typical element in this behaviour is a design obtained from the fax —that lambda-calculus would simply note $\lambda x.x$. Hence

$\lambda x.x \in \exists X \forall Y (AX \Rightarrow AY)$, but we cannot find any witness for X, i.e. the existence property fails. Concretely this means that we cannot remove the biorthogonal in the definition of the behaviour $\exists X \forall Y (AX \multimap AY)$, in other terms that this behaviour is incomplete. This is not a surprise, Gödel's theorem —or its version *ante litteram*, Cantor's theorem— forbids such a thing, but with a heavy argument based on diagonalisation : the paragon of incompleteness is Gödel's formula, which is Π_1^0, i.e., Σ^1. Nothing of the like here : there is no witness, period ; this is quite different from « I cannot name the witness ». So ludics does not enjoy the existence property... But is it a deadly sin ? Indeed the existence property is required for numerical quantifiers, i.e., quantifiers restricted to natural numbers —and of course it holds in that case.

4.3.4. *First order quantification*

First order quantification is a sort of big conjunction, maybe uniform in some sense : this is the viewpoint of model-theory, of German style proof-theory, etc. This is also the viewpoint we must adopt if we want to extend completeness to predicate calculus. It seems that the extant tools in ludics —esp. *uniformity*, a topic we avoided in this survey— are sharp enough to make it. But do we really need this, i.e., is there a need for a « constructive » —procedural— interpretation of predicate calculus ? To tell the truth, I never saw any such interpretation. In fact first order quantification has always been used in the particular case of *numerical* quantification, i.e., in contexts $\forall x \ (x \in \mathbb{N} \Rightarrow \ \cdot \ )$ and $\exists x \ (x \in \mathbb{N} \wedge \ \cdot \ )$. For instance coming back to realisability, one defines

$$c \circledR \forall n A \Leftrightarrow (\forall n \ \{c\}n \circledR A[n/x]) \wedge \ldots \,^{32} \qquad (29)$$

But I never saw any *real* definition of $c \circledR \forall x A$.

It is therefore legitimate to question the interest of the predicate part of Heyting's logic, which has been unimaginatively modelled on classical predicate calculus. What would for instance be the result of treating first order quantification in the locative spirit, i.e., by means of equations (22) and (23) ? Obviously more formulas would be validated, with an incompleteness of the existential quantifier, typically $\exists x \forall y (Ax \Rightarrow Ay)$. But with no consequence of the form $\exists m \forall n (Am \Rightarrow An)$: this would not alter our beloved existence property for *numerical* quantification : $x \in \mathbb{N}$ translates

---

[32] Some nonsense to fix the leakage.

into the usual Dedekind formulation

$$\forall X(\forall z(z \in X \Rightarrow z + 1 \in X) \Rightarrow (0 \in X \Rightarrow x \in X)) \qquad (30)$$

which admits a forgetful propositional image in system $\mathbb{F}$ as

$$\forall X((X \Rightarrow X) \Rightarrow (X \Rightarrow X)) \qquad (31)$$

a Π^1 formula. To make the long story short, the completeness of the type $\forall X((X \Rightarrow X) \Rightarrow (X \Rightarrow X))$ provides the required witness —even if we interpret the first order quantification in a locative way.

5. The challenge of exponentials

5.1. ON INTEGERS

5.1.1. *Kronecker*
God created the integers, all else is the work of man ; the sentence is the best illustration of the intrinsic difficulty of foundations : how can we proceed beyond integers ? There is no way of making the natural number series lose its absoluteness : for instance model theory has considered *non-standard* integers, but the very choice of the expression betrays the existence of standard ones. Our approach to natural numbers must therefore be *oblique*, not because we want a warped notion, but because we meet a blind spot of our intuition. The possibility of an oblique approach is backed by quantum physics —by the way the major scientific achievement of last century— : one has been able to speak of non-realistic artifacts, in complete opposition to our « fundamental intuitions » concerning position, momentum, ... The short story of quantum mechanics begins with the discovery by Planck of small *cracks* in this impressive realistic building, *thermodynamics*. Our only hope will be in the discovery of a « crack » in the definition of integers : even a small crack may do it.

5.1.2. *Rates of growth*
But it is quite desperate to seek a crack in this desperately flat $\{0, 1, 2, \dots\}$, as long as we see it as a set. It becomes different if we see it as a « process », since we can play on the *rate of growth*. This is backed by the development of computational complexity, and paradoxically by the impossibility of proving any non-immediate separation theorem between the various classes : something about integers could be hidden there. The « rate of growth » could concern the possible functions involving integers ; more likely, there could be several rates of growth (polytime, logspace, ...) corresponding to different notions of integers.

5.1.3. *Norms*

Can we imagine different notions of integers ? Again, let us try an analogy : there is no problem with the finite dimensional space $\mathbb{C}^n$, corresponding to the basis $\{e_1, \ldots, e_n\}$; but $\mathbb{C}^\omega$, which corresponds to the basis $\{e_1, \ldots, e_n, \ldots\}$ has no meaning at all —some quantitative information should be added, typically a norm, usually the ℓ^1-norm or the ℓ^2-norm. Could we find something like a « norm » (or norms) in the case of integers ?

5.1.4. *Dedekind*

There is no reason to criticise Dedekind's definition of integers (30), and little reasons to refuse its simplified version (31) : we shall try to plug in real numbers in (31). Fortunately, $\forall X((X \Rightarrow X) \Rightarrow (X \Rightarrow X))$ is not flat like $\{0, 1, 2, \ldots\}$, it offers two asperities, one is $\forall X$: we could play on the possible X —i.e., restrict comprehension. This has been the main activity of an *ism* —predicative mathematics— : nothing never came nor can be expected from this approach, which is pure bondage. Remains the other entrance, the connective $\Rightarrow$.

5.2. ON IMPLICATION

5.2.1. *The input of Scott domains*

If we consider implication, there is a real crack in the building, not a very recent one, since it dates back to 1969, to Scott domains. This work was the final point to a problem illustrated by the names of Kleene, Kreisel, Gandy, ... on higher order computation[33]. The question was to find a natural topology on function spaces, in modern terms, to build a CCC of topological spaces. Scott (and independently Ershov) solved the question beyond doubt, but there is something puzzling : this is achieved by a restriction to queer spaces whose topology is not Hausdorff —only $\mathcal{T}_0$, the only separation property which costs nothing. In other terms, the Scott topology succeeds in keeping the cardinality of functionals quite low ; but it is is cheap topology. This is our crack : the logical rules for implication contradict usual topology : we cannot interpret them with spaces like $\mathbb{R}$.

Let us be more precise : there is no canonical topology on a function space. Depending on the context, we are likely to use simple or uniform convergence[34]. When we come to the point of interpreting lambda-calculus in a continuous way, we observe that $f, g \rightsquigarrow f \circ g$ can only be separately

---

[33] I have written somewhere that beyond second-order, higher order has only be useful to Ph D's ; this was not quite true at that time.

[34] I am excluding things like the *compact-open* topology : I am interpreting something natural, logic, into something natural, topology, and I must use the same spaces as mathematicians.

continuous in f, g, whatever topology we put on the function space ; in particular, $f \rightsquigarrow f \circ f$ is not continuous. Scott domains are indeed bleak topological spaces with no uniformity, e.g., no metrics : one cannot distinguish between « simple » and « uniform », and as a « reward » separately continuous functions become continuous.

There are two possible interpretations of the absence of a convincing « continuous » explanation : the usual way is to try to fix it, and this is what everybody has so far done. But there is another way out, maybe our logical rules, typically composition $f, g \rightsquigarrow f \circ g$ are *wrong*. Removing —more likely simply tampering with— composition may change a lot of things, and the kind of modification I am seeking is likely to have a different fate from intuitionism whose novelty was tamed by Gödel's translation : the « enemy » was still present —it only took the $\neg\neg$-disguise. We must accept the idea that perhaps the exponential $n \rightsquigarrow 2^n$ —which is a typical product of composition— may disappear once for all... This is shocking, but do we want something new, or are we happy with metaturtles ?

5.2.2. *The input of linear logic*
Linear logic started with a decomposition of implication :

$$A \Rightarrow B \ = \ !A \multimap B \tag{32}$$

into a more basic *linear implication* and a repetition operation, the *exponential* $!A$ which mainly takes care of contraction. The exponential-free part is something peculiar, extremely basic —so to speak foundationally neutral ; this is the fragment so far interpreted by ludics. Exponentials were excluded from the present version of ludics—not because of any essential impossibility— but because of a slight *malaise*, in particular as to their precise locativity : by the way, this malaise is part of the crack we are seeking.

Using linear logic, it was possible to revisit continuous interpretations :

▶ Formulas become Banach spaces[35], proofs linear maps of norm $\leqslant 1$.
▶ Positive connectives are interpreted by means of ℓ^1-norms ; typically the connective $\oplus$ corresponds to a direct sum equipped with $\|x \oplus y\| = \|x\| + \|y\|$.
▶ Negative connectives are interpreted by means of ℓ^∞-norms ; typically the connective $\&$ corresponds to a direct sum equipped with $\|x \oplus y\| = \sup(\|x\|, \|y\|)$.
▶ Usual implication $\mathbf{E} \Rightarrow \mathbf{F}$ corresponds to *analytical* functions from the open unit ball of $\mathbf{E}$ to the closed unit ball of $\mathbf{F}$.

---

[35] Indeed *coherent* Banach spaces, in which the « dual space » is specified ; this accounts for the want of reflexivity of Banach spaces like ℓ^1, ℓ^∞.

The crack already noticed in Scott domains becomes more conspicuous : what prevents us from composing functions is the difference between an open ball and a closed ball —more precisely the impossibility of extending an analytic function to the closed ball in any reasonable way.

5.3. OBJECTIVE : EXPONENTIALS

5.3.1. *On infinity*
Methodologically speaking, the introduction of exponentials induces a schizophrenia between labile operations —the basic linear connectives— and « stable » ones, the exponentials !, ?. One of the basic theses of linear logic was about infinity : infinity is not in some external reality, but in the possibility of *reuse*, i.e., contraction, i.e., exponentials :

Infinity as Eternity

For instance, no diagonalisation argument is possible in the absence of exponentials ; typically, Prawitz's naive set-theory, rewritten in a (basic) linear framework is consistent : this naive set theory normalises in logspace.

5.3.2. *Light logics*
In 1986 when linear logic was created, there was no question of departing from usual logic, to create yet another *Broccoli* logic. This is why the standard rules of exponentials have been chosen so as to respect intuitionistic logic through the translation (32). Infinity is concentrated in the exponentials, hence any tampering with exponentials will alter the properties of infinity. This is the oblique approach I mentioned : in this way, we can expect accessing to something beyond our realistic intuitions. Several systems with « light exponentials » have been produced ; my favourite being **LLL**, *light linear logic*, which has a polytime normalisation algorithm and can harbour all polytime functions.
Unfortunately these systems are good for nothing, they all come from bondage : artificial restrictions on the rules which achieve certain effects, but are not justified by use, not even by some natural « semantic » considerations.

5.3.3. *Some science-fiction*
Let us put things together. In the mismatch logic/topology, my thesis is that logic is wrong, not topology : so let us modify logic. Using linear logic, the modification must take into account the mismatch open ball/closed ball, and one can imagine several ways out —e.g., changing the diameters of balls— which all would have the effect of plugging real parameters in

logical definitions such as Dedekind's.

However we are not yet in position to do so : the Banach space thing is only a *semantics*, i.e., a badly leaking interpretation. We have to import parameters —say complex— in ludics and accommodate exponentials in a continuous setting. We should eventually get a norm (rather several norms, none of them distinguished), not quite on the integers, but on a wider space... But the best programs are those written after their fulfilment.

References

1. J.-Y. Girard. **Locus Solum**. *Mathematical Structures in Computer Science*, 11:301 – 506, 2001. [36]

---

[36] This paper contains a very comprehensive bibliography, this is why we give no other reference here.

NAÏVE COMPUTATIONAL TYPE THEORY

ROBERT L. CONSTABLE
Cornell University
Ithaca, New York, U.S.A.
rc@cs.cornell.edu

Preface

In 1960 the mathematician Paul Halmos wrote a 104-page book called *Naïve Set Theory* that made the subject accessible to generations of mathematicians. This article interprets that book in an abbreviated manner to introduce type theory and the approach to algorithms and computational complexity theory inherent in computational type theory. I start by paraphrasing the preface to Halmos' book. The sections of this article follow his chapters closely.

Every computer scientist agrees that every computer scientist must know some type theory; the disagreement begins in trying to decide how much is some. This article contains my partial answer to that question. The purpose of the article is to tell the beginning student of advanced computer science the basic type theoretic facts of life, and to do so with a minimum of philosophical discourse and logical formalism. The point throughout is that of a prospective computer scientist eager to study programming languages, or database systems, or computational complexity theory, or distributed systems or information discovery.

In type theory, "naïve" and "formal" are contrasting words. The present treatment might best be described as informal type theory from a naïve point of view. The concepts are very general and very abstract; therefore they may take some getting used to. It is a mathematical truism, however, that the more generally a theorem applies, the less deep it is. The student's task in learning type theory is to steep himself or herself in unfamiliar but essentially shallow generalities until they become so familiar that they can be used with almost no conscious effort.

Type theory has been well exposited in articles by N. G. de Bruijn and the Automath group; the writings of Per Martin-Löf, the originator of many

H. Schwichtenberg and R. Steinbrüggen (eds.), Proof and System-Reliability, 213–259.
© 2002 *Kluwer Academic Publishers. Printed in the Netherlands.*

of the basic ideas; the writings of Jean-Yves Girard, another originator; the writings of the Coq group, the Cornell group, and the Gothenberg group; and the writings of others who have collectively expanded and applied type theory.

What is new in this account is treatment of classes and computational complexity theory along lines that seem very natural. This approach to complexity theory raises many new questions as can be seen by comparison to the lectures of Niel Jones at the summer school.

Table of Contents

1. Types and equality

Section 1 of *Naïve Set Theory* notes that the book will not define sets. Instead an intuitive idea (perhaps erroneous) will be delineated by saying what can be correctly done with sets. Halmos notes already on page one that for the purposes of mathematics one can assume that the only members of sets are other sets. *It is believed that all mathematical concepts can be coded as sets.* This simplifying approach makes set theory superficially very different from type theory.

Likewise, we delineate an intuition (possibly erroneous) about types: that intuition might be acquired from programming languages or databases or from set theory. We explain what can be correctly done with types, and how they are used.

Beginning students of computer science might believe that the only members of types need to be *bits* since all implementations ultimately reduce any data to bits. But this level of abstraction is too low to allow a good mathematical theory. The right abstraction is at the "user level" where distinctions are made between numbers, characters, booleans and other basic kinds of data. Type theory starts in the middle, axiomatizing the user-level building blocks for computation and information. From these it is possible to define realizations as bits — going "downward" to the machine — and to define abstractions to classes, going "upward" to systems.

ELEMENTS

Types are collections of elements, possibly empty. Elements are the data. When a type is defined, the structure of the elements is specified. A type definition says how to construct elements and how to take them apart. The basic way to do this is to provide a construction pattern. On top of these, a more abstract characterization can then be given in terms of operations called *constructors* and *destructors*. But in order to define these computationally, we need at least one concrete symbolic representation of the basic data. The representation must be concrete like bits but general enough to naturally describe the objects we manipulate mentally when we calculate.

The structure of data in type theory is given by *abstract syntax*. This follows in the tradition of Lisp with its S-expressions as the basic data. To define a type, we specify the form of its data elements. To say that we have a data element is precisely to say that we have it in a specific predetermined format. This is what is critical about all implementations of data types: we must know the exact format. We call this exact format the *canonical form* of the data. To have an element is to have access to its canonical form.

Let's illustrate these ideas for the type of *bits*. The data elements we normally have in mind are 0 and 1, but let us analyze what this choice means. We could also think of bits as boolean values; then we might use *true* and *false*. We might prefer the constants of some programming language such as $0B, 1B$.

The notion we have in mind is that there are two *distinct symbols* that represent the bits, say 0, 1, and a bit is formed from them. We can say that $bit\{0\}$, $bit\{1\}$ are the data formats, or $0B, 1B$ are.

We intend that there are precisely two bits, given in distinct data formats. To say this precisely we need a criterion for when two canonical data elements are equal. For instance, we need to agree that we are using $bit\{0\}$, not $0B$, or agree that $bit\{0\}$ and $0B$ are equal. Defining a type means settling these matters.

Let us agree that $bit\{0\}$ and $bit\{1\}$ are the data elements formed from the distinct symbols 0 and 1. The only equalities are $bit\{0\}=bit\{0\}$ and

$bit\{1\}=bit\{1\}$. These are required since equality must be an equivalence relation — hence reflexive. How do we say that $bit\{0\}\neq bit\{1\}$?

The answer comes from understanding what it means to take the data elements apart or *use them*. Key to our notion of the type is that 0 and 1 are *distinct* characters. This is part of the type definition process, and moreover we mean that we can *effectively* distinguish the two symbols. For example, the operation to check for equality of bits could be to access the parts inside $\{\}$ and compare them. This depends on our presumed ability to distinguish these characters. We can make the computation more abstract by creating a computational form inside the type theory rather than relying on our computational intuition about symbols so directly.

The way we accomplish the abstraction is to postulate an effective operation on the data formats. The operation is to decide whether the data is $bit\{0\}$ or $bit\{1\}$. We represent this operation by a computation rule on a syntactic form created explicitly to make decisions. Let us call the form a *conditional expression* written as

$$\textit{if } b \textit{ then } s \textit{ else } t \textit{ fi.}$$

The only meaning given to the form at this stage is in terms of the computation rules.

$$\textit{if } bit\{0\} \textit{ then } s \textit{ else } t \textit{ fi} \quad \text{reduces to } s$$
$$\textit{if } bit\{1\} \textit{ then } s \textit{ else } t \textit{ fi} \quad \text{reduces to } t$$

With this form we can distinguish $bit\{0\}$ and $bit\{1\}$ as long as we can distinguish anything. That is, suppose we know that $s\neq t$. Then we can conclude that $bit\{0\}\neq bit\{1\}$ as long as if-then-else-fi *respects equality*. That is, $e_1 = e_2$ should guarantee that

$$\textit{if } e_1 \textit{ then } s \textit{ else } t \textit{ fi} = \textit{if } e_2 \textit{ then } s \textit{ else } t \textit{ fi.}$$

Given this, if $bit\{0\}=bit\{1\}$, then s=t. Let's take 0 for s and 1 for t. Then since $0\neq 1$ as characters, $bit\{0\}\neq bit\{1\}$.

Why don't we just postulate that $bit\{0\}\neq bit\{1\}$? One reason is that this can be derived from the more fundamental computational fact about if-then-else-fi. This computational fact must be expressed one way or another. We'll see later that a second reason arises when we analyze what it means to know a proposition; knowing $0\neq 1$ is a special case.

EQUALITY

In a sense the equality relation defines the type. We can see this clearly in the case of examples such as the integers with respect to different equalities. Recall that "the integers modulo k," $\mathbb{Z}_k$, are defined in terms of this equivalence relation, $mod(k)$, defined

$$x = y \ mod(m) \text{ iff } (x - y) = k \cdot m \text{ for some } m.$$

$\mathbb{Z}_2$ equates all even numbers, and in set theory we think of $\mathbb{Z}_2$ as the two equivalence classes (or residue classes)

$$\{0, \pm 2, \pm 4, \ldots\} \text{ and } \{\pm 1, \pm 3, \pm 5, \ldots\}.$$

It is easy to prove that $x = y \ mod(k)$ is an equivalence relation. What makes $\mathbb{Z}_k$ so interesting is that these equivalence relations are also congruences on the algebraic structure of $\mathbb{Z}$ with respect to addition $(+)$, subtraction $(-)$, and multiplication $(*)$.

In type theory we do not use equivalence classes to define $\mathbb{Z}_k$. Instead we define $\mathbb{Z}_k$ to be $\mathbb{Z}$ with a new equality relation. We say

$$\mathbb{Z}_k == \mathbb{Z}//mod(k).$$

The official syntax is

$$\mathbb{Z}_k == quotient(\mathbb{Z}; x, y.x = y \ mod(k)).$$

This syntax treats $x, y.x + y \ mod(k)$ as a *binding construct*. The binding variables x and y have as scope the expression following the dot. In general, if A is a type and E is an equivalence relation on A, then $A//E$ is a new type such that $x = y \ in \ A//E$ iff xEy.

Recall that an equivalence relation written xEy instead of the usual relation notation $E(x, y)$ satisfies:

1. xEx reflexivity
2. xEy implies yEx commutativity
3. xEy and yEz implies xEz transitivity

The official syntax is quotient$(A; \ x, y. \ xEy)$.

These types, $A//E$, are called *quotient types*, and they reveal quite clearly the fact that a type is characterized by its equality. We see that $\mathbb{Z}$ and $\mathbb{Z}_2$ *have the same elements but different equalities.*

We extend the notion of membership from the canonical data to any expression in this way. If expression a' reduces to expression a, and a is a canonical element of A, then a' is an element of A by definition. For example, if $a \in A$, then *if bit$\{0\}$ then a else b fi* belongs to A as well.

If A is a type, and if a, b are elements of A, then $a = b \ in \ A$ denotes the equality on A. We require that equality respect computation. If $a \in A$ and a' reduces to a, then $a' = a \ in \ A$.

In set theory the membership proposition, "x is a member of y," is written $x \in y$. This proposition is specified axiomatically. It is not presented as a relation defined on the collection of all sets. Because $x \in y$ is a proposition, it makes sense to talk about *not* $x \in y$, symbolically $\neg(x \in y)$ or $x \notin y$.

In type theory, *membership is not a proposition* and not a relation on a predefined collection of all objects that divides objects into types. We write *a in A* to mean that a is an object of type A; this is a basic judgment. It tells us what the form of object a is. If you like, A is a specification for how to build a data object of type A. To judge *a in A* is to judge that indeed a is constructed in this way.

It does not make sense to regard *a in B* as a proposition; for instance, it is not clear what the negation of *a in B* would mean. To write *a in B* we must first know that a is an object. So first we would need to know *a in A* for some type A. That will establish what a is. Then if B is a type, we can ask whether elements of A are indeed also elements of B.

We could define a relative membership relation which is a proposition, say $x \in B$ *wrt* A, read "x belongs to B with respect to A." This means that given x is of type A, this relation is true exactly when x is also in the type B. We will have little occasion to use this predicate except as a comparison to set theory. The reasons for its scarce use are discussed when we talk about universes and open-endedness in Section 6. But we see next a proposition that addresses some useful examples of relative membership.

2. Subtypes and set types

We say that A is a *subtype* of B (symbolically, $A \sqsubseteq B$) if and only if $a = a'$ in A implies that $a = a'$ in B. Clearly $A \sqsubseteq A$, and if $A \sqsubseteq B$ and $B \sqsubseteq C$, then $A \sqsubseteq C$. For the empty type, *void*, *void* $\sqsubseteq A$ for any A.

The subtype relation induces an *equivalence relation* on types as follows. Define $A \equiv B$ if and only if $A \sqsubseteq B$ and $B \sqsubseteq A$. Clearly $A \equiv B$ is an equivalence relation, i.e. $A \equiv A$, if $A \equiv B$ then $B \equiv A$ and if $A \equiv B$ and $B \equiv C$, then $A \equiv C$. This equivalence relation is called *extensional equality*. It means that A and B have the same elements; moreover, the equality relations on A and on B are the same.

In set theory, two sets S_1, S_2 are equal iff they have the same elements. This is called *extensional equality*. Halmos writes this on page 2 as an axiom. In type theory it is a definition not an axiom, and furthermore, it is not the only equality on types. There is a more primitive equality that is *structural* (or *intensional*); we will encounter it soon.

The subtype relation allows us to talk about relative membership. Given *a in A* and given that B is a type, we can ask whether $A \sqsubseteq B$. If $A \sqsubseteq B$, then we know that *a in A* implies *a in B*, so talking about the "B-like structure of a" makes sense.

We now introduce a type that is familiar from set theory. Halmos takes it up in his section 2 under the title "axiom of specification." The axiom is also called separation. He says that:

> To every set A and to every condition $S(x)$, there corresponds a set B whose elements are exactly those elements x of A for which $S(x)$ holds.

This set is written $\{x : A \mid S(x)\}$. Halmos goes on to argue that given the set $B = \{x : A \mid x \notin x\}$, we have the curious relation

$$(*) \quad x \in B \text{ iff } x \in A \text{ and } x \notin x.$$

If we assume that either $B \in B$ or $B \notin B$, then we can prove that $B \notin A$. Since A is arbitrary, this shows that either there is no set of all sets or else the law of excluded middle, P or $\neg P$, does not hold on sets. The assumption that there is a universal set and that the law of excluded middle is true leads to the contradiction known as *Russell's paradox*.

In type theory the set type is defined just as in set theory. To every type A and relation S on A there is a type whose elements are exactly those elements of A that satisfy S. The type is denoted $\{x : A \mid S(x)\}$. But unlike in set theory, we cannot form $B = \{x : A \mid x \in x\}$ because $x \in x$ is not a relation of type theory. It is a judgment. The closest proposition would be relative membership: $x \in y$ *wrt* A (but y must be a type).

Also unlike set theory, we will not assume the law of excluded middle, for reasons to be discussed later. Nevertheless, we cannot have a type of all types. The reasons are deeper, and are discussed in Section 6.

Notice that we do know this:

$$\{x : A \mid S(x)\} \sqsubseteq A$$

Also we can define an empty type, *void*. We use $\emptyset$ as its display.

$$\emptyset = \{x : A \mid x \neq x \text{ in } A\}.$$

We know that for any type B

$$\emptyset \sqsubseteq B.$$

In type theory, we distinguish $\{x : A \mid x = x \text{ in } A\}$ from A itself. While $A \equiv \{x : A \mid x = x \text{ in } A\}$, we say that $\{x : A \mid S(x)\} = \{x : A' \mid S'(x)\}$ iff $A = A'$ and $S(x) = S'(x)$ for all x.

Here is another interesting fact about subtyping.

Theorem 1 *For all types A and equivalence relations E on A, $A \sqsubseteq A//E$.*

This is true because the elements of A and $A//E$ are the same, and since E is an equivalence relation over A, we know that E must respect the equality on A, that is

$$x = x' \text{ in } A \text{ and } y = y' \text{ in } A \text{ implies that } xEy \text{ iff } x'Ey'.$$

Thus, since xEx, then $x = x'$ in A implies xEx'.

3. Pairs

Halmos devotes one chapter to *unordered pairs* and another one to *ordered pairs*. In set theory ordered pairs are built from unordered ones using a clever "trick;" in type theory the ordered pair is primitive. Just as in programming, ordered pairing is a basic building block. Ordered pairs are the quintessential data elements. But unordered pairs are not usually treated as distinct kinds of elements.

Given a type A, an unordered pair of elements can be defined as:

$$\{x : A \mid x = a \vee x = b\}.$$

This is a type. We might also write it as $\{a, b\}_A$. It is unordered because $\{a, b\}_A \equiv \{b, a\}_A$. We'll see that this notion does not play an interesting role in type theory. It does not behave well as a data element, as we see later.

Given types A, B, their *Cartesian product, $A \times B$*, is the type of ordered pairs, $pair(a; b)$. We abbreviate this as $< a, b >$. Given a in A, b in B, then $< a, b >$ in $A \times B$. The constructor $pair(a; b)$ structures the data. The obvious destructors are operations that pick out the first and second elements:

$$1of(< a, b >) = a \qquad 2of(< a, b >) = b.$$

These can be defined in terms of a single operator, *spread()*, which splits a pair into its parts. The syntax of *spread()* involves the idea of *binding variables*. They are used as a pattern to describe the components. Here is the full syntax and the rules for computing with it.

If p is a pair, then $spread(p; u, v.g)$ describes an operator g for decomposing it as follows:

$spread(< a, b >; u, v.g)$ reduces in one step to $g[a/u, b/v]$

where $g[a/u, b/v]$ is the result of substituting a for the variable u, and b for the variable v.

Define

$$1of(p) == spread(p; u, v.u)$$
$$2of(p) == spread(p; u, v.v).$$

Notice that

$$spread(< a, b >; u, v.u) \text{ reduces to } a$$
$$spread(< a, b >; u, v.v) \text{ reduces to } b.$$

Is there a way to treat $\{a, b\}$ as a data element analogous to $< a, b >$? Can we create a type $Pair(A; B)$ such that

$$\{a, b\} \text{ in } Pair(A; B)$$
$$\{a, b\} = \{b, a\} \text{ in } Pair(A; B) \text{ and } Pair(A; B) = Pair(B; A)?$$

If we had a union type, $A \cup B$, we might define the pairs as $\{a, b\} = \{x : A \cup B | (x = a \text{ in } A) \text{ or } (x = b \text{ in } B)\}$, and let $Pair(A; B)$ be the collection of all such pairs. We exploe this collection later, in Sections 4 and 6.

How would we use $\{a, b\}$? The best we can do is pick elements from the pair, say $pick(p; u, v.g)$ and allow either reduction:

$pick(\{a, b\}; u, v.g)$ reduces either to $g[a/u, b/v]$ or to $g[b/u, a/v]$.

4. Union and intersection

Everyone is familiar with taking unions and intersections of sets and writing them in the standard cup and cap notations respectively, as in the union $X \cup Y$ and intersection $X \cap Y$. These operations are used in type theory as well, with the same notations. But the meanings go beyond the set idea, because the definitions must take account of equalities.

Intersection is the easier idea. If A and B are types, then equality holds on $A \cap B$ when it holds in A and in B; that is,

$$a = b \text{ in } A \cap B \quad \text{iff} \quad a = b \text{ in } A \quad \text{and} \quad a = b \text{ in } B.$$

In particular,

$$a = a \text{ in } A \cap B \quad \text{iff} \quad a = a \text{ in } A \quad \text{and} \quad a = a \text{ in } B.$$

For example, $\mathbb{Z}_2 \cap \mathbb{Z}_3$ has elements such as 0, since $0 = 0$ in $\mathbb{Z}_2$ and $0 = 0$ in $\mathbb{Z}_3$. And $0 = 6$ holds in $\mathbb{Z}_2 \cap \mathbb{Z}_3$ since $0 = 6$ in $\mathbb{Z}_2$ and $0 = 6$ in $\mathbb{Z}_3$. In fact, $\mathbb{Z}_2 \cap \mathbb{Z}_3 = \mathbb{Z}_6$.

Intersections can be extended to families of types. Suppose $B(x)$ is a type for every x in A. Then $\cap x : A.\ B(x)$ is the type such that

$$b = b'\ in\ \cap x : A.\ B(x) \text{ iff } b = b'\ in\ B(x) \text{ for all } x \text{ in } A.$$

For example, if $\mathbb{N}_k = \{0, \dots, k-1\}$ and $\mathbb{N}^+ = \{1, 2, \dots\}$ then $\cap x : \mathbb{N}^+.\mathbb{N}_x$ has only 0 in it.

It is interesting to see what belongs to $\cap x : A.\ B(x)$ if A is empty. We can show that there is precisely one element, and any closed expression of type theory denotes that element. We give this type a special name because of this interesting property.

Definition $Top == \cap x : void.\ x$, for $void$, the empty type.

Theorem 2 *If A, A' are types and $B(x)$ is a family of types over A, then*

1. $A \cap A' \sqsubseteq A \quad A \cap A' \sqsubseteq A'$
2. $\cap x : A.\ B(x) \sqsubseteq B(a)$ *for all a in A*
3. $A \sqsubseteq Top$
4. *If $C \sqsubseteq A$ and $C \sqsubseteq A'$, then $C \sqsubseteq A \cap A'$*

If A and B are disjoint types, say $A \cap B = void$, then their *union*, $A \cup B$, is a simple idea, namely

$$a = b \text{ in } A \cup B \text{ iff } a = b \text{ in } A \text{ or } a = b \text{ in } B.$$

In general we must consider equality on elements that A and B have in common. The natural thing to do is extend the equality so that if $a = a'$ in A and $a' = b$ in B, then $a = b$ in $A \cup B$. Thus the equality of $A \cup B$ is the *transitive closure* of the two equality relations, i.e.

$$a = b\ in\ A \cup B \text{ iff} \quad \begin{array}{l} a = b\ in\ A \text{ or } a = b\ in\ B \quad or \\ \exists c : A \cup B \quad a = c\ in\ A \cup B \text{ and } c = b\ in\ A \cup B. \end{array}$$

Note in $\mathbb{Z}_2 \cup \mathbb{Z}_3$ all elements are equal, and $\mathbb{Z}_4 \cup \mathbb{Z}_6 = \mathbb{Z}_2$.

Exercise: What is the general rule for membership in $\mathbb{Z}_m \cup \mathbb{Z}_n$?

Exercise: Unions can be extended to families of types. Give the definition of $\cup x : A.\ B(x)$. See the Nuprl Web page under basic concepts, unions, for the answer.

Theorem 3 *If A and A' are types and $B(x)$ is a family of types over A, then*

1. $A \sqsubseteq A \cup A' \quad A' \sqsubseteq A \cup A'$

2. $B(a) \sqsubseteq \cup x : A.\ B(x)$ *for all a in A*

3. *If* $A \sqsubseteq C$ *and* $A' \sqsubseteq C$, *then* $A \cup A' \sqsubseteq C$.

In set theory the disjoint union $A \oplus B$ is defined by using tags on the elements to force disjointness. We could use the tags *inl* and *inr* for the left and right disjuncts respectively. The definition is

$$A \oplus B = \{< inl, a >, < inr, b > \mid a \in A, b \in B\}.$$

The official definition is a union:

$$(\{inl\} \times A) \cup (\{inr\} \times B).$$

We could define a disjoint union in type theory in a similar way. Another approach that is more common is to take disjoint union as a new primitive type constructor, $A + B$.

If A and B are types, then so is $A + B$, called their disjoint union. The elements are $inl(a)$ for *a in A* and $inr(b)$ for *b in B*.

The destructor is $decide(d; u.g_1; v.g_2)$, which reduces as follows:

$decide(inl(a); u.g_1; v.g_2)$ reduces to $g_1[a/u]$ in one step

$decide(inr(b); u.g_1; v.g_2)$ reduces to $g_2[b/v]$ in one step.

Sometimes we write just $decide(d; g_1; g_2)$, if g_1 and g_2 do not depend on the elements of A and B, but only on the tag. In this way we can build a type isomorphic to *Bit* by forming $Top + Top$.

5. Functions and relations

We now come to the heart of type theory, an abstract account of *computable functions over all types*. This part of the theory tells us what it means to have an effectively computable function on natural numbers, lists, rational numbers, real numbers, complex numbers, differentiable manifolds, tensor algebras, streams, on any two types whatsoever. It is the most comprehensive theory of effective computability in the sense of deterministic sequential computation. *Functions are the main characters of type theory in the same way that sets are the main characters of set theory and relations are the main characters of logic.* Types arise because they characterize the domains on which an effective procedure terminates.

The computational model is more abstract than machine models, say Turing machines or random access machines or networks of such machines. Of all the early models of computability, this account is closest to the idea of a high-level programming language or the lambda calculus. A distinguishing feature is that all of the computation rules are independent of the type of the data, e.g., they are *polymorphic*. Another distinguishing feature is that the untyped computation system is *universal* in the sense that it captures at least all effective sequential procedures — terminating and non-terminating.

FUNCTIONS

If A and B are types, then there is a type of the effectively computable functions from A to B, and it is denoted $A \to B$. Functions are also data elements, but the operations on functions do not expose their internal structure. We say that the canonical form of this data is $\lambda(x.\ b)$. The symbol lambda indicates that the object is a function. (Having any particular canonical form is more a matter of convenience and tradition, not an essential feature.) The variable x is the formal input, or *argument*. It is a binding variable that is used in the *body* of the function, b, to designate the input value. The *body* b is the *scope* of this binding variable x. Of course the exact name of the variable is immaterial, so $\lambda(x.\ b)$ and $\lambda(y.\ b[y/x])$ are equal canonical functions.

The important feature of the function notation is that the body b is an expression of type theory which is *known to produce a value of type B after finitely many reduction steps* provided an element of type A, say a, is input to the function. The precise number of reduction steps on input a is the *time complexity* of $\lambda(x.\ b)$ on input a. We write this as

$$b[a/x] \downarrow b' \text{ in } n \text{ steps}$$

Recall that $b[a/x]$ denotes the expression b with term a substituted for all free occurrences of x. A free occurrence of x in b is one that does not occur in any subexpression c which is in the scope of a binding occurrence of x, i.e. not in $\lambda(x.\ c)$ or $spread(p; u, v.c)$, where one of u, v is x. The notion of binding occurrence will expand as we add more expressions to the theory. The above definition applies, of course, to all future extensions of the notion of a binding occurrence.

The simplest example of a computable function of type $A \to A$ is the identity function $\lambda(x.\ x)$. Clearly if $a \in A$, then $x[a/x]$ is $a \in A$.

If a_0 is a constant of type A, then $\lambda(x.\ a_0)$ is the constant function whose value is a_0 on any input. Over the natural numbers, we will have functions such as $\lambda(x.\ x + 1)$, the successor, $\lambda(x.\ 2 * x)$, doubling, etc.

We say that two functions, f and g, are *equal on* $A \to B$ when they produce the same values on the same inputs.

$$(f = g \text{ in } A \to B) \text{ iff } f(a) = g(a) \text{ in } B \text{ for all } a \text{ in } A.$$

This relation is called *extensional equality*. It is not the only sensible equality, but it is the one commonly used in mathematics.

We might find a tighter (finer) notion of equality more useful in computing, but no widely agreed-upon concept has emerged. If we try to look closely at the structure of the body b, then it is hard to find the right "focal length." Do we want the details of the syntax to come into focus or only some coarse features of it, say the combinator structure?

Functions are not meant to be "taken apart" as we do with pairs, nor do we directly use them to make distinctions, as we use the bits 0, 1. Functions are encapsulations of computing procedures, and the principal way to use them is to *apply them*. If f is a function, we usually display its application as $f(a)$ or fa. To conform to our uniform syntax, we write application primitively as $ap(f; a)$ and display this as $f(a)$ or fa when no confusion results.

The computation rule for $ap(f; a)$ is to first reduce the expression for f to canonical form, say $\lambda(x.\ b)$, and then to reduce $ap(\lambda(x.\ b); a)$. One way to continue is to reduce this to $b[a/x]$ and continue. Another way is to reduce a, say to a', and then continue by reducing $b[a'/x]$. The former reduction method is called *call by name* and the latter is *call by value*. We will use both kinds, writing $apv(f; a)$ for call by value.

Notice that when we use call by name evaluation, the constant function $\lambda(x.\ a_0)$ maps B into A for *any* type B, even the void type. So if $a_0 \in A$, then $\lambda(x.\ a_0) \in B \to A$ for any type B.

The function $\lambda(x.\ \lambda(y.x))$ has the type $A \to (B \to A)$ for any types A and B, regardless of whether they are empty or not. We can see this as follows. If we assume that z is of type A, then $ap(\lambda(x.\ \lambda(y.\ x)); z)$ reduces in one step to $\lambda(y.\ z)$. By what we just said about constant functions, this belongs to $(B \to A)$ for any type B. Sometimes we call the function $\lambda(x.\ \lambda(y.x))$ the K *combinator*. This stresses its polymorphic nature and indicates a connection to an alternative theory of functions based on combinators.

Exercise: What is the general type of these functions?

1. $\lambda(x.\ \lambda(y.\ <x, y>))$
2. $\lambda(f.\ \lambda(g.\ \lambda(x.\ g(x)(f(x)))))$

The function in (1) could be called the *Currying combinator*, and Curry called the function in (2) the S *combinator;* it is a form of composition.

The function $\lambda(f.\lambda(g.\lambda(x.g(f(x)))))$ belongs to the type $(A \to B) \to ((B \to C) \to (A \to C))$ because for $x \in A$, $f(x) \in B$ and $g(f(x)) \in C$. This is a *composition combinator, Comp.* It shows clearly the polymorphic nature of our theory. We can express this well with intersection types:

$$Comp\ in\ (\cap A, B, C : Type.\ (A \to B) \to ((B \to C) \to (A \to C))).$$

We will need to discuss the notion $A : Type$ etc. in Section 6 before this is entirely precise.

There are polymorphic λ-terms that denote sensible computations but which cannot be directly typed in the theory we have presented thus far. One of the most important examples is the so-called *Y-combinator* discovered by Curry,

$$\lambda(f.\ \lambda(x.f(xx))\lambda(x.f(xx))).$$

The subtyping relation on $A \rightarrow B$ behaves like this:

Theorem 4 *For all types* $A \sqsubseteq A'$, $B \sqsubseteq B'$

$$A' \rightarrow B \sqsubseteq A \rightarrow B'.$$

To see this, let $f = g \in A' \rightarrow B$. We prove that $f = g$ *in* $A \rightarrow B'$. First notice that for $a \in A$, we know $a \in A'$, thus $f(a')$ and $g(a')$ are defined, and $f(a') = g(a')$ in B. But $B \sqsubseteq B'$, so $f(a') = g(a')$ in B'.

This argument depends on the polymorphic behavior of the functions; thus, if $f(a)$ terminates on any input in A', it will, as a special case, terminate for any element of a smaller type. We say that $\sqsubseteq$ is *co-variant* in the domain.

In set theory, functions are defined as *single-valued relations,* and relations are defined as sets of ordered pairs. This reduces both relations and functions to sets, and in the process the reduction obliterates any direct connection between functions and algorithms.

A function $f \in A \rightarrow B$ does generate a type of ordered pairs called its *graph,* namely

$$graph(f) = \{x : A \times B | f(1of(x)) = 2of(x)\ in\ B\}.$$

Clearly $graph(f) \sqsubseteq A \times B$, somewhat as in set theory. If we said that a relation R on $A \times B$ is any subtype of $A \times B$, then we would know that $R \sqsubseteq A \times B$. But we will see that in type theory we cannot collect all such R in a single "power type." Let us see how relations are defined in type theory.

RELATIONS

Since functions are the central objects of type theory, we define a *relation* as a certain kind of function, a logical function in essence. This is Frege's idea of a relation. It depends on having a type of propositions in type theory. For now we denote this type as $Prop$, but this is amended later to be $Prop_i$ for fundamental reasons.

A relation on A is a *propositional function* on A, that is, a function $A \to Prop$. Here are propositional functions we have already encountered:

$$x = y \; in \; A$$

is an atomic proposition of the theory. From it we can define two propositional functions:

$$\lambda(x. \; \lambda(y. \; x = y \; in \; A)) \; in \; A \to (A \to Prop)$$

$$\lambda(p. \; 1of(p) = 2of(p) \; in \; A) \; in \; A \times A \to Prop.$$

We also discussed the proposition

$$x \in B \; wrt \; A,$$

from which we can define the propositional function

$$\lambda(x. \; x \in B \; wrt \; A) \; in \; A \to Prop \text{ for a fixed } B.$$

This propositional function is well-defined iff

$$x = y \; in \; B \text{ implies that } (x \in B \; wrt \; A)$$

is equal to the proposition

$$(y \in B \; wrt \; A).$$

The type $Prop$ includes propositions built using the *logical operators* & (and), $\vee$ (or), $\Rightarrow$ (implies), and $\neg$ (not), as well as the *typed quantifiers* $\exists x : A$ (there is an x of type A) and $\forall x : A$ (for all x of type A). We mean these in the *constructive sense* (see Section 9).

6. Universes, powers and openness

In set theory we freely treat sets as objects, and we freely quantify over them. For example, the ordered pair $< x, y >$ in set theory is $\{x, \{x, y\}\}$. We freely form nested sets, as in the sequence starting with the empty set: $\emptyset, \{\emptyset\}, \{\{\emptyset\}\}, \{\{\{\emptyset\}\}\}, \ldots$ All of these are distinct sets, and the process can go on indefinitely in "infinitely many stages." For example, we can collect all of these sets together

$$\{\emptyset, \{\emptyset\}, \{\{\emptyset\}\}, \{\{\{\emptyset\}\}\}, \ldots\}$$

and then continue $\{\{\emptyset, \{\emptyset\}, \ldots\}\}$, $\{\{\{\emptyset, \{\emptyset\}, \ldots\}\}\}$. There is a precise notion called the *rank* of a set saying how often we do this.

In set theory we are licensed to build these more deeply nested sets by various axioms. The axiom to justify $\{\emptyset, \{\emptyset\}, \{\{\emptyset\}\}, \{\{\{\emptyset\}\}\}, \ldots\}$ is the *axiom of infinity*. Another critical axiom for nesting sets is called the *power set* axiom. It says that for any set x, the set of all its subsets exists, called $\mathcal{P}(x)$, the power set of x. One of its subsets is $\{x\}$; another is $\emptyset$.

In type theory we treat types as objects, and we can form $\{\emptyset\}, \{\{\emptyset\}\}, \ldots$ But the licensing mechanism is different, and types are not much used as objects of computation, because more efficient data is available and is almost always a better choice.

The licensing mechanism for building large sets is critical because paradoxes arise if sets are allowed to be "too large," as with a set of all sets, called a *universe*. A key idea from set theory is used to justify large types. The idea is this: if we allow a fixed number of ways of building sets safely, such as unions, separation, and power sets, then we can form a safer kind of universe defined by allowing the iteration of all these operations indefinitely. This is the idea of a universe in type theory.

A universe is a collection of types closed under the type-building operations of pairing, union, intersection, function space and other operations soon to be defined. But it is not closed under forming universes (though a controlled way of doing this is possible), and we keep careful track of how high we have gone in this process. As a start, we index the universes as $\mathbb{U}_1, \mathbb{U}_2, \mathbb{U}_3, \ldots$ The elements are codes that tell us precisely how to build types. We tend to think of these indices as the types themselves.

Here is how we can use a universe to define $\{\emptyset\}$. Recall that $\emptyset$, the void type, could be defined as $\{x : \mathbb{B} \mid x \neq x \text{ in } \mathbb{B}\}$. This type belongs to $\mathbb{U}_i$ for all i. Now define $\{\emptyset\}$ as $\{x : \mathbb{U}_1 \mid x = \emptyset \text{ in } \mathbb{U}_1\}$. In our type theory, we cannot define this set without mentioning $\mathbb{U}_1$. Thus we are forced to keep track of the universe through which a type of this sort is admitted.

With universes we can define a *limited power type*, namely

$$\mathcal{P}_i(A) = \{x : \mathbb{U}_i \mid x \sqsubseteq A\}.$$

These might be seen as "pieces" of some ideal $\mathcal{P}(A)$ for any type A. But we do not know a provably safe way to define the ideal $\mathcal{P}(A)$.

The universe construction forms a *cumulative hierarchy*, $\mathbb{U}_1 \in \mathbb{U}_2, \mathbb{U}_2 \in \mathbb{U}_3, \ldots$, and $\mathbb{U}_i \sqsubseteq \mathbb{U}_{i+1}$. Note that $Top \text{ in } \mathbb{U}_i$ for all i, and yet $\mathbb{U}_i \sqsubseteq Top$. There is nothing like Top in set theory.

7. Families

Unions, intersections, products and function spaces can all be naturally extended from pairs of types to whole families. We saw this already in the case of unions. Halmos devotes an entire chapter to families.

Let us consider the disjoint union of a family of types indexed by a type A, that is, for every x in A, there is a type $B(x)$ effectively associated to it. The disjoint union in both set theory and type theory is denoted by $\Sigma x : A.\ B(x)$, and the elements are pairs $< a, b >$ such that $b\ in\ B(a)$. In type theory this is a new primitive type constructor, but it uses the pairing operator associated with products. It also uses $spread(p; u, v.g)$ to decompose pairs. Unlike pairing for products where $A \times B$ is empty if either A or B is, the disjoint union is not empty unless all of the $B(x)$ for each x in A are empty.

In set theory the disjoint union of a family of types is defined in terms of $A \times \cup x : A.\ B(x)$,

$$\Sigma x : A.\ B(x) = \{p : A \times \cup x : A.\ B(x) | 1of(p) \in A\ \&\ 2of(p) \in B(1of(p))\}$$

and where the ordinary union $\cup x : A.\ B(x)$ is defined as the union of the range of the function B from A into sets, e.g. $\cup\{B(x) | x \in A\}$.

The type $\Sigma x : A.\ B(x)$ behaves both like a union and a product. It is a product because its elements are pairs, so we expect the type to be related to $A \times B$. But it is not like a product in that it can be nonempty even if one of the $B(x)$ types is empty.

One reason for computer scientists to think of this as a product is that it is called a *variant record* in the literature of programming languages, and records are treated as products. It is a "variant" record because the initial componenets (leftmost) can influence the type of the later ones, causing them to "vary."

As a product, we call this type a *dependent product*, and we employ notation reminiscent of products, namely

$$x : A \times B(x)$$

where x is a binding variable, bound to be of type A and having *scope* $B(x)$. So

$$y : A \times B(y)$$

is the same type as (equal to) $x : A \times B(x)$.

Here is an example of this type. We'll consider which alternative view is most natural. Suppose the index set (first component) is in $\mathbb{N}^2 + \mathbb{R}^2$; thus it is either a pair of natural numbers or a pair of computable reals. Suppose the second component will be a number representing the area of a rectangle defined by the first pair. Its type will be $\mathbb{N}$ or $\mathbb{R}$, depending on the first value. To define this, let $Area(x)$ be defined as

$$Area(x) = decide(x; \mathbb{N}; \mathbb{R})$$

(or, more informally, $Area(x) = if\ is\_left(x)\ then\ \mathbb{N}\ else\ \mathbb{R}\ fi$). The type we want is $\Sigma x : (\mathbb{N}^2 + \mathbb{R}^2). Area(x)$, or equivalently $x : (\mathbb{N}^2 + \mathbb{R}^2) \times Area(x)$. How do we think of this? Is it a union of two types $(\mathbb{N}^2 \times \mathbb{N})$ and $(\mathbb{R}^2 \times \mathbb{R})$, or does it describe data that looks like $< inl(< nat, nat >), nat >$ or $< inr(< real, real >), real >$?

There are examples where the contrast is clearer. One of the best is the definition of dates as $< month, day >$, where

$$
\begin{aligned}
Month &= \{1, \ldots, 12\} \text{ and} \\
Day(2) &= \{1, \ldots, 28\} \\
Day(i) &= \{1, \ldots, 30\} \text{ for } i = 3, 6, 9, 11 \\
Day(i) &= \{1, \ldots, 31\} \text{ otherwise.}
\end{aligned}
$$

$Date = m : Month \times Day(m)$ seems like the most natural description of the data we use to represent dates, and the idea behind it is clearly a pair whose second component depends on the first.

We can also extend the function space constructor to families, forming a *dependent function space*. Given the family $B(x)$ indexed by A, we form the type

$$x : A \to B(x).$$

The elements are functions f such that for each $a \in A$, $f(a)$ *in* $B(a)$.

In set theory the same type is called an *infinite product* of a family, written $\Pi x : A. B(x)$, and defined as

$$\{f : A \to \cup x : A.\ B(x) | \ \forall x : A.\ f(x) \in B(x)\}.$$

An example of such a function takes as input a month, and produces the maximum day of the month — call it *maxday*. It belongs to

$$m : Month \to Day(m).$$

The intersection type also extends naturally to families, yet this notion was only recently discovered and exploited by Kopylov, in ways that we illustrate later. Given type A and family B over A, define

$$x : A \cap B(x)$$

as the collection of elements x of A such that x is also in $B(x)$. If $a = b$ in A and $a = b$ in $B(a)$, then $a = b$ in $x : A \cap B(x)$.

8. Lists and numbers

The type constructors we have examined so far all build finite types from finite types. The list constructor is not this way. The type $List(A)$ is limitless or *infinite* if there is at least one element in A. Moreover, the type

$List(A)$ is *inductive*. From $List(A)$ we can build the natural numbers, another inductive type. These types are an excellent basis for a computational understanding of the infinite and induction.

In set theory induction is also important, but it is not as explicitly primitive; it is somewhat hidden in the other axioms, such as the axiom of infinity—which explicitly provides $\mathbb{N}$—and the axiom of regularity, which provides for (transfinite) induction on sets by asserting that every ε-chain $x_1 \in x_0,\ x_2 \in x_1,\ x_3 \in x_2, \ldots$ must terminate.

If A is a type, then so is $List(A)$. The canonical data of $List(A)$ is either the empty list, *nil*, or it is a list built by the basic *list constructor*, or *cons* for short. If L is a list, and a *in* A, then $cons(a; L)$ is a list. The standard way to show such an inductive construction uses the pattern of a rule,

$$\frac{a \, in \, A \quad L \, in \, List(A)}{cons(a; L) \, in \, List(A).}$$

Here are some lists built using elements $a_1, a_2, a_3, \ldots$ The list *nil* is in any type $List(A)$ for any type A, empty or not. *So $List(A)$ is always nonempty.* Next, $cons(a_1; nil)$ is a list, and so are

$$cons(a_1; nil), cons(a_1; cons(a_1; nil)), cons(a_1; cons(a_2; nil)),$$

and so forth. We typically write these as

$$[a_1], [a_1, a_2], [a_1, a_2, a_3].$$

If A is empty, then $List(A)$ has only *nil* as a member.

Here is a particularly clear list type. Let $\mathbf{1}$ be the type with exactly one element, 1. Then the list elements can be enumerated in order, *nil*, $cons(1; nil), cons(1; cons(1; nil)), cons(1; cons(1; cons(1; nil))), \ldots$ We will define the natural numbers to be this type $List(\mathbf{1})$. The type $List(Top)$ is isomorphic to this since Top has just one element, but that element has a limitless number of canonical names.

The method of destructing a list must make a distinction between *nil* and $cons(a; L)$. So we might imagine an operator like *spread*, say $dcons(L; g_1; u, t.g_2)$ where $dcons(nil; g_1; u, v.g_1)$ reduces to g_1 in one step and $dcons(cons(a; L); g_1, u, t.g_2)$ reduces to $g_2[a/u, L/t]$ in one step. This allows us to disassemble one element. The power of lists comes from the *inductive pattern* of construction,

$$\frac{a \, in \, A \quad L \, in \, List(A)}{cons(a; L) \, in \, List(A).}$$

This pattern distinguishes $List(A)$ from all the other types we built, making it infinite. We need a destructor which recognizes this inductive character, letting us apply *dcons* over and over until the list is eventually *nil*.

How can we build an inductive list destructor? We need an inductive definition for it corresponding to the inductive pattern of the elements. But just decomposing a list will be useless. We want to leave some trace as we work down into this list. That trace can be the corresponding construction of another object, perhaps piece by piece from the inner elements of the list.

Let's imagine that $build(L)$ is constructing something from L as it is decomposed. Then the inductive pattern for building something in B is just this:

$$\text{build } b_0 \text{ in } B \qquad \frac{a \text{ in } A, \text{assume } b \text{ in } B \text{ is built from } L}{\text{combine } a, L, \text{ and } b \text{ to build } g(a, L, b) \text{ in } B.}$$

This pattern can be expressed in a recursive computation on a list

$$build(nil) = b_0 \qquad build(cons(a; L)) = g(a, L, build(L)).$$

This pattern can be written as a simple recursive function if we use $dcons$ as follows:
$$f(L) = dcons(L; b_0; u, t.g(u, t, f(t)).$$

There is a more compact form of this expression that avoids the equational form. We extend $dcons$ to keep track of the value being built up. The form is $list\_ind(L; b_0; u, t, v.g(u, t, v))$, where v keeps track of the value.

We say that $list\_ind(L; b_0; u, t, v.g(u, t, v))$ in B, provided that b_0 in B and assuming that $u \in A, t \in List(A)$ and v in B, then $g(u, t, v)$ in B.

The reduction rule for $list\_ind$ is just this:

$$list\_ind(nil; b_0; u, t, v.g(u, t, v)) \qquad \text{reduces to } b_0$$
$$list\_ind(cons(a; L); b_0; u, t, v.g(u, t, v)) \quad \text{reduces to } g(a, L, v_0)$$

where $v_0 = list\_ind(L; b_0; u, t, v.g(u, t, v))$.

There are several notations that are commonly adopted in discussing lists. First, we write $cons(a; l)$ as $a.l$. Next we notice that

$$list\_ind(l; a; u, t, v.b)$$

where v does not occur in b is actually just a case split on whether l is nil followed by a decomposition of the cons case. We write this as:

$$\text{case of } l; \ nil \to b; \ a.t \to b$$

Here are some of the basic facts about lists along with a sketch of how we prove them.

Fact: $nil \neq cons(a; L)$ in $List(A)$ for any a or L.

This is because if $nil = cons(a; L)$ in $List(A)$, then $list\_ind(x; 0; u, t, v.1)$ would reduce in such a way that $0 = 1$, which is a contradiction. The basic fact is that all expressions in the theory respect equality.

One of the most basic operations on lists is *appending* one onto another, say $[a_1, a_2, a_3]@[a_4, a_5] = [a_1, a_2, a_3, a_4, a_5]$. Here is a recursive definition:

$$x@y = \text{case of } x; nil \to y; a.t \to a.(t@y).$$

This abbreviates

$$list\_ind(x; \lambda(y.y); a, t, v.\lambda(y.a.v(y))).$$

Fact: For all x, y, z in $List(A)$, $(x@y)@z = x@(y@z)$.

We prove this by induction on x. The base case is $(nil@x)@y = nil@(x@y)$. This follows immediately from the nil case of the definition of @.

Assuming that $(t@x)@y = t@(x@y)$, then $((u.t)@y)@z = u.t@(y@z)$ again follows by the definition of @ in the cons case. This ends the proof.

If f is a function from A to B, then here is an operation that applies it to every element of l in $List(A)$:

$$map(f; l) = \text{case of } l; nil \to nil; a.t \to f(a).map(f; t).$$

Fact: If $f : A \to B$ and $g : B \to C$ and $l : List(A)$, then $map(g; map(f; l))$ $= map(\lambda(x.g(f(x))); l)$ in $List(C)$.

Fact: If $f : A \to B$ and $x, y : List(A)$, then $map(f; x@y) = map(f; x)$ $@map(f; y)$ in $List(B)$.

Fact: $x@y = nil$ in $List(A)$ iff $x = nil$ and $y = nil$ in $List(A)$.

Exercise: What would $List(\{nil\})$ look like? Do all the theorems still work in this case? How does this compare to $List(\emptyset)$?

9. Logic and the Peano axioms

Following the example of Halmos, we have avoided attention to logical matters, but there is something especially noteworthy about logic and type theory. *We use logical language in a way that is sensitive to computational meaning.* For instance, when we say $\exists x : A. \ B(x)$, read as "we can find an x of type A such that $B(x)$," we mean that to know that this proposition is true is to be able to exhibit an object a of type A, called the *witness*, and *evidence* that $B(a)$ is true; let this evidence be $b(a)$. It turns out that we can think of a proposition P as *the type of evidence for its truth*. We require of the evidence only that it carry the *computational content* of the sense of the proposition. So in the case of a proposition of the form $\exists x : A. \ B(x)$, the evidence must contain a witness a and the evidence for $B(a)$. We can think of this as a pair, $< a, b(a) >$.

When we give the type of evidence for a proposition, we are specifying the computational content. Here is another example. When we know P implies Q for propositions P and Q, we have an effective procedure for taking evidence for P into evidence for Q. So the computational content for P *implies* Q is the function space $P \to Q$, where we take P and Q to be the types of their evidence.

This computational understanding of logic diverges from the "classical" interpretation. This is especially noticeable for statements involving *or*. To know $(P \text{ or } Q)$ is to either know P or know Q, and to know which. The rules of evidence for $(P \text{ or } Q)$ behave just as the rules for elements of the disjoint union type, $P + Q$.

The computational meaning of $\forall x : A.\ B(x)$ is that we can exhibit an effective method for taking elements of A, a, to evidence for $B(a)$. If $b(a)$ is evidence for $B(a)$ given any element of a in A, then $\lambda(x.\ b(x))$ is computational evidence for $\forall x : A.\ B(x)$. So the evidence type behaves just as $x : A \to B(x)$.

For atomic propositions like $a = b$ in B, there is no interesting computational content beyond knowing that when a and b are reduced to canonical form, they will be identical, e.g. $bit\{0\} = bit\{0\}$ or $bit\{1\} = bit\{1\}$. For the types we will discuss in this article, there will be no interesting computational content in the equality proposition, even in $f = g \text{ in } A \to B$. We say that the computational content of $a = b \text{ in } A$ is *trivial*. We will only be concerned with whether there is evidence. So we take some atomic object, say *is_true*, to be the evidence for any true equality assertion.

What is remarkably elegant in this account of computational logic is that the rules for the evidence types are precisely the expected rules for the logical operators. Consider, for example, this computational interpretation of $\exists x : A.\neg Q(x) \text{ implies } (\forall x : A.\ (P(x) \lor Q(x)) \text{ implies } \exists x : A.\ P(x).$

To prove $\exists x : A.\ P(x)$, let a be the element of A such that $\neg Q(a)$. We interpret $\neg S$ for any proposition S as meaning S implies *False*, and the evidence type for *False* is empty. Taking a for x in $\forall x : A.(P(x) \lor Q(x))$ we know $P(a) \lor Q(a)$. So there is some evidence $b(a)$ which is either evidence for $P(a)$ or for $Q(a)$. We can analyze $b(a)$ using *decide* since $P(a) \lor Q(a)$ is like $P(a) + Q(a)$. If $b(a)$ is in $P(a)$ then we have the evidence needed. If $b(a)$ is in $Q(a)$, then since we know $Q(a) \to False$, there is a method, call it f, taking evidence for $Q(a)$ into *False*.

To finish the argument, we look at the computational meaning of the assertion $(False \text{ implies } S)$ for any proposition S. The idea is that *False* is empty, so there is no element that is evidence for *False*. This means that if we assume that x is evidence for *False*, we should be able to provide *evidence for any proposition whatsoever.*

To say this computationally, we introduce a form $any(x)$ with the typing rule that if x is of type *False*, then $any(x)$ is of type S for any proposition S.

Now we continue the argument. Suppose $b(a)$ is in $Q(a)$. Then $f(b(a))$ is in *False*, so $any(f(b(a)))$ is in $P(a)$. Thus in either case of $P(a)$ or $Q(a)$ we can prove the $\exists x : A.\ P(a)$. Here is the term that provides the computation we just built:

$$\lambda(e.\lambda(all.decide(all(1of(e)));$$
$$p.p;$$
$$q.any(f(q)))))).$$

Note that e is evidence for $\exists x : A.\neg Q(x)$, so $1of(e)$ belongs to A. The function all produces the evidence for either $P(a)$ or $Q(a)$, so $all(1of(e))$ is what we called $b(a)$. We use the *decide* form to determine the kind of evidence $b(a)$ is; in one case we call it p and in the other q. So $any(f(q))$ is precisely what we called $any(f(b(a)))$ in the discussion.

Now we turn to using this logic. In 1889 Peano provided axioms for the natural numbers that have become a standard reference point for our understanding of the properties of natural numbers. We have been using $\mathbb{N}$ to denote these numbers $(0, 1, 2, \dots)$.

Set theory establishes the adequacy of its treatment of numbers by showing that the set of natural numbers, ω, satisfies the five Peano axioms. These axioms are usually presented as follows, where $s(n)$ is the *successor* of n.

1. 0 is a natural number, $0 \in \mathbb{N}$.
2. If $n \in \mathbb{N}$, then $s(n) \in \mathbb{N}$.
3. $s(n) = s(m)$ implies $n = m$.
4. Zero has no predecessor, $\neg(s(n) = 0)$.
5. The induction axiom, if $P(0)$, and if $P(n)$ implies $P(s(n))$, then P holds for all natural numbers.

In set theory axioms 1 and 2 are part of the definition of the axiom of infinity; axiom 3 is a general property of the successor of a set defined as $s(x) = x \cup \{x\}$. Induction comes from the definition of ω as the least inductive subset of the postulated infinite set.

In type theory we also deal with axioms 1 and 2 as part of the definition. One way to treat $\mathbb{N}$ is to define it as a new type whose canonical members are $0, s(0), s(s(0))$, and so forth, say using these rules:

$$0 \ in \ \mathbb{N} \qquad \frac{n \ in \ \mathbb{N}}{s(n) \ in \ \mathbb{N}.}$$

Another approach is to define ℕ using lists. We can take ℕ as *List*(**1**), with *nil* as 0 and *cons*(1; *n*) as the successor operation. Then the induction principle follows as a special case of list induction.

In this definition of ℕ, the addition operation $x + y$ corresponds exactly to $x@y$, e.g. $2 + 3$ is just $[1, 1]@[1, 1, 1]$.

Exercise: Show how to define multiplication on lists.

10. Structures, records and classes

Bourbaki's encyclopedic account of mathematics begins with set theory, and then treats the general concept of a *structure*. Structures are used to define algebraic structures such as monoids, groups, rings, fields, vector spaces, and so forth. They are also used to define topological structures and order structures, and then these are combined to provide a modular basis for real analysis and complex analysis.

STRUCTURES

The idea of a structure is also important in computer science; they are the basis for modules in programming languages and for classes and objects in object-oriented programming. Also, just as topological, order-theoretic, and algebraic structures are studied separately and then combined to explain *aspects* of analysis, so also in computing, we try to understand separate aspects of a complex system and then combine them to understand the whole.

The definition of structures in set theory is similar to their definition in type theory, as we next illustrate; later we look at key differences. In algebra we define a *monoid* as a structure $< M, op, id >$, where M is a set, *op* is an associative binary operation on M, and *id* is a constant of M that behaves as an identity, e.g.

$$x \ op \ id = x \ \ \text{and} \ \ id \ op \ x = x.$$

A *group* is a structure $< G, op, id, inv >$ which extends a monoid by including an inverse operator, e.g.

$$x \ op \ inv(x) = id \ \text{and} \ inv(x) \ op \ x = id.$$

In algebra, a group is considered to be a monoid with additional structure. These ideas are naturally captured in type theory in nearly the same way. We start with a structure with almost no form,

$$A : \mathbb{U}_i \times Top.$$

This provides a carrier A and a "slot" for its extension. We can extend by refining Top to have structure, for example, $B \times Top$. Notice that $B \times Top \sqsubseteq Top$, and

$$A \times (B \times Top) \sqsubseteq A \times Top, \quad \text{since}$$
$$A \sqsubseteq A \text{ and } B \times Top \sqsubseteq Top.$$

Generally,

$$A \times (B \times (C \times Top)) \sqsubseteq A \times (B \times Top).$$

We can define a monoid as a dependent product:

$$M : \mathbb{U}_i \times ((M \to (M \to M)) \times (M \times Top)).$$

An element has the form $< M, < op, < id, \bullet >>>$ where:

$$M \text{ in } \mathbb{U}_i, \quad op \text{ in } M \to (M \to M), \quad id \in M, \text{ and } \bullet \text{ in } Top.$$

Call this structure $Monoid$.

A *group* is an extension of a monoid which includes an inverse operator, $inv : G \to G$. So we can define the type $Group$ as

$$G : \mathbb{U}_i \times (G \to (G \to G) \times (G \times ((G \to G) \times Top))).$$

It is easier to compare these dependent structures if we require that the type components are related. So we define *parameterized* structures. We specify the *carrier* C of a monoid or group, etc., and define

$$Monoid(C) = C \to (C \to C) \times (C \times Top)$$
$$Group(C) = C \to (C \to C) \times (C \times (C \to C \times Top)).$$

Then we know

Fact: If $M \in \mathbb{U}_i$, then $Group(M) \sqsubseteq Monoid(M)$.

Exercise: Notice that $G \sqsubseteq M$ need not imply that

$$Group(G) \sqsubseteq Monoid(M).$$

These subtyping relationships can be extended to richer structures such as rings and fields, though not completely naturally. For example, a ring consists of two related structures: a group part (the *additive* group) and a monoid part (the *multiplicative* monoid). We can combine the structures as follows, putting the group first:

$$Ring(R) = R \to (R \to R) \times ((R \times (R \to R) \times (R \to (R \to R) \times (R \times Top)))).$$

We can say that $Ring(R) \sqsubseteq Group(R)$, but it is not true that the multiplicative structure is directly a substructure of $Monoid(R)$. We need to project it off,

$$Mult\_Ring(R) = (R \to (R \to R) \times (R \times Top)).$$

Then we can say $Mult\_Ring(R) \sqsubseteq Monoid(R)$. Given a ring

238

$$Rng \ in \ Ring(R),$$

$$< R, < add\_op, < add\_id, < add\_inv, < mulop, < mulid, \ \bullet >>>>>>,$$

we can project out the multiplicative part and the additive part:

$$add(Rng) \ = \ < add\_op, < add\_id, < add\_inv, - >>>$$
$$mul(Rng) \ = \ < mulop, < mulid, \ \bullet >>,$$

and we know that $add(Rng) \ in \ Group(R)$ and $mul(Rng) \ in \ Monoid(R)$.

RECORDS

Our account of algebraic structure so far is less convenient than the informal one on which it is based. One reason is that we must adhere to a particular ordering of the components and access them in this order. Programming notations deal with this inconvenience by associating names with the components and accessing them by name. The programming construct is called a *record type*; the elements are *records*. A common notation for a record type is this:

$$\{a : A; \ b : B; \ c : C\}.$$

Here A, B, C are types and a, b, c are names called *field selectors*. If r is a record, the notations $r.a, r.b, r.c$ select the component with that name. The order is irrelevant, so $\{b : B; \ c : C; \ a : A\}$ is the same type, and we know that $r.a \in A, r.b \in B, r.c \in C$.

In this notation, one type for a group over G is

$$\{add\_op : G \to (G \to G); \ add\_id : G; \ add\_inv : G \to G\}.$$

The field selectors in this example come from a type called *Label*, but more generally we can say that a family of these group types abstracted over the field selectors, say $GroupType(G; x, y, z)$, is

$$\{x : G \to (G \to G); \ y : G; \ z : G \to G\}.$$

We can combine $MonoidType(G; u, v) = \{u : G \to (G \to G); v : G\}$ with the $GroupType$ to form a $RingType(R; x, y, z, u, v) =$

$$\{x : G \to (G \to G); \ y : G; \ z : G \to G; \ u : G \to (G \to G); v : G\}.$$

Record types can be defined as function spaces over an index type such as *Label*. First we associate with each x in *Lable* a type, say $T : Label \to \mathbb{U}_i$. If we have in mind a type such as group, parameterized by G, then we map *add_op* to $G \to (G \to G)$, *add_id* to G, *add_inv* to $G \to G$, and all other lables to *Top*. Call this map $Grp(x)$. Then the group type is

$$x : Label \to Grp(x).$$

An element of this type, g, is a function such that $g(add\_op) \in G \to (G \to G)$, $g(add\_id) \in G$, $g(add\_inv) \in G \to G$, and $g(z) \in Top$ for all other z in *Label*.

The record types we use have the property that only finitely many labels are mapped to types other than *Top*. For example, $\{a : A;\ b : B;\ c : C\}$ is given by a map $T : Label \to \mathbb{U}_i$ such that $T(a) = A$, $T(b) = B$, $T(c) = C$, and $T(x) = Top$ for x not in $\{a, b, c\}$. Thus

$$\{a : A;\ b : B;\ c : C\} = x : Label \to T(x).$$

We say that such records have *finite support* of $\{a, b, c\}$ in *Label*. If I is the finite support for T, we sometimes write the record as $x : I \to T(x)$.

It is very interesting that for two records having finite support of I_1 and I_2 such that $I_1 \subseteq I_2$, and such that $T : Label \to \mathbb{U}_i$ agree on I_2, we know that

$$x : I_2 \to T(x) \sqsubseteq x : I_1 \to T(x).$$

For example, consider $\{a, b, c\} \subseteq \{a, b, c, d\}$, with records $R_1 = \{a : A; b : B; c : C\}$ and $R_2 = \{a : A;\ b : B;\ c : C;\ d : D\}$. Then $R_2 \sqsubseteq R_1$. This natural definition conforms to programming language practice and mathematical practice. We will see that this definition, while perfectly natural in type theory, is not sensible in set theory.

If we use standard labels for algebraic operations on monoids, groups, and rings, say

$$Alg : \{add\_op, add\_id, add\_inv, mul\_op, mul\_id\} \to \mathbb{U}_1$$

then

$$
\begin{aligned}
Add\_Monoid(G) &= i : \{add\_op, add\_id\} \to Alg(i) \\
Group(G) &= i : \{add\_op, add\_id, add\_in\} \to Alg(i) \\
Mul\_Monoid(G) &= i : \{mul\_op, mul\_id\} \to Alg(i) \\
Ring(G) &= i : \{add\_op, add\_id, add\_inv, mul\_op, mul\_id\} \to Alg(i)
\end{aligned}
$$

and we have

$$
\begin{aligned}
Group(G) &\sqsubseteq Add\_Monoid(G) \\
Ring(G) &\sqsubseteq Group(G) \sqsubseteq Add\_Monoid(G) \\
Ring(G) &\sqsubseteq Mul\_Monoid(G).
\end{aligned}
$$

The reason that these definitions don't work in set theory is that the subtyping relation on function spaces is not valid in set theory. Recall that the relation is

$$\frac{A \subseteq A' \quad B \subseteq B'}{A' \to B \subseteq A \to B'}.$$

This is true in type theory because functions are *polymorphic*. If $f \in A' \to B$, then given $a \in A$, the function f applies to a; so $f(a)$ is well-defined and the result is in B, hence in B'. In set theory, the function f is a set of ordered pairs, e.g. $f = \{< a, b > \in A' \times B | f(a) = b\}$. This set of ordered pairs can be larger than $\{< a, b > \in A \times B' | f(a) = b\}$, so $A' \to B \not\subseteq A \to B'$. The difference in the notion of function is fundamental, and it is not clear how to reconcile them.

DEPENDENT RECORDS

A full account of algebraic structures must include the axioms about the operators. For a monoid we need to say that *op* is associative, say

(1) $Assoc\,(M, op)$ is

$\forall x, y, z : M.\ (x\ op\ y)op\ z = x\ op(y\ op\ z)\ in\ M.$

and we say that *id* is a two-sided identity:

(2) $Id(M, op, id)$ is

$\forall x : M.\ (x\ op\ id = x\ in\ M)$ and $(id\ op\ x = x\ in\ M).$

For the group inverse the axiom is

(3) $Inv(M, op, id, inv)$ is

$\forall x : M.\ (x\ op\ inv(x) = id\ in\ M)$ and $(inv(x)op\ x = id\ in\ M).$

In set theory, these axioms are not included inside the algebraic structure because the axioms are propositions, which are "logical objects," not sets. But as we have seen, propositions can be considered as types. So we can imagine an account of a *full-monoid* over M that looks like this:

$\{op : G \to (G \to G);\ id : G;\ ax1 : Assoc(G, op);\ ax2 : Id(G, op, id)\}.$

If g is a full-monoid, then $g(op)$ is the operator and $g(ax1)$ is the computational evidence that op is associative; that is, $g(ax1)$ is a mathematical object in the type $Assoc(G, op)$.

Does type theory support these kinds of records? We call them *dependent records*, since the type $Assoc(G, op)$ depends on the object $g(op)$. Type theory does allow us to define them.

The object we define is the general dependent record of the form

$$\{x_1 : A_1; x_2 : A_1(x_1); \ldots ; x_n : A_{n-1}(x_1, \ldots, x_{n-1})\}.$$

In Automath these are called *telescopes,* and they are primitive concepts. We define them from the dependent intersection following Kopylov. Other researchers have added these as new primitive types, and Jason Hickey defined them from a new primitive type called the very-dependent function space. (See Bibliographic Notes for this section.)

Recall that the dependent intersection, $x : A \cap B(x)$ is the collection of those elements a of A which are also in $B(a)$.

We define $\{x : A; y : B(x)\}$ as the type

$$f : \{x : A\} \cap \{y : B(f.x)\}.$$

The elements of $\{x : A\}$ are the functions $\{x\} \to A$; that is, functions in $i : Label \to A(i)$ where $A(x) = A$ and $A(i) = Top$ for $i \neq x$. The singleton label, $\{x\}$, is the finite support. The elements of the intersection are those functions f in $\{x\} \to A$ such that on input y from $Label$, $f(y)$ in $B(f(x))$.

To define $\{x : A; y : B(x); z : C(x, y)\}$, we have the choice of associating the dependent intersection to the right or left; we chose to the left.

$$*g : (f : \{x : A\} \cap \{y : B(f(x))\}) \cap \{z : C(g(x), g(y))\}.$$

The function g must agree with f on label x. We can see that the outermost binding, g, satisfies the inner constraints as well. So as we intersect in more properties, we impose more constraints on the function.

The value of associating to the left is that we can think of building up the dependent record by progressively adding constraints. It is intuitively like this:

$$(\{x : A\} \cap \{y : B(x)\}) \cap \{z : C(x, y)\}.$$

This kind of notational simplicity can be seen as an abbreviation of

$$** \quad s : (s : \{x : A\} \cap \{y : B(s(x))\}) \cap \{z : C(s(x), s(y))\},$$

because the scoping rules for binding operators tell us that * and ** are equal types.

In programming languages such as ML, modules and classes are used to modularize code and to make the code more abstract. For example, our treatment of natural numbers so far has been particularly concrete. We introduced them as lists of a single atomic object such as 1. Systems like HOL take the natural numbers ($\mathbb{N}$) as primitive, and Nuprl takes the integers ($\mathbb{Z}$) as primitive, defining $\mathbb{N}$ as $\{z : \mathbb{Z} | 0 \leq z\}$. All these approaches

242

can be subsumed using a class to axiomatize an abstract structure. We define numbers, say integers, abstractly as a class over some type D which we axiomatize as an *ordered discrete integral domain*, say $Domain(D)$. The class is a dependent record with structure. We examine this in more detail here.

First we define the stucture without induction and then add both recursive definition and induction. We will assume *display forms* for the various field selectors when we write axioms. Here is a table of displays:

Field Selector Name	Display (inside the class and outside)
add	$+$
zero	0
minus	-
mult	$*$
one	1
div	$\div$
mod	*mod infix*
less_eq	$\leq$

The binary operators are given by the type $BinaryOp(D)$. This can be the "curried style" type $D \rightarrow (D \rightarrow D)$ that we used previously, or the more "first-order" style, $D \times D \rightarrow D$, or it can even be the "Lisp-style," where we allow any list of arguments from D, say $List(D) \rightarrow D$. We can define monoids, groups, etc. using the same abstraction, which we select only at the time of implementation. Likewise, we can abstract the notion of a binary relation to $BinaryRel(D)$. We can use $D \rightarrow (D \rightarrow Prop_i)$ or $D \times D \rightarrow Prop_i$ or $List(D) \rightarrow Prop_i$.

Definition For D a type in $\mathbb{U}_i$ the class $Domain(D)$ is:

$\{ add \qquad : BinaryOp\,(D)$; $assoc\_add : Assoc\,(D, add)$;

$zero \qquad : D$; $identity\_zero : Identity\,(D, add, zero)$;

$minus \qquad : D \rightarrow D$; $inverse\_minus : Inverse\,(D, add, zero, minus)$;

$mult \qquad : BinaryOp\,(D)$; $assoc\_mult : Assoc\,(D, mult)$;

$one \qquad : D$; $identity\_one : Identity\,(D, mult, one)$;

$div \qquad : BinaryOp\,(D)$;

$rm \qquad : BinaryOp\,(D)$; $\forall x, y : D.(x = y * div(x, y) + rm(x, y)\,in\,D)$;

$discrete : \forall x, y : D.(x = y\,in\,D\,or\,x \neq y\,in\,D)$;

$less\_eq \quad : BinaryRel\,(D)$; $porder : PartialOrder\,(D, less\_eq)$;

$trichot \quad : \forall x, y : D.(x \leq y\,or\,y \leq x)$
$$\qquad\qquad\qquad and\,(x \leq y\,\&\,y \leq x\,implies\,x = y\,in\,D);$$

$cong \qquad : \forall x, y, z : D.(x \leq y\,implies\,x + z \leq y + z)\,and$
$$\qquad\qquad (x \leq y\,and\,z \geq 0\,implies\,x * z \leq y * z)\,and$$
$$\qquad\qquad (x \leq y\,and\,z < 0\,implies\,y * z \leq x * z).\ \}$$

The domain can be made *inductive* if we add an induction principle such as

$$ind : \forall P : D \rightarrow Prop_i.\ (P(0)\ \&\ \forall z : \{x : D | x \geq 0\}.P(z) \quad implies \quad P(z + 1)$$

$$\&\ \forall z : \{x : D | x < 0\}.\ P(z + 1)\,implies\,P(z))\,implies\,\forall x : D.\ P(x).$$

In type theory it is also easy to allow a kind of *primitive recursive definition* over D by generalizing the induction to

$$ind : \forall A : D \rightarrow \mathbb{U}_i.A(0) \rightarrow (z : \{y : D | y \geq 0\} \rightarrow A(z) \rightarrow A(z + 1)) \rightarrow$$

$$(z : \{y : D | y < 0\} \rightarrow A(z + 1) \rightarrow A(z))\forall x : D.A(x).$$

$$ind\_eq : \forall b : A(0).\forall f : (z : \{y : D | y \geq 0\} \rightarrow A(z) \rightarrow A(z + 1)).$$

$$\forall g : z : (\{y : D | y < 0\} \rightarrow A(z + 1) \rightarrow A(z)).$$

$$ind(b)(f)(g)(0) = b\,in\,A(0)\,and\,\forall y : D(y \geq 0\,implies$$

$$ind(b)(f)(g)(y) = f(y)(ind(b)(f)(g)(y - 1))\,in\,A(y))$$

$$and\,\forall y : D.(y < 0\,implies$$

$$ind(b)(f)(g)(y) = g(y)(ind(b)(f)(g)(y + 1))\,in\,A(y)).$$

If we take A to be the type $(D \rightarrow D)$, then the induction constructor allows us to define functions $D \rightarrow (D \rightarrow D)$ by induction. For example, here is a definition of factorial over D. We take $A(x) = D$ for all x, so we are defining a function from D to D. On input 0, we build 1; in the case for $z > 0$ and element u in $A(z)$, we build the element $z * u$ as result. For $z < 0$, we take 0 as the result. The form of definition is

$$ind(\lambda(x.D))(1)(\lambda(z.\lambda(u.z * u)))(\lambda(z.\lambda(u.0))).$$

11. The axiom of choice

Halmos notes that it has been important to examine each consequence of the axiom of choice to see "the extent to which the axiom is needed in the proof." He said, "an alternative proof without the axiom of choice spelled victory." It was thought that results without the axiom of choice were safer. But in fact, in the computational mathematics used here, which is very safe, one form of the axiom of choice is provable! We start this section with a proof of its simplest form.

If we know that $\forall x : A.\ \exists y : B.\ R(x, y)$, then the axiom of choice tells us that there is a function f from A to B such that $R(x, f(x))$ for all x in A. We can state this symbolically as follows (using $\Rightarrow$ for implication):

Axiom of Choice $\forall x : A.\ \exists y : B.\ R(x, y) \Rightarrow \exists f : A \to B.\ \forall x : A.\ R(x, f(x))$.

Here is a proof of the axiom. We assume $\forall x : A.\ \exists y : B.\ R(x, y)$. According to the computational meaning of $\forall x : A$, we know that there is a function g from A to evidence for $\exists y : B.\ R(x, y)$. The evidence for $\exists y : B.\ R(x, y)$ is a pair of a witness, say $b(x)$, and evidence for $R(x, b(x))$; call that evidence $r(x)$. Thus the evidence is the pair $< b(x), r(x) >$.

So now we know that on input x from A, g produces $< b(x), r(x) >$. We can define f to be $\lambda(x.\ b(x))$. We know that f in $A \to B$ since $b(x)$ in B for any x in A. So the witness to $\exists f : A \to B$ is now known. Can we also prove $\forall x : A.\ R(x, f(x))$? For this we need a function, say h, which on input x produces a proof of $R(x, f(x))$. We know that $\lambda(x.r(x))$ is precisely this function. So the pair we need for the conclusion is

$$< \lambda(x.b(x)), \lambda(x.r(x)) >$$

where $b(x) = 1of(g(x))$ and $r(x) = 2of(g(x))$. Thus the implication is exhibited as

$$\lambda(g.\ < \lambda(x.1of(g(x))), \lambda(x.2of(g(x))) >).$$

This is the computational content of the axiom of choice.

In set theory the corresponding statement of the axiom of choice is the statement that the product of a family of sets $B(x)$ indexed by A is nonempty if and only if each $B(x)$ is. That is:

$$\Pi x : A.\ B(x) \text{ is inhabited}$$
$$\text{iff}$$
$$\text{for each } x \text{ in } A,\ B(x) \text{ is inhabited.}$$

We can state this as a special case of our axiom by taking $R(x, y)$ to be trivial, say *True*.

$$\exists f : (x : A \to B(x)).\ True \text{ iff } \forall x : A.\ \exists y : B(x).\ True.$$

Another formulation of the axiom in set theory is that for any collection C of nonempty subsets of a set A, there is a function f taking $x \in C$ to an element $f(x)$ in x. Halmos states this as

$$\exists f. \forall x : (\mathcal{P}(A) - \{\emptyset\}). \; f(x) \in x$$

We can almost say this in type theory as: there is an element of the type

$$x : \{Y : \mathcal{P}(A) \,|\, Y \text{ is nonempty}\} \to x.$$

One problem with this formulation is that $\mathcal{P}(A)$, the type of all subsets of A, does not exist. The best we can do, as we discussed in the section on power sets, is define

$$\mathcal{P}_i(A) = \{x : \mathbb{U}_i \,|\, x \sqsubseteq A\}.$$

If we state the result as the claim that

$$x : (Y : \mathcal{P}_i(A) \times Y) \to x$$

is inhabited, then it is trivially true since the choice function f takes as input a type Y, that is, a subtype of A and an element $y \in Y$, and it produces the y, e.g.

$$f(< Y, y >) = y$$

so $f(x) = 2 of(x)$.

But this simply shows that we can make the problem trivial. In a sense our statements of the axiom of choice have made it too easy.

Another formulation of the axiom of choice would be this:

$$\text{Set Choice} \quad x : \{y : \mathcal{P}_i(A)|y\} \to x.$$

Recall that the evidence for $\{x : A|B\}$ is simply an object a *in* A; the evidence for B is suppressed. That is, we needed it to show that a is in $\{x : A|B\}$, but then by using the subtype, we agree not to reveal this evidence. Set Choice says that if we have an arbitrary subtype x of A which is inhabited but we do not know the inhabitant, then we can recover the inhabitant uniformly.

This Set Choice is quite unlikely to be true in type theory. We can indeed make a recursive model of type theory in which it is false. This is perhaps the closest we can get in type theory to stating the axiom of choice in its classical sense, and this axiom is totally implausible. Consider this version:

$$P : \{p : Prop_i|p \vee \neg p\} \to P \vee \neg P.$$

We might call this "propositional choice." It is surely quite implausible in computational logic.

12. Computational complexity

As Turing and Church showed, computability is an abstract mathematical concept that does not depend on physical machines (although its practical value and its large intellectual impact do depend very much on physical machines, and on the sustained steep exponential growth in their power). Computability can be axiomatized as done in these notes by underpinning mathematical objects with data having explicit structure, and by providing reduction rules for destructor operations on data.

Hartmanis and Stearns showed that the cost of computation can also be treated mathematically, even though it might at first seem that this would depend essentially on characteristics of physical machines, such as how much *time* an operation required, or how much circuitry or how much electrical power or how much bandwidth, etc. It turns out that our abstract internal characterizations of resource expenditure during computation is correlated in a meaningful way with actual physical costs. We call the various internal measures *computational complexity measures*.

Although we can define computational complexity mathematically, it is totally unlike all the ideas we have seen in the first eleven sections. Moreover, there is no comparably general account of computational complexity in set theory, since not all operations of set theory have computational meaning.

Let's start our technical story by looking at a simple example. Consider the problem of computing the *integer square root* of a natural number, say $root(0) = 0$, $root(2) = 1$, $root(4) = 2$, $root(35) = 5$, $root(36) = 6$, etc. We can state the problem as a theorem to be proved in our computational logic:

Root Theorem $\forall n : \mathbb{N}. \ \exists r : \mathbb{N}. \ r^2 \leq n < (r+1)^2.$

We prove this theorem by induction on n. If n is zero, then taking r to be 0 satisfies the theorem.

Suppose now that we have the root for n; this is our induction hypothesis, namely

$$* \ \exists r : \mathbb{N}. \ r^2 \leq n < (r+1)^2.$$

Let r_0 be this root. Our goal is the find the root for $n+1$. As in the case of $root(34) = 5$ and $root(35) = 5$ and $root(36) = 6$, the decision about whether the root of $n+1$ is r_0 or $r_0 + 1$ depends precisely on whether $(r_0 + 1)^2 \leq n + 1$. So we consider these two cases:

1. Case $n + 1 < (r_0 + 1)^2$; then since $r_0{}^2 \leq n$, we have:
$$r_0{}^2 \leq n + 1 < (r_0 + 1)^2.$$

Hence, r_0 is the root of $n + 1$.

2. Case $(r_0 + 1)^2 \leq n + 1$. Notice that since $n < (r_0 + 1)^2$, it follows that:
$$n + 1 < (r_0 + 1)^2 + 1 \text{ and } ((r_0 + 1) + 1)^2 > (r_0 + 1)^2 + 1.$$

Hence $r_0 + 1$ is the root of $n + 1$. This ends the proof.

By the axiom of choice, there is a function $root \in \mathbb{N} \to \mathbb{N}$ such that
$$\forall n{:}\mathbb{N}.\ root(n)^2 \leq n < (root(n) + 1)^2.$$

Indeed, we know the code for this function because it is derived from the computational meaning of the proof. It is this recursive function:

$$
\begin{aligned}
root(0) &= 0. \\
root(n + 1) &= \textit{let } r_0 = root(n) \\
&\quad \textit{in if } n + 1 < (r_0 + 1)^2 \textit{ then } r_0 \\
&\quad \textit{else } r_0 + 1 \\
&\quad \textit{end}.
\end{aligned}
$$

It's easy to determine the number of computation steps needed to find the root of n. Basically it requires $4 \cdot root(n)$. This is a rather inefficient computation. It is basically the same as the cost of the program:

$$
\begin{aligned}
&r := 0 \\
&\textit{while } r^2 \leq n \textit{ do} \\
&\quad r := r + 1 \\
&\textit{end}.
\end{aligned}
$$

There are worse computations that look similar, such as

$$
\begin{aligned}
slow\_root(n) = &\textit{ if } n < 0 \textit{ then } 0 \\
&\textit{else if } n < (slow\_root(n - 1) + 1)^2 \\
&\textit{then } slow\_root(n - 1) \\
&\textit{else } slow\_root(n - 1) + 1.
\end{aligned}
$$

This computation takes $4 \cdot 2^{root(n)}$.

We might call $slow\_root$ an "exponential algorithm," but usually we measure computational complexity in terms of the length of the input, which is essentially $\log(n)$. So even $4 \cdot root(n)$ is an exponential algorithm. It is possible to compute $root(n)$ in time proportional to $\log(n)$ if we take large steps as we search for the root. Instead of computing $root(n - 1)$, we look at $root(n \div 4)$. This algorithm will call $root$ at most $\log(n)$ times. Here is the algorithm:

$$
\begin{aligned}
sqrt(x) = &\textit{ if } x = 0 \textit{ then } 0 \\
&\textit{else let } r = sqrt(x \div 4) \\
&\textit{in if } x < (2 * r + 1)^2 \textit{ then } 2 * r \\
&\textit{else } 2 * r + 1 \\
&\textit{end}.
\end{aligned}
$$

This algorithm comes from the following proof. It uses the following *efficient-induction* principle:

If $P(0)$ and if for y *in* N, $P(y \div 4)$ implies $P(y)$, then $\forall x : \text{N. } P(x)$.

Fast Root Theorem $\forall x : \text{N. } \exists r : \text{N. } r^2 \leq x < (r+1)^2$.

Proceed by efficient induction. When $x = 0$, take $r = 0$. Otherwise, assume $\exists r : \text{N.} r^2 \leq x \div 4 < (r+1)^2$. Now let r_0 be the root assumed to exist and compare $(2 * r_0 + 1)^2$ with x.

1. Case $x < (w * r_0 + 1)^2$; then since $r_0{}^2 \leq x \div 4$, we know that
$$(2 * r_0)^2 \leq x.$$

   So $2 \cdot r$ is the root of x.

2. Case $(2 * r_0 + 1)^2$. Then we know that:
$$(2 * r_0 + 2)^2 = (2 \cdot (r_1))^2 = 4 \cdot (r+1)^2 > x,$$
   since
$$x \div 4 < (r_0 + 1)^2 \text{ and } 4 \cdot (x \div 4) < 4 \cdot (r_0 + 1)^2.$$

So $2 \cdot r_0 + 1$ is the root of x. This ends the proof.

   If we use binary (or decimal) notation for natural numbers and implement the basic operations of addition, subtraction, multiplication and integer division efficiently, then we know that this algorithm operates in number of steps $0(\log(x))$. This is a reasonably efficient algorithm.

   The question we want to explore is how to express this basic fact about runtime of *sqrt* inside the logic. Our observation that the runtime is $0(\log(x))$ is made in the *metalogic*, where we have access to the computation rules and the syntax of the algorithm. Inside the logic we do not have access to these aspects, and we cannot easily extend our rules to include them because these rules conflict with other more basic decisions. Let's look at this situation more carefully.

   In the logic, the three algorithms *root*, *slow_root*, and *sqrt* are functions from N to N, and as functions they are equal, because *slow_root*$(x) =$ *sqrt*(x) for all x *in* N. Thus *slow_root* = *sqrt in* N $\to$ N. This fact about equality means that we cannot have a function *Time* such as

$$Time(slow\_root)(x) = 4 * 2 \, root(x)$$

and

$$Time(sqrt)(x) = 0(\log(x)),$$

because a function such as *Time* must respect equality; so if $f = g$ then $Time(f) = Time(g)$.

At the metalevel we are able to look at *slow_root* and *sqrt* as terms rather than as functions. This is the nature of the metalogic; it has access to syntax and rules. So our exploration leads us to ask whether we can somehow express facts about the metalogic inside the object logic. Gödel showed one way to do this, by "encoding" terms and rules as numbers by the mechanism of *Gödel numbering*.

We propose to use a mechanism more natural than Gödel numbering; we will add the appropriate metalogical types and rules into the logic itself. We are interested in these components of the metalogic:

Metalogic	Object Logic
term	Term
x *evalsto* y in m steps	x *EvalsTo* y in m Steps
$eval(x)$	$Eval(x)$

The idea is that we create a new type, called $Term$. We show in the metalogic that $Term$ represents $term_2$ in that for each $t \in term$, there is an element $rep(t)\,in\,Term$. If $t = t'\,in\,term$, then $rep(t) = rep(t')\,in\,Term$.

If t evaluates to $t'\,in\,term$, then $rep(t)\,EvalsTo\,rep(t')\,in\,Term$, and if $eval(t) = t'$, then $Eval(rep(t)) = rep(t')$.

We will also introduce a function, $ref(t)$, which provides the meaning of elements of $Term$. So for a closed $term\,t\,in\,Term$, $ref(t)$ will provide its meaning as an element of a type A of the theory, if it has such a meaning. For each type A, there will be the collection of $Terms$ that represent elements of A, denoted $[A] = \{x : Term | \exists y : A.ref(x) = y\,in\,A\}$. The relation $ref(x)\,in\,A$ is a proposition $Term \times A \to Prop$.

For each type A we can define a collection of $Terms$ that denote elements of A. Let

$$[A] = \{x : Term | \exists y : A.ref(x) = y\,in\,A\}.$$

The relation $ref(x)\,is\_in\,A$ is defined as

$$\exists y : A.\,ref(x) = y\,in\,A.$$

This is a propositional function on $Term \times A$.

Now given an element of $[A]$, we can measure the number of steps that it takes to reduce it to canonical form. This is done precisely as in the meta theory, using the relation $e\,EvalsTo\,e'$ in n steps. Having $Term$ as a type makes it possible to carry over into the type theory the evaluation relation and the step counting measure that is part of it. We can also define other resource measures as well, such as the amount of *space* used during a computation. For this purpose we could use the size of a term as the basic building block for a space measure.

Once we have complexity measures defined on $Term$, we can define the concept of a complexity class, as follows.

The evaluation relation on $Term$ provides the basis for defining computational complexity measures such as time and space. These measures allow us to express traditional results about complexity classes as well as recent results concerning complexity in higher types. The basic measure of time is the *number of evaluation steps to canonical form*. Here is a definition of the notion that e runs within time t:

$$Time(e,t) \text{ iff } \exists n{:}[0 \cdots t].\, \exists f{:}[0 \cdots n] \to Term.\, f(0) = e \text{ in } Term \, \wedge$$
$$iscanon(f(n)) = true \text{ in } \mathbb{B} \, \wedge$$
$$\forall i{:}[0 \cdots n-1].$$
$$f(i) \ EvalsTo \ f(i+1).$$

We may define a notion of space in a similar manner. First, we may easilynoindent define a function $size$ with type $Term \to \mathbb{N}$ which computes the number of operators in a term. Then we define the predicate $Space(e,s)$, that states that e runs in space at most s:

$$Space(e,s) \text{ iff } \exists n{:}\mathbb{N}.\, \exists f{:}[0 \cdots n] \to$$
$$Term.\, f(0) = e \text{ in } Term \text{ and}$$
$$iscanon(f(n)) = true \text{ in } \mathbb{B} \text{ and}$$
$$\forall i{:}[0 \cdots n-1].\, f(i) \ EvalsTo \ f(i+1) \text{ and}$$
$$\forall i{:}[0 \cdots n].\, size(f(i)) \leq s.$$

Using these, we may define the resource-indexed type $[T]_s^t$ of terms that evaluate (to a member of T) within time t and space s:

$$[T]_s^t \stackrel{\text{def}}{=} \{e : [T] \mid Time(e,t) \text{ and } Space(e,s)\}$$

One interesting application of the resource-indexed types is to define types like Parikh's *feasible numbers*, numbers that may be computed in a "reasonable" time. Benzinger shows another application.

With time complexity measures defined above, we may define complexity classes of functions. Complexity classes are expressed as function types whose members are required to fit within complexity constraints. We call such types *complexity-constrained* function types. For example, the quadratic time, polynomial time, and polynomial space computable functions may be defined as follows:

$$Quad(x{:}A \longrightarrow B) =$$
$$\{f : [x{:}A \to B] \mid \exists c{:}\mathbb{N}.\, \forall a{:}[A]^0.\, Time(\langle\!\langle app^*, f, a \rangle\!\rangle, c \cdot size(a)^2)\}$$

$$Poly(x{:}A \longrightarrow B) =$$

$$\{f : [x{:}A \to B] \mid \exists c, c'{:}\mathbb{N}. \, \forall a{:}[A]^0. \, Time(\langle\!\langle app^*, f, a \rangle\!\rangle, c \cdot size(a)^{c'})\}$$

$$PSpace(x{:}A \longrightarrow B) =$$

$$\{f : [x{:}A \to B] \mid \exists c, c'{:}\mathbb{N}. \, \forall a{:}[A]^0. \, Space(\langle\!\langle app^*, f, a \rangle\!\rangle, c \cdot size(a)^{c'})\}$$

One of the advantages of constructive logic is that when the existence of an object is proven, that object may be constructed, as we saw in our discussion of the axiom of choice, where a computable function is constructed from a proof. However, there is no guarantee that such functions may feasibly be executed. This has been a serious problem in practice, as well as in principle.

Using the complexity-constrained functions, we may define a *resource-bounded logic* that solves this problem. As we noted in Section 9 under the propositions-as-types principle, the universal statement $\forall x : A. \, B$ corresponds to the function space $x : A \to B$. By using the complexity-constrained function space instead, we obtain a *resource-bounded* universal quantifier. For example, let us denote the quantifier corresponding to the polynomial-time computable functions by $\forall_{poly} x{:}A. \, B$. By proving the statement $\forall_{poly} x{:}A. \, \exists y{:}B. \, P(x, y)$, we guarantee that the appropriate y may actually be feasibly computed from a given x.

The following is a proposition expressing the requirement for a *feasible integer square root*:

$$\forall_{poly} x : \mathbb{N}. \, \exists r : \mathbb{N}. \, \{r^2 \leq x < (r + 1^2)\}.$$

Bibliographic notes

SECTION 1 — TYPES AND EQUALITY

The Halmos book [57] does not cite the literature since his account is of the most basic concepts. I will not give extensive references either, but I will cite sources that provide addtional references.

One of the best books about basic set theory, in my opinion, is still *Foundations of Set Theory*, by Fraenkel, Bar-Hillel and Levy [50].

There are a few text books on basic type theory. The 1986 book by the PRL group [37] is still relevant, and it is now freely available at the Nuprl Web site (www.cs.cornell.edu/Info/Projects/NuPrl/). Two other basic texts are *Type Theory and Functional Programming*, by Thompson [103], and *Programming in Martin-Löf's Type Theory* [86], by Nordstrom, Petersson and Smith. A recent book by Ranta, *Type-theoretical Grammar* [93],

252

has a good general account of type theory. Martin-Löf type theory is presented in his *Intuitionistic Type Theory* [80] and *Constructive Mathematics and Computer Programming* [79].

SECTION 2 — SUBTYPES AND SET TYPES

The notion of subtype is not very thoroughly presented in the literature. There is the article by Constable and Hickey [39] which cites the basic literature. Another key paper is by Pierce and Turner [91]. The PhD theses from Cornell and Edinburgh deal with this subject, especially Crary [46], Hickey [63], and Hofmann [65].

SECTION 3 — PAIRS

Cartesian products are standard, even in the earliest type theories, such as Curry [47] and deBruijn [49].

SECTION 4 — UNION AND INTERSECTION

The intersection type is deeply studied in the lambda calculus. See the papers of Coppo and Dezani-Ciancaglini [42], Compagnoni [32], and Pierce [92].

The logical role of intersection is discussed by Caldwell [25, 26] and earlier by Allen [4]. The newest results are from Kopylov [71] and Girard [53].

SECTION 5 — FUNCTIONS AND RELATIONS

This section is the heart of the untyped and typed lambda calculus. See Barendregt [7, 8], Stenlund [102], and Church [29]. The treatment of relations goes back to Frege and Russell and is covered well in Church [30].

SECTION 6 — UNIVERSES, POWERS AND OPENNESS

The account of universes is from Per Martin-Löf [79] and is informed by Allen [4] and Palmgren [88]. Insights about power sets can be found in Fraenkel et al. [50], Beeson [11], and Troelstra [104].

SECTION 7 — FAMILIES

Families are important in set theory, and accounts such as Bourbaki [18] inform Martin-Löf's approach [79].

SECTION 8 — LISTS AND NUMBERS

List theory is the basis of McCarthy's theory of computing [82]. The Boyer-Moore prover was used to create an extensive formal theory [19], and the Nuprl libraries provide a constructive theory.

SECTION 9 — LOGIC AND THE PEANO AXIOMS

Our approach to logic comes from Brouwer as formalized in Heyting [61]. One of the most influential accounts historically is Howard [66] and also deBruijn [48, 49] for Automath. The Automath papers are collected in [85].

The connection between propositions and types has found its analogue in set theory as well. The set theory of Anthony P. Morse from 1986 equated sets and propositions. He "asserted a set" by the claim that it was nonempty. Morse believed "that every (mathematical) thing is a set." For him, conjunction is intersection, disjunction is union, negation is complementation. Quantification is the extension of these operations to families.

"Each set is either true or false, and each sentence is a name for a set."

SECTION 10 — STRUCTURES, RECORDS AND CLASSES

The approach to records and classes developed here is based entirely on the work of Constable, Hickey and Crary. The basic papers are [39, 46]. The account by Betarte and Tasistro [15] is related. There is an etensive literature cited in the books of Gunter and Mitchell [56]. The treatment of inductive classes is based on Basin and Constable [9].

SECTION 11 — THE AXIOM OF CHOICE

There are many books about the axiom of choice. One of the best is Fraenkel et al.[50] Another is Gregory Moore's *Zermelo's Axiom of Choice : Its Origins, Development, and Influence* [83]. Our account is based on Martin-Löf [79].

SECTION 12 — COMPUTATIONAL COMPLEXITY

The fundamental concepts and methods of computational complexity theory were laid down in the seminal paper of Hartmanis and Stearns, *On the Computational Complexity of Algorithms* [60]. Many textbooks cover this material, for example [75]. The extension of this theory to higher-order objects is an active field [99], and the study of feasible computation is another active area related to this article [12, 69, 72, 73, 74]. These topics are covered also in Schwichtenberg [13], and in the articles of Jones [70], Schwichtenberg [98], and Wainer [87] in this book.

The work reported here is new and based largely on Constable and Crary [38] and Benzinger [14], as well as examples from Kreitz and Pientka [90].

One interesting application of the resource-indexed types is to define types like Parikh's feasible numbers [89], numbers that may be computed in a "reasonable" time. Benzinger [14] shows another application.

Acknowledgements

I would like to thank Juanita Heyerman for preparing the manuscript so carefully and for contributing numerous suggestions to improve its form. I also want to thank the excellent students at the Marktoberdorf summer school for encouraging me to write from this point of view.

References

1. Abadi, M. and L. Cardelli: 1996, *A Theory of Objects*. Springer.
2. Abadi, M. and L. Lamport: 1995, 'Conjoining Specifications'. *ACM Toplas* **17**(3).
3. Aczel, P.: 1999, 'Notes on the Simply Typed Lambda Calculus'. In: U. B. et al. (ed.): *Computational Logic. Proceedings of the NATO ASI, Marktoberdorf, Germany, July 29–August 10, 1997*, Vol. 165 of *NATO ASI Ser., Ser. F, Comput. Syst. Sci.* Berlin, pp. 57–97.
4. Allen, S. F.: 1987, 'A Non-Type-Theoretic Semantics for Type-Theoretic Language'. Ph.D. thesis, Cornell University.
5. Backhouse, R. C.: 1989, 'Constructive Type Theory–an introduction'. In: M. Broy (ed.): *Constructive Methods in Computer Science*, NATO ASI Series, Vol. F55: *Computer & System Sciences*. Springer-Verlag, New York, pp. 6–92.
6. Backhouse, R. C., P. Chisholm, G. Malcolm, and E. Saaman: 1989, 'Do-it-yourself Type theory (part I)'. *Formal Aspects of Computing* **1**, 19–84.
7. Barendregt, H. P.: 1977, 'The Typed Lambda Calculus'. In: J. Barwise (ed.): *Handbook of Mathematical Logic*. North-Holland, NY, pp. 1091–1132.
8. Barendregt, H. P.: 1992, *Handbook of Logic in Computer Science*, Vol. 2, Chapt. Lambda Calculi with Types, pp. 118–310. Oxford University Press.
9. Basin, D. A. and R. L. Constable: 1993, 'Metalogical Frameworks'. In: G. Huet and G. Plotkin (eds.): *Logical Environments*. Great Britain: Cambridge University Press, Chapt. 1, pp. 1–29.
10. Bates, J. L. and R. L. Constable: 1985, 'Proofs as Programs'. *ACM Transactions on Programming Languages and Systems* **7**(1), 53–71.
11. Beeson, M. J.: 1985, *Foundations of Constructive Mathematics*. Springer-Verlag.
12. Bellantoni, S. and S. Cook: 1992, 'A New Recursion-Theoretic Characterization of the Poly-time Functions'. *Computational Complexity* **2**, 97–110.
13. Bellantoni, S. J., K.-H. Niggl, and H. Schwichtenberg: 2000, 'Higher type recursion, ramification and polynomial time'. *Annals of Pure Applied Logic* **104**(1-3), 17–30.
14. Benzinger, R.: 2000, 'Automated Complexity Analysis of Nuprl Extracted Programs'. *Journal of Functional Programming*. To Appear.
15. Betarte, G. and A. Tasistro: 1999, 'Extension of Martin Löf's Type Theory with Record Types and Subtyping. Chapter 2, pages 21–39, In Twenty-Five Years of Constructive Type Theory'. Oxford Science Publications.
16. Bickford, M. and J. Hickey: 1998, 'An Object-Oriented Approach to Verifying Group Communication Systems'. Department of Computer Science, Cornell University, Unpublished.

17. Bourbaki, N.: 1968a, *Elements of Mathematics, Algebra, Volume 1*. Reading, MA: Addison-Wesley.
18. Bourbaki, N.: 1968b, *Elements of Mathematics, Theory of Sets*. Addison-Wesley, Reading, MA.
19. Boyer, R. S. and J. S. Moore: 1979, *A Computational Logic*. Academic Press, New York.
20. Boyer, R. S. and J. S. Moore: 1981, 'Metafunctions: Proving Them Correct and Using Them Efficiently as New Proof Procedures'. In: *The Correctness Problem in Computer Science*. pp. 103–84.
21. Brouwer, L. E. J.: 1975, *Collected Works* A. Heyting, ed., Vol. 1. North-Holland, Amsterdam.
22. Bruce, K. and J. Mitchell: 1992, 'PER models of subtyping, recursive types and higher–order polymorphism'. *Proceedings of the Nineteenth ACM Symposium on Principles of Programming Lanuages* pp. 316–327.
23. Bruce, K. B. and G. Longo: 1994, 'A Modest Model of Records, Inheritance, and Bounded Quantification'. In: J. C. M. C. A. Gunter (ed.): *Theoretical Aspects of Object-Oriented Programming, Types, Semantics and Language Design*. Cambridge, MA: MIT Press, Chapt. III, pp. 151–196.
24. Buss, S.: 1986, 'The polynomial hierarchy and intuitionistic bounded arithmetic'. In: *Structure in Complexity Theory*, Lecture Notes in Computer Science. 223. pp. 77–103.
25. Caldwell, J.: 1997, 'Moving Proofs-as-Programs into Practice'. In: *Proceedings of the 12th IEEE International Conference on Automated Software Engineering*.
26. Caldwell, J., I. Gent, and J. Underwood: 2000, 'Search Algorithms in Type Theory'. *Theoretical Computer Science* **232**(1–2), 55–90.
27. Cardelli, L.: 1994, 'Extensible Records in a Pure Calculus of Subtyping'. In: C. A. Gunter and J. C. Mitchell (eds.): *Theoretical Aspects of Object-Oriented Programming: Types, Semantics and Language Design*. MIT Press. A preliminary version appeared as SRC Research Report No. 81, 1992.
28. Cardelli, L. and J. Mitchell: 1994, 'Operations on records'. In: J. C. M. C. A. Gunter (ed.): *Theoretical Aspects of Object-Oriented Programming, Types, Semantics and Language Design*. Cambridge, MA: MIT Press, Chapt. IV, pp. 295–350.
29. Church, A.: 1951, *The Calculi of Lambda-Conversion*, Vol. 6 of *Annals of Mathematical Studies*. Princeton: Princeton University Press.
30. Church, A.: 1956, *Introduction to Mathematical Logic, Vol. I*. Princeton University Press.
31. Cobham, A.: 1965, 'The Intrinsic Computational Difficulty of Functions'. In: Y. Bar-Hillel (ed.): *Proceedings of the 1964 International Conference for Logic, Methodology, and Philosophy of Science*. pp. 24–30.
32. Compagnoni, A. B.: 1995, 'Higher-Order Subtyping with Intersection Types'. Ph.D. thesis, Katholieke Universiteit Nijmegen.
33. Compagnoni, A. B. and B. C. Pierce: 1995, 'Higher-Order Intersection Types and Multiple Inheritance'. *Math. Struct. in Comp. Science* **11**, 1–000.
34. Constable, R. L.: 1994, 'Using Reflection to Explain and Enhance Type Theory'. In: H. Schwichtenberg (ed.): *Proof and Computation*, Vol. 139 of *NATO Advanced Study Institute, International Summer School held in Marktoberdorf, Germany, July 20-August 1, NATO Series F*. Berlin: Springer, pp. 65–100.
35. Constable, R. L.: 1995, 'Experience Using Type Theory as a Foundation for Computer Science'. In: *Proceedings of the Tenth Annual IEEE Symposium on Logic in Computer Science*. pp. 266–279.
36. Constable, R. L.: 1998, 'Types in Logic, Mathematics and Programming'. In: S. R. Buss (ed.): *Handbook of Proof Theory*. Elsevier Science B.V., Chapt. X, pp. 684–786.
37. Constable, R. L., S. F. Allen, H. M. Bromley, W. R. Cleaveland, J. F. Cremer, R. W. Harper, D. J. Howe, T. B. Knoblock, N. P. Mendler, P. Panangaden, J. T. Sasaki, and S. F. Smith: 1986, *Implementing Mathematics with the Nuprl Development*

256

System. NJ: Prentice-Hall.

38. Constable, R. L. and K. Crary: 2001, 'Computational Complexity and Induction for Partial Computable Functions in Type Theory'. In: W. Sieg, R. Sommer, and C. Talcott (eds.): *Reflections on the Foundations of Mathematics: Essays in Honor of Solomon Feferman*, Lecture Notes in Logic. Association for Symbolic Logic, pp. 166–183. 2001.

39. Constable, R. L. and J. Hickey: 2000, 'Nuprl's Class Theory and its Applications'. In: F. L. Bauer and R. Steinbrueggen (eds.): *Foundations of Secure Computation*. pp. 91–116.

40. Constable, R. L. and S. F. Smith: 1993, 'Computational Foundations of Basic Recursive Function Theory'. *Theoretical Computer Science* **121**, 89–112.

41. Constable, R. L. a. P. B. J., P. Naumov, and J. Uribe: 2000, 'Constructively Formalizing Automata'. In: *Proof, Language and Interaction: Essays in Honour of Robin Milner*. MIT Press, Cambridge, pp. 213–238.

42. Coppo, M. and M. Dezani-Ciancaglini: 1980, 'An Extension of the Basic Functionality Theory for the λ–Calculus'. *Notre-Dame Journal of Formal Logic* **21**(4), 685–693.

43. Coquand, T. and G. Huet: 1988, 'The Calculus of Constructions'. *Information and Computation* **76**, 95–120.

44. Crary, K.: 1997, 'Foundations for the Implementation of Higher-Order Subtyping'. In: *1997 ACM SIGPLAN International Conference on Functional Programming*. Amsterdam. to appear.

45. Crary, K.: 1998a, 'Simple, efficient object encoding using intersection types.'. Technical Report TR98-1675, Department of Computer Science, Cornell University.

46. Crary, K.: 1998b, 'Type-Theoretic Methodology for Practical Programming Languages'. Ph.D. thesis, Cornell University, Ithaca, NY.

47. Curry, H. B., R. Feys, and W. Craig: 1958, *Combinatory Logic*, Studies in Logic and the Foundations of Mathematics. Amsterdam: North-Holland.

48. deBruijn, N. G.: 1975, 'Set Theory with Type Restrictions'. In: A. Jahnal, R. Rado, and V. T. Sos (eds.): *Infinite and Finite Sets*, Vol. I of *Collections of the Mathematics Society*. J. Bolyai 10, pp. 205–314.

49. deBruijn, N. G.: 1980, 'A Survey of the Project Automath'. In: J. P. Seldin and J. R. Hindley (eds.): *To H. B. Curry: Essays in Combinatory Logic, Lambda Calculus, and Formalism*. pp. 589–606.

50. Fraenkel, A. A., Y. Bar-Hillel, and A. Levy: 1984, *Foundations of Set Theory*, Vol. 67 of *Studies in Logic and the Foundations of Mathematics*. Amsterdam: North-Holland, 2nd edition.

51. Girard, J.-Y.: 1999, 'On the Meaning of Logical Rules. I: Syntax Versus Semantics'. In: U. B. et al. (ed.): *Computational Logic. Proceedings of the NATO ASI, Marktoberdorf, Germany, July 29–August 10, 1997*, Vol. 165 of *NATO ASI Ser., Ser. F, Comput. Syst. Sci.* Berlin, pp. 215–272.

52. Girard, J.-Y.: 2000, 'On the Meaning of Logical Rules II: Multiplicatives and Additives'. In: F. L. Bauer and R. Steinbrueggen (eds.): *Foundations of Secure Computing*. Berlin: IOS Press, pp. 183–212.

53. Girard, J.-Y.: 2002, 'Ludics'. In: *this volume*. Kluwer.

54. Girard, J.-Y., P. Taylor, and Y. Lafont: 1989, *Proofs and Types*, Cambridge Tracts in Computer Science, Vol. 7. Cambridge University Press.

55. Giunchiglia, F. and A. Smaill: 1989, 'Reflection in constructive and non-constructive automated reasoning'. In: H. Abramson and M. H. Rogers (eds.): *Meta-Programming in Logic Programming*. MIT Press, Cambridge, Mass., pp. 123–140.

56. Gunter, C. A. and J. C. Mitchell (eds.): 1994, *Theoretical Aspects of Object-Oriented Programming, Types, Semantics and Language Design*, Types, Semantics, and Language Design. Cambridge, MA: MIT Press.

57. Halmos, P. R.: 1974, *Naive Set Theory*. New York: Springer-Verlag.

58. Harper, R. and M. Lillibridge: 1994, 'A type-theoretic approach to higher-order

modules with sharing.'. *Twenty-First ACM SIGACT-SIGPLAN Symposium on Principles of Programming Languages* pp. 123–137.

59. Hartmanis, J.: 1978, *Feasible Computations and Provable Complexity Properties.* Philadelphia, PA: SIAM.

60. Hartmanis, J. and R. Stearns: 1965, 'On the Computational Complexity of Algorithms'. *Transactions of the American Mathematics Society* **117**, 285–306.

61. Heyting, A.: 1971, *Intuitionism.* Amsterdam: North-Holland.

62. Hickey, J.: 1996, 'Formal Objects in Type Theory Using Very Dependent Types'. *Foundations of Object-Oriented Languages* **3**.

63. Hickey, J.: 2000, 'The Metaprl Logical Programming Environment'. Ph.D. thesis, Cornell University, Ithaca, NY.

64. Hoare, C. A. R.: 1972, 'Notes on data structuring'. In: *Structured Programming.* Academic Press, New York.

65. Hofmann, M.: 1995, 'Extensional Concepts in Intensional Type Theory'. Ph.D. thesis, University of Edinburgh.

66. Howard, W.: 1980, 'The Formulas-as-Types Notion of Construction'. In: *To H.B. Curry: Essays on Combinatory Logic, Lambda-Calculus and Formalism.* pp. 479–490.

67. Howe, D. J.: 1996a, 'Importing Mathematics from HOL into Nuprl'. In: J. von Wright, J. Grundy, and J. Harrison (eds.): *Theorem Proving in Higher Order Logics,* Vol. 1125, of Lecture Notes in Computer Science. Berlin: Springer-Verlag, pp. 267–282.

68. Howe, D. J.: 1996b, 'Semantic Foundations for Embedding HOL in Nuprl'. In: M. Wirsing and M. Nivat (eds.): *Algebraic Methodology and Software Technology,* Vol. 1101 of Lecture Notes in Computer Science. Berlin: Springer-Verlag, pp. 85–101.

69. Immerman, N.: 1987, 'Languages Which Capture Complexity Classes'. *SIAM Journal of Computing* **16**, 760–778.

70. Jones, N. D.: 2002, 'Computability and Complexity from a Programming Perspective'. In: *this volume.* Kluwer.

71. Kopylov, A.: 2000, 'Dependent Intersection: A New Way of Defining Records in Type Theory'. Technical Report TR2000–1809, Cornell University, Ithaca, New York.

72. Leivant, D.: 1991, 'Finitely Stratified Polymorphism'. *Information and Computation* **93**(1), 93–113.

73. Leivant, D.: 1994, 'Predicative recurrence in finite Type'. In: A. Nerode and Y. V. Matiyasevich (eds.): *Logical Foundations of Computer Science, Lecture Notes in Computer Science 813.* Berlin: Springer-Verlag, pp. 227–239.

74. Leivant, D.: 1995, 'Ramified Recurrence and Computational Complexity I: Word Recurrence and Polynomial Time'. In: P. Clote and J. Remmel (eds.): *Feasible Mathematics II, Perspectives in Computer Science.* Birkhäuser.

75. Lewis, H. R. and C. H. Papadimitriou: 1994, *Elements of the Theory of Computation.* Englewood Cliffs, New Jersey: Prentice-Hall.

76. Liu, X., C. Kreitz, R. van Renesse, J. H. ckey, M. Hayden, K. Birman, and R. Constable: 1999, 'Building Reliable, High-Performance Communication Systems from Compone nts'. *Operating Systems Review* **34**(5), 80–92. Presented at the 17th ACM Symposium on Operating Systems Principles (SOSP'99).

77. MacLane, S. and G. Birkhoff: 1967, *Algebra.* MacMillan.

78. Martin-Löf, P.: 1973, 'An Intuitionistic Theory of Types: Predicative Part'. In: *Logic Colloquium '73.* pp. 73–118.

79. Martin-Löf, P.: 1982, 'Constructive Mathematics and Computer Programming'. In: *Proceedings of the Sixth International Congress for Logic, Methodology, and Philosophy of Science.* Amsterdam, pp. 153–175.

80. Martin-Löf, P.: 1984, *Intuitionistic Type Theory. Notes by Giovanni Sambin of a Series of Lectures given in Padua, June 1980,* No. 1 in Studies in Proof Theory, Lecture Notes. Napoli: Bibliopolis.

81. Martin-Löf, P.: 1998, 'An Intuitionistic Theory of Types'. In: G. Sambin and J. Smith (eds.): *Twenty-Five Years of Constructive Type Theory*, Vol. 36 of *Oxford Logic Guides*. Oxford: Clarendon Press, pp. 127–172.

82. McCarthy, J.: 1963, 'A Basis for a Mathematical Theory of Computation'. In: P. Braffort and D. Hirschberg (eds.): *Computer Programming and Formal Systems*. Amsterdam: North-Holland, pp. 33–70.

83. Moore, G. H.: 1982, *Zermelo's Axiom of Choice : Its Origins, Development, and Influence*, Oxford Science Publications. New York: Springer-Verlag.

84. Naumov, P.: 1998, 'Formalizing Reference Types in NuPRL'. Ph.D. thesis, Cornell University.

85. Nederpelt, R. P., J. H. Geuvers, and R. C. D. Vrijer: 1994, *Selected Papers in Automath*, Vol. 133 of *Studies in Logic and The Foundations of Mathematics*. Amsterdam: Elsevier.

86. Nordstrom, B., K. Petersson, and J. Smith: 1990, *Programming in Martin-Löf's Type Theory*. Oxford: Oxford Sciences Publication.

87. Ostrin, G. E. and S. S. Wainer: 2002, 'Proof Theory and Complexity'. In: *this volume*. Kluwer.

88. Palmgren, E.: 1998, 'On Universes in Type Theory'. In: G. Sambin and J. Smith (eds.): *Twenty-Five Years of Constructive Type Theory*. Oxford: Clarendon Press, pp. 191–204.

89. Parikh, R.: 1971, 'Existence and Feasibility in Arithmetic'. *Jour. Assoc. Symbolic Logic* **36**, 494–508.

90. Pientka, B. and C. Kreitz: 1998, 'Instantiation of Existentially Quantified Variables in Inductive Specification Proofs'. In: *Fourth International Conference on Artificial Intelligence and Symbolic Computation (AISC'98)*. Kaiserslauten, Germany, pp. 247–258.

91. Pierce, B. and D. Turner: 1994, 'Simple Type-theoretic Foundations for Object-oriented Programming'. *Journal of Functional Programming* **4**(2).

92. Pierce, B. C.: 1991, 'Programming with Intersection Types, Union Types, and Polymorphism'. Technical Report CMU-CS-91-106, Carnegie Mellon University.

93. Ranta, A.: 1994, *Type-theoretical Grammar*, Oxford Science Publications. Oxford, England: Clarendon Press.

94. Reynolds, J. C.: 1974, 'Towards a theory of type structure'. In: *Proceedings Colloque sur, la Programmation*, Lecture Notes in Computer Science, Vol. 19. pp. 408–23.

95. Reynolds, J. C.: 1984, 'Polymorphism is not set-theoretic'. In: *Lecture Notes in Computer Science*. Berlin, pp. 145–156. vol. 173.

96. Russell, B.: 1908, 'Mathematical Logic as Based on a Theory of Types'. *Am. J. Math.* **30**, 222–62.

97. Sazonov, V. Y.: 1989, 'On Feasible Numbers'. In: *Proceedings of the ASL Meeting*. West Berlin.

98. Schwichtenberg, H.: 2002, 'Feasible Computation with Higher Types'. In: *this volume*. Kluwer.

99. Seth, A.: 1994, 'Turing Machine Characterizations of Feasible Functionals of All Finite Types'. In: P. Clote and J. Remmel (eds.): *Proceedings of MSI Workshop on Feasible Mathematics*.

100. Smith, D. R.: 1993, 'Constructing Specification Morphisms'. *Journal of Symbolic Computation, Special Issue on Automatic Programming* **16**(5-6), 571–606.

101. Smith, D. R.: 1996, 'Toward a Classification Approach to Design'. *Proceedings of the Fiftieth International Conference on Algebraic Methodology and Software Technology, AMAST'96, LNCS* pp. 62–84. Springer Verlag.

102. Stenlund, S.: 1972, *Combinators, λ-Terms, and Proof Theory*. D. Reidel, Dordrechte.

103. Thompson, S.: 1991, *Type Theory and Functional Programming*. Addison-Wesley.

104. Troelstra, A.: 1998, 'Realizability'. In: S. Buss (ed.): *Handbook of Proof Theory*, Vol. 137 of *Studies in Logic and the Foundations of Mathematics*. Elsevier, pp.

407–473.

105. Troelstra, A. S. and H. Schwichtenberg: 1996, *Basic Proof Theory*. Cambridge University Press.

PROOF-CARRYING CODE. DESIGN AND IMPLEMENTATION

GEORGE C. NECULA

Department of Electrical Engineering and Computer Sciences
University of California, Berkeley, CA 94720, USA
(necula@cs.berkeley.edu)

Abstract. Proof-Carrying Code (PCC) is a general mechanism for verifying that a code fragment can be executed safely on a host system. The key technical detail that makes PCC simple yet very powerful is that the code fragment is required to be accompanied by a detailed and precise explanation of why it satisfies the safety policy. This leaves the code receiver with the simple task of verifying that the explanation is correct and that it matches the code in question.

In this paper we explore the basic design and the implementation of a system using Proof-Carrying Code. We consider two possible representations for the proofs carried with the code, one using Logical Frameworks and the other using hints for guiding a non-deterministic proof reconstructor. In the second part of the paper we discuss issues related to generating the required proofs, which is done through cooperation between a compiler and a theorem prover. We will conclude with a presentation of experimental results in the context of verifying that the machine-code output of a Java compiler is type safe.

1. Introduction

More and more software systems are designed and build to be extensible and configurable dynamically. The proportion of extensions can range from a software upgrade, to a third-party add-on or component, to an applet. On another dimension, the trust relationship with the producer of the extension code can range from completely trusted, to believed-not-malicious, to completely unknown and untrusted. In such a diverse environment there is a need for a general mechanism that can be used to allow even untrusted system extensions to be integrated into an existing software system without compromising the stability and security of the host system.

H. Schwichtenberg and R. Steinbrüggen (eds.), Proof and System-Reliability, 261–288.
© 2002 *Kluwer Academic Publishers. Printed in the Netherlands.*

Proof-Carrying Code (PCC) [8, 10] was designed to be a *general* mechanism that allows the receiver of code (referred to as the **host**) to check *quickly* and *easily* that the code (referred to as the **agent**) has certain safety properties. The key technical detail that makes PCC very general yet very simple is a requirement that the agent producer cooperates with the host by attaching to the agent code an "explanation" of why the code complies with the safety policy. Then all that the host has to do to ensure the safe execution of the agent is to define a framework in which the "explanation" must be conducted, along with a simple yet sufficiently strong mechanism for checking that (a) the explanation is acceptable (i.e., is within the established framework), that (b) the explanation pertains to the safety policy that the host wishes to enforce, and (c) that the explanation matches the actual code of the agent.

Below is a list of the most important ways in which PCC improves over other existing techniques for enforcing safe execution of untrusted code:

– PCC operates at **load time** before the agent code is installed in the host system. This is in contrast with techniques that enforce the safety policy by relying on extensive run-time checking [14] or even interpretation [4, 5]. As a result PCC agents run at native-code speed, which can be ten times faster than interpreted agents (written for example using Java bytecode) or 30% faster than agents whose memory operations are checked at run time.

  Additionally, by doing the checking at load time it becomes possible to enforce certain safety policies that are hard or impossible to enforce at run time. For example, by examining the code of the agent and the associated "explanation" PCC can verify that a certain interrupt routine terminates within a given number of instructions executed or that a video frame rendering agent can keep up with a given frame rate. Run-time enforcement of timing properties of such fine granularity is hard.

– The PCC trusted infrastructure is **small**. The PCC checker is simple and small because it has to do a relatively simple task. In particular, PCC does not have to discover on its own whether and why the agent meets the safety policy.

– For the same reason, PCC can operate even on agents expressed in native-code form. And because PCC can verify the code after compilation and optimization, the checked code is ready to run without needing an additional interpreter or compiler on the host. This has serious software engineering advantages since it reduces the amount of security-critical code and it is also a benefit when the host environment is too small to contain an interpreter or a compiler, such as is the case for many embedded software systems.

- PCC is **general.** All PCC has to do is to verify safety explanations and to match them with the code and the safety policy. By standardizing a language for expressing the explanations and a formalism for expressing the safety policies it is possible to implement a single algorithm that can perform the required check, for any agent code, any valid explanation and a large class of safety policies. In this sense a single implementation of PCC can be used for checking a variety of safety policies.

The combination of benefits that PCC offers is unique among the techniques for safe execution of untrusted code. Previously one had to sacrifice one or more of these benefits because it is impossible to achieve them all in a system that examines just the agent code and has to discover on its own why the code is safe.

The PCC infrastructure is designed to complement a cryptographic authentication infrastructure [6]. While cryptographic techniques such as digital signatures can be used by the host to verify external properties of the agent program, such as freshness and authenticity, or the author's identity, the PCC infrastructure checks internal semantic properties of the code such as what the code does and what it does not do. This enables the host to prevent safety breaches due to either malicious intent (for agents originating from untrusted sources) or due to programming errors (for agents originating from trusted sources).

The description of PCC so far has been in abstract terms without referring to a particular form of safety explanations. There are a number of possible forms of explanations each with its own advantages and disadvantages. Safety explanations must be precise and comprehensive, just like formal proofs. In fact, in the first realization of a PCC architecture [9] the explanations were precisely formal proofs represented as terms in a variant of the dependently-typed λ-calculus called the Edinburgh Logical Framework (LF) [3]. The major advantage of this approach is that proof checking can be accomplished using a relatively simple and well understood LF type checker.

The proof-based realization of PCC is very useful for gaining initial experience with the PCC technique in various applications. However, the size of proofs represented this way can grow out of control very quickly. In an attempt to address this issue we first devised a simple extension of the LF type checker that enables it to reconstruct certain small but numerous fragments of proofs while performing type checking [11]. This variant of LF, which we call LF_i (for implicit LF), allows us to omit those

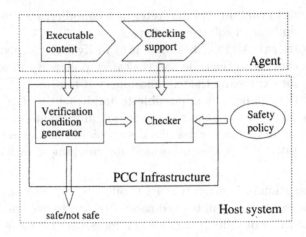

Figure 1.   The high-level structure of the PCC architecture.

fragments from the representation of proofs with the major benefit that proof sizes and the checking times are now growing linearly with the size of the program. (This is in the context of checking that certain assembly language programs are type safe.) This allows us to process examples up to several thousands of line of code with proof sizes averaging 2.5 to 5 times the size of the code. But even LF_i-based proof representations are too large for the hundred-thousand line examples that we want to process.

In this paper we describe a series of improvements to the PCC architecture that together allow us to process even large examples. These improvements range from major changes in the representation and checking of proofs to changes in the way different components of the PCC system communicate. We start in Section 2 with an overview presentation of a PCC system in the context of an example. For this example, we will discuss both the LF-based proof representation (in Section 3) and the oracle-based proof representation (in Section 4). Although most of this document focuses on the code-receiving end, we observed that for better scalability the code producer must also cooperate. We describe in Section 5 one way in which a compiler that produces PCC agents can speed up the checking process. Finally, in Section 6 we discuss experimental evidence that substantiates the claim that this combination of techniques does indeed scale to very large examples.

```
bool forall(bool *data, int dlast) {
  int i;
  for(i=dlast; i >= 0; i --)
    if(! data[i]) return false;
  return true;
}
```

Figure 2. The C code for a function that computes the conjunction of all elements in an array of booleans.

2. Overview and an extended example

A high-level view of the architecture of a PCC system is shown in Figure 1. The untrusted code is first inspected by a verification condition generator (VCGen) that checks simple syntactic conditions (e.g. that direct jumps are within the code boundary). Each time VCGen encounters an instruction whose execution could violate the safety policy, it asks the Checker module to verify that in the current context the instruction actually behaves safely. For this purpose, VCGen encodes the property to be checked as a simple symbolic formula. The implementation of the Checker module is intrinsically linked to the choice of the proof representation. We shall describe two implementations of the Checker, one corresponding to each proof representation strategy.

In the rest of this section we describe the operation of the PCC checking infrastructure in the context of verifying a simple type-safety policy for an agent written in a generic assembly language. To keep the presentation concise we consider here only a trivial agent consisting of a function that computes the boolean value representing the conjunction of all elements in an array of booleans. The C source code for the function is shown in Figure 2. The inputs consist of a pointer to the start of an array of booleans along with the index of the last element. To simplify somewhat the assembly-language generated from this program the array is scanned backwards.

In Figure 3 we show one possible compilation of the forall function into a generic assembly language. The resulting program uses four registers, r_d and r_l to hold respectively the start address and the index of the last element of the array, a register r_i to hold the value of the local variable i and a temporary register r_t. The instruction "$r_t = \text{ge } r_i, 0$" stores in r_t either 1 or 0 depending on whether the contents of register r_i is greater-or-equal to 0 or not.

```
1                              /* r_d=data, r_l=dlast */
2        r_i = r_l
3  L_0:
4        r_t = ge r_i, 0
5        jfalse r_t, L_1
6        r_t = add r_d, r_i
7        r_t = load r_t
8        jfalse r_t, L_2
9        r_i = sub r_i, 1
10       jump L_0
11 L_1:   ret 1
12 L_2:   ret 0
```

Figure 3.   The assembly language code for the function shown in Figure 2.

2.1. THE SAFETY POLICY

Before the code producer and the host can exchange agents they must agree on a safety policy. Setting up the safety policy is mostly the host's responsibility. For our example, the safety policy is a variant of type safety and requires that all memory accesses be contained in a memory range whose starting value and last-element index are being passed by the host in the input argument registers r_d and r_l. Furthermore, only boolean values may be written to that memory range and the agent may assume that the values read from that memory range are themselves boolean values. The safety policy further specifies that the function's return value must be a boolean value. Finally, the safety policy specifies that only the values 0 and 1 are valid boolean values.

This safety policy is stated as a pair of a function precondition and postcondition for the agent code. These formulas are constructed using a number of type constructors defined by the host's safety policy. For our example the host requires the following specification:

$$\text{Pre}_{\text{forall}} = r_d: \text{array}(\text{bool}, r_l)$$
$$\text{Post}_{\text{forall}} = \text{res}: \text{bool}$$

This specification uses the infix binary predicate constructor ":" to denote that the expression on the left of the operator has a certain type, along with two type constructors, **array** and **bool**. Note that the **array** type constructor declares the index of the last element of the array along with the type of the array elements. In the postcondition, the name **res** refers to the result value.

$$\frac{A:\texttt{array}(T,\ L) \quad I \geq 0 \quad L \geq I}{\texttt{saferd}(\texttt{add}(A,\ I))}\ \texttt{rd} \qquad \frac{A:\texttt{array}(T,\ L) \quad I \geq 0 \quad L \geq I}{\texttt{mem}(\texttt{add}(A,\ I)):T}\ \texttt{mem}$$

$$\frac{A:\texttt{array}(T,\ L) \quad I \geq 0 \quad L \geq I \quad E:T}{\texttt{safewr}(\texttt{add}(A,\ I),\ E)}\ \texttt{wr}$$

$$\frac{}{0:\texttt{bool}}\ \texttt{bool0} \qquad \frac{}{1:\texttt{bool}}\ \texttt{bool1}$$

$$\frac{}{E=E}\ \texttt{eqid} \quad \frac{}{E \geq E}\ \texttt{geqid} \quad \frac{E \geq I \quad \texttt{ge}(I,\ 0)}{E \geq \texttt{sub}(I,\ 1)}\ \texttt{dec} \quad \frac{\texttt{ge}(E,E')}{E \geq E'}\ \texttt{geq}$$

Figure 4. The rules of inference that constitute the safety policy for our example.

These predicate and type constructors are declared as part of the trusted safety policy along with inference rules (or typing rules) that specify how the type constructors can be used. We show in Figure 4 the rules of inference that are necessary for our example. In order to understand these rules we preview briefly the operation of the VCGen. We do so here with only enough details to motivate the form of the rules. More details on VCGen follow in Section 2.3.

When VCGen encounters a memory read at an address E_a, it asks the Checker module to verify that the predicate $\texttt{saferd}(E_a)$ holds according to the trusted rules of inference. The first rule of inference from Figure 4 says that one instance in which a memory read is considered safe is when it falls within the boundaries of an array. The rule mem goes even further and says that the value read from an array has the same type as the declared array element type.[1] Similarly, the predicate $\texttt{safewr}(E_a, E)$ is generated by VCGen to request a verification that a memory write of value E at address E_a is safe. The third rule of inference in our safety policy can prove such a predicate if E_a falls inside an array and if E has the array element type. Notice that in all of these rule the function add has been used instead of the usual mathematical function "+". This is because we want to preserve the distinction between the mathematical functions and their approximate implementation in a processor. More precisely, we shall assume that the

---

[1] Since the contents of memory does not change in our example we can introduce a constructor mem such that $\texttt{mem}(E)$ denotes the contents of memory at address E. In general, a different mechanism must be used to keep track of the memory contents in the presence of memory writes. The reader is invited to consult [9] for details.

domain and ranges of the expression constructors consist of the subset of integers that are representable on the underlying machine.

The rules `bool0` and `bool1` specify the representation of booleans. The last row of inference rules in Figure 4 are a sample of the rules that encode the semantics of arithmetic and logic machine instructions. Consider for example the `dec` rule. Here `ge` and `sub` are the predicate constructors that encode the result of the machine instructions with the same name. VCGen uses such predicate constructors to encode the result of the corresponding instructions, leaving their interpretation to the safety policy. The rule `dec` says that the result of performing a machine subtraction of 1 from the value I is less or equal to some other value E if I itself is known to be less-or-equal to E and also if the test $ge(I, 0)$ is successful. This rule deserves an explanation. If we could assume that the machine version of subtraction is identical in behavior to the standard subtraction operation then the rule would be sound even without the second hypothesis. However, the second hypothesis must be added to prevent the case when I is the smallest representable integer value and the `sub` operation underflows. (A weaker hypothesis would also work but this one is sufficient for our example.) The rule `geq` say that the meaning of the machine `ge` operation is the same as that of the mathematical greater-or-equal comparison.

The rules of Figure 4 constitute a representative subset of a realistic safety policy. The safety policy used for the experiments discussed in Section 6 for the type system of the Java language consists of about 140 such inference rules.

2.2. THE ROLE OF PROGRAM ANNOTATIONS

The VCGen module attempts to execute the untrusted program symbolically in order to signal all potentially unsafe operations. To make this execution possible in finite time and without the need for conservative approximations on the part of VCGen, we require that the program be annotated with invariant predicates. At least one such invariant must be specified for each cycle in the program's control-flow graph. To further simplify the work of VCGen each invariant must also declare those registers that are live at the invariant point and are guaranteed to preserve their values between two consecutive times when the execution reaches the invariant point. We call these registers the *invariant registers*.

For our example, at least one invariant must be specified for some point inside the loop that starts at label L_0. We place the following invariant at label L_0:

$$L_0 : \quad \text{INV} = r_l \geq r_i \quad \text{REGS} = \{r_d, r_l\}$$

The invariant annotation says that whenever the execution reaches the label L_0 the contents of register r_i is less-or-equal to the contents of register r_l and also that in between two consecutive hits of the program point the registers r_d and r_l are the only ones among the live registers that are guaranteed to preserve their values.

A valid question at this point is who discovers this annotation and how. There are several possibilities. First, annotations can be inserted by hand by the programmer. This is the only alternative when the agent code is programmed in assembly language or when the programmer wants to hand-optimize the output of a compiler. It is true that this method does not scale well, but it is nevertheless a feature of PCC that the checker does not care whether the code is produced by a trusted compiler and will gladly accept code that was written or optimized by hand.

Another possibility is that the annotations can be produced automatically by a certifying compiler. Such a compiler first inserts bounds checks for all array accesses. In the presence of these checks the invariant predicates must include only type declarations for those live registers that are not declared invariant. An exception are those registers that contain integers, for which no declaration is necessary or useful since the integer data type contains all representable values. In our example, the reader can verify that in the presence of bounds checks preceding the memory read the invariant predicate **true** is sufficient. In general, the invariant predicates that are required in the absence of optimizations are very easy to generate by a compiler.

An optimizing compiler might analyze the program and discover that the lower bound check is entailed by the loop termination test and the upper bound check is entailed by a loop invariant "$r_l \geq r_i$". With this information the optimizing compiler can eliminate the bounds checks but it must communicate through the invariant predicate what are the loop invariants that it discovered and used in optimization. This is the process by which the code and annotations of Figure 3 could have been produced automatically. The reader can consult [9] for the detailed description of a certifying compiler for a safe subset of C and [1] for the description of a certifying compiler for Java.

Finally, note that the invariant annotations are required but cannot be trusted to be correct as they originate from the same possibly untrusted source as the code itself. Nevertheless, VCGen can still use them safely, as described in the next section.

2.3. THE VERIFICATION CONDITION GENERATOR

The verification condition generator (VCGen) is implemented as a symbolic evaluator for the program being checked. It scans the program in a forward direction and at each program point it maintains a symbolic value for each register in the program. These symbolic values are then used at certain program points (e.g. memory operations, function calls and returns) to formulate checking goals for the Checker module. To assist the checker in verifying these goals VCGen also records for each goal that is submitted to the Checker a number of assumptions that the Checker is allowed to make. These assumptions are generated from the control-flow instructions (essentially informing the Checker about the location of the current program point) or from the program-supplied invariants. Figure 5 shows a summary of the sequence of actions performed by VCGen for our example.

First, VCGen initializes the symbolic values of all four registers with fresh new symbolic values to denote unknown initial values in all registers. But the initial values of the registers r_d and r_l are constrained by the precondition. To account for this, VCGen takes the specified precondition and, without trying to interpret its meaning, substitutes in it the current symbolic values of the registers. The result is the symbolic predicate formula "d_0 : array(bool, l_0)" that is added to a stack of assumptions. (These assumptions are shown underlined in Figure 5 and with an indentation level that corresponds to their position in the stack.)

After these initial steps VCGen proceeds to consider each instruction in turn. The assignment instruction of line 2 is modeled as a assignment of symbolic values. On line 3, VCGen encounters an invariant. Since the invariant is not trusted VCGen asks the Checker to verify first that the invariant predicate holds at this point. To do this VCGen substitutes the current symbolic register values in the invariant predicate and the resulting predicate "$l_0 \geq l_0$" is submitted to the Checker for verification. (Such verification goals are shown in Figure 5 right-justified and boxed.)

Then, VCGen simulates symbolically an *arbitrary* iteration through the loop. To achieve this VCGen generates fresh new symbolic values for all registers, except the invariant ones. Next VCGen adds to the assumption stack the predicate "$l_0 \geq i_1$" obtained from the invariant predicate after substitution of the new symbolic values for registers.

When VCGen encounters a conditional branch it simulates both possible outcomes. First, the branch is assumed to be taken and a corresponding assumption is placed on the assumption stack. When that symbolic execution of that branch terminates (e.g. when encountering a return instruction or an invariant for the second time), VCGen restores the state of the assumption stack and processes in a similar way the case when the branch is not taken.

1: Generate fresh values $r_d = d_0$, $r_l = l_0$, $r_i = i_0$ and $r_t = t_0$

1: Assume Precondition $\underline{d_0 \; : \; \texttt{array(bool, } l_0\texttt{)}}$

2: Set $r_i = l_0$

3: Invariant (first hit) $\boxed{l_0 \geq l_0}$

3: Generate fresh values $r_i = i_1$, $r_t = t_1$

3: Assume invariant $\underline{l_0 \geq i_1}$

4: Set $r_t = \texttt{ge}(i_1, 0)$

5: Branch 5 taken $\underline{\texttt{not } (\texttt{ge}(i_1, \; 0))}$

11: Check postcondition $\boxed{\texttt{1 : bool}}$

5: Branch 5 not taken $\texttt{ge}(i_1, \; 0)$

6: Set $r_t = \texttt{add}(d_0, i_1)$

7: Check load $\boxed{\texttt{saferd(add}(d_0, \; i_1)\texttt{))}}$

7: Set $r_t = \texttt{mem}(\texttt{add}(d_0, i_1))$

8: Branch 8 taken $\underline{\texttt{not } (\texttt{mem}(\texttt{add}(d_0, \; i_1)))}$

12: Check postcondition $\boxed{\texttt{0 : bool}}$

8: Branch 8 not-taken $\texttt{mem}(\texttt{add}(d_0, \; i_1))$

9: Set $r_i = \texttt{sub}(i_1, 1)$

3: Invariant (second hit) $\boxed{l_0 \geq \texttt{sub}(i_1, \; 1)\texttt{, } d_0 = d_0\texttt{, } l_0 = l_0}$

Figure 5. The sequence of actions taken by VCGen. We show on the left the program points and a brief description of each action. Some actions result in extending the stack of assumptions that the Checker is allowed to make. These assumptions are shown underlined and with an indentation level that encodes the position in the stack of each assumption. Thus an assumption at a given indentation level implicitly discards all previously occurring assumptions at the same or larger indentation level. Finally, we show right-justified and boxed the checking goals submitted to the Checker.

Consider for example the branch of line 5. In the case when the branch is taken, VCGen pushes the assumption "not $(\texttt{ge}(i_1, 0))$" and continues the execution from line 11. There a return instruction is encountered and VCGen asks the Checker to verify that the postcondition is verified. The precise verification goal is produced by substituting the actual symbolic return value (the literal 1 in this case) for the name res in the postcondition Post$_{\texttt{forall}}$. Once the Checker module verifies this goal, VCGen restores the symbolic values of registers and the state of the assumption stack to their states from before the branch and then simulates the case when the branch is not taken. In this case, the memory read instruction is encountered and VCGen produces a saferd predicate using the symbolic value of the registers to construct a symbolic representation of the address being read.

The final notable item in Figure 5 is what happens when VCGen encounters the invariant for the second time. In this case VCGen instructs the Checker to verify that the invariant predicate still holds. At this point VCGen also asks the Checker to verify that the symbolic values of those registers that were declared invariant are equal to their symbolic values at the start of the arbitrary iteration thorough the loop. In our case, since the registers declared invariant were not assigned at all, their symbolic values before and after the iteration are not only equal but identical.

This completes our simplified account of the operation of VCGen. For details on how VCGen deals with more complex issues such as function calls and memory updates, the reader is invited to consult [9]. Note that in the VCGen used in previous versions of PCC (such as that described in [9]) the result of the verification condition generator is a single large predicate that encodes all of the goals (using conjunctions) and all of the assumptions (using implications). This means that the VCGen runs first to completion and produces this predicate which is then consumed by the Checker. This approach, while natural, turned out to be too wasteful. For large examples it became quite common for this monolithic predicate to require hundreds of megabytes for storage slowing down the checking process considerably. In some of the largest examples not even a virtual address space of 1Gb could hold the whole predicate. In the architecture that we propose here VCGen produces one small goal at a time and then passes the control to the Checker. Once a goal is validated by the Checker, it is discarded and the symbolic evaluation continues. This optimization might not seem interesting from a scientific point of view but it is illustrative of a number of purely engineering details that we had to address to make PCC scalable.

3. The Checker module. The LF-based representation of proofs

In a Proof-Carrying Code system, as well as in any application involving the explicit manipulation of proofs, it is of utmost importance that the proof representation is compact and proof checking is efficient. In this section we present a logical framework derived from the Edinburgh Logical Framework [3], along with associated proof representation and proof checking algorithms, that have the following desirable properties:

- The framework can be used to encode judgments and derivations from a wide variety of logics, including first-order and higher-order logics.
- The implementation of the proof checker is parameterized by a high-level description of the logic. This allows a unique implementation of the proof checker to be be used with many logics and safety policies.
- The proof checker performs a directed, one-pass inspection of the proof object, without having to perform search. This leads to a simple im-

plementation of the proof checker that is easy to trust and install in existing extensible systems.

– Even though the proof representation is detailed, it is also compact.

We chose the Edinburgh Logical Framework (LF) as the starting point in our quest for efficient proof manipulation algorithms because it alone scores very high on the first three of the four desirable properties listed above. In this respect, our design can also be viewed as a testimony to the usefulness of LF for practical systems applications, such as Proof-Carrying Code.

3.1. THE EDINBURGH LOGICAL FRAMEWORK

The Edinburgh Logical Framework (also referred to as LF) has been introduced by Harper, Honsell and Plotkin [3] as a metalanguage for high-level specification of logics. LF provides natural support for the management of binding operators and of the hypothetical and schematic judgments through LF bound variables. This is a crucial factor for the succinct formalization of proofs.

The LF type theory is a language with entities at three levels, namely objects, types and kinds, whose abstract syntax is shown below:

$$
\begin{array}{lll}
\text{Kinds} & K ::= \textbf{Type} \mid \Pi x\!:\!A.K \\
\text{Types} & A ::= a \mid A\,M \mid \Pi x\!:\!A_1.A_2 \\
\text{Objects} & M ::= x \mid c \mid M_1 M_2 \mid \lambda x\!:\!A.M
\end{array}
$$

The notation $A_1 \to A_2$ is sometimes used instead of $\Pi x\!:\!A_1.A_2$ when x is not free in A_2.

The encoding of a logic in LF consists of an LF signature Σ that contains declarations for a set of constants corresponding to the syntactic formula constructors and to the proof rules. A fragment of the signature that defines the first-order predicate logic extended as needed by our example is shown in Figure 6. The top section of the figure contains declarations of the type constructors ι and o corresponding respectively to individuals and predicates, and of the type family **pf** indexed by predicates. If P is the representation of a predicate, then "**pf** P" is the type of all valid proofs of P. The rest of Figure 6 contains the declarations of a few syntactic constructors followed by a few constants corresponding to proof rules of first-order logic. Note the use of higher-order features of LF for the succinct representation of the hypothetical and parametric judgments that characterize the implication elimination (**impi**), the universal quantifier introduction (**alli**) and elimination (**alle**) rules.

We write $\ulcorner P \urcorner$ to denote the LF representation of the predicate P. As an example of an LF representation of a proof we show in Figure 7 the

$$\iota \qquad : \text{Type}$$
$$o \qquad : \text{Type}$$
$$\text{pf} \qquad : o \to \text{Type}$$

$$\text{true} \quad : o$$
$$\text{and} \quad : o \to o \to o$$
$$\text{imp} \quad : o \to o \to o$$
$$\text{all} \quad : (\iota \to o) \to o$$

$$\text{zero} \quad : \iota.$$
$$\text{bool} \quad : \iota.$$
$$\text{array} \quad : \iota \to \iota \to \iota.$$

$$\text{hastype} : \iota \to \iota \to o.$$
$$\text{ge} \qquad : \iota \to \iota \to o.$$
$$\text{saferd} : \iota \to o.$$

$$\text{rd} \qquad : \Pi A : \iota.\Pi T : \iota.\Pi L : \iota.\Pi I : \iota.$$
$$\text{pf (of } A \text{ (array } T\ L)) \to \text{pf (ge } I \text{ zero)} \to$$
$$\text{pf (ge } L\ I) \to \text{pf (saferd (add } A\ I)).$$

$$\text{truei} \quad : \text{pf true}$$
$$\text{andi} \quad : \Pi P : o.\Pi R : o.\text{pf } P \to \text{pf } R \to \text{pf (and } P\ R)$$
$$\text{andel} \quad : \Pi P : o.\Pi R : o.\text{pf (and } P\ R) \to \text{pf } P$$
$$\text{impi} \quad : \Pi P : o.\Pi R : o.(\text{pf } P \to \text{pf } R) \to \text{pf (imp } P\ R)$$
$$\text{alli} \quad : \Pi P : \iota \to o.(\Pi X : \iota.\text{pf } (P\ X)) \to \text{pf (all } P)$$
$$\text{alle} \quad : \Pi P : \iota \to o.\Pi E : \iota.\text{pf (all } P) \to \text{pf } (P\ E)$$

Figure 6. Fragment of the LF signature corresponding to first-order predicate logic.

representation of a proof of the predicate $P \Rightarrow (P \wedge P)$, for some predicate P.

At this point the reader might have noticed that the formulas produced by the VCGen on our example (the framed formulas of Figure 5) are always atomic, yet we have introduced LF constants for proof rules like implication and quantification introduction. First, such proof rules might be useful in situations where the annotations can be arbitrary formulas of first-order logic. In such cases some of the formulas that VCGen produces can be non-atomic. But more importantly the proof rules for quantification and implication introduction demonstrate how LF can encode parametric and hypothetical proofs, that is, proofs of facts parameterized by some individual names and proofs that are allowed to use certain hypotheses.

$$M = \text{impi } \ulcorner P \urcorner \ (\text{and } \ulcorner P \urcorner \ulcorner P \urcorner)$$
$$(\lambda x \text{:pf } \ulcorner P \urcorner.\text{andi } \ulcorner P \urcorner \ulcorner P \urcorner \ x \ x)$$

Figure 7. The LF representation of the proof by implication introduction followed by conjunction introduction of the predicate $P \Rightarrow (P \wedge P)$.

If we now look back at Figure 5 we notice that in addition to the framed formulas, which are facts to be proved, VCGen also communicates to the Checker some assumptions, which are represented underlined in Figure 5. Also, the new symbolic names that VCGen creates while evaluating the code are nothing else than parameters. Thus, the **saferd** checking goal can be represented entirely in first-order logic as:

$$\forall d_0 l_0 i_1.d_0 : \text{array}(\text{bool}, l_0) \Rightarrow (l_0 \geq i_1 \Rightarrow (\text{ge}(i_1, 0) \Rightarrow \text{saferd}(\text{add}(d_0, i_1)))))$$

or equivalently as the LF term:

```
all (λD:ι.
    all (λL:ι.
        (all (λI:ι.
            imp (of D (array bool L))
            (imp (ge L I)
                (imp (ge I zero) (saferd (add D I)))))))))
```

An important benefit of using LF for representing proofs is that we can use LF type checking to verify that a proof is valid and also that it proves the required predicate. The LF type-checking judgment is written as $\Gamma \vdash^{LF} M : A$, where Γ is a type environment for the free variables of M and A.[2] A definition of LF type checking, along with a proof of adequacy of using LF type checking for proof validation for first-order and higher-order logics, can be found in [3]. For example, the validation of the proof representation M shown in Figure 7 can be done by verifying that M has the LF type "pf (imp $\ulcorner P \urcorner$ (and $\ulcorner P \urcorner \ulcorner P \urcorner$))".

Owing to the simplicity of LF and of the LF type system, the implementation of the type checker is simple and easy to trust. Furthermore, because all of the dependencies on the particular object logic are segregated in the signature, the implementation of the type checker can be reused directly for proof checking in various first-order or higher-order logics. The only logic-dependent component of the proof-checker is the signature, which is usually easy to verify by visual inspection.

---

[2] The LF typing judgment and all the other typing judgments discussed in this document depend also on a signature Σ, which is henceforth omitted in order to simplify the notation.

276

$$\text{impi } *_1 \ *_2 \ (\lambda x \colon *_3.\text{andi } *_4 \ *_5 \ x \ x)$$

Figure 8. The LF$_i$ representation of a proof of $P \Rightarrow P \wedge P$.

Unfortunately, the above-mentioned advantages of LF representation of proofs come at a high price. The typical LF representation of a proof is large, due to a significant amount of redundancy. This fact can already be seen in the proof representation shown in Figure 7, where there are six copies of $\ulcorner P \urcorner$ as opposed to only three in the predicate to be proved. The effect of redundancy observed in practice increases non-linearly with the size of the proofs. Consider for example, the representation of the proof of the n-way conjunction $P \wedge \ldots \wedge P$. This representation contains about n^2 redundant copies of $\ulcorner P \urcorner$ and n occurrences of **andi**. The redundancy of representation is not only a space problem but also leads to inefficient proof checking, because all of the redundant copies have to be type checked and then checked for equivalence with the copies of P from the predicate.

The proof representation and checking framework presented in the remainder of this section is based on the observation that it is possible to retain only the skeleton of an LF representation of a proof and use a modified LF type-checking algorithm to reconstruct on the fly the missing parts. The resulting *implicit LF* representation inherits the advantages of the LF representation (i.e., small and logic-independent implementation of the proof checker) without the disadvantages (i.e., large proof sizes and slow proof checking).

3.2. IMPLICIT LF

The implicit LF representation of a proof is similar to the corresponding LF representation, with the exception that select parts of the representation are missing. For expository purposes, the positions in the representation where subterms are missing are marked with placeholders, written as $*$. To exemplify the use of placeholders we show in Figure 8 an LF$_i$ representation of the proof shown in Figure 7.

The major difficulty in dealing with placeholders is in type checking. In particular, the type checking algorithm must be able to reconstruct on the fly the missing parts of a proof. That is why we refer to the LF$_i$ type checking algorithm as the *reconstruction algorithm*.

The first step in establishing a reconstruction algorithm for LF$_i$ is to introduce the LF$_i$ type system, through a typing judgment $\Gamma \vdash^i M : A$, whose definition is shown in Figure 9. The LF$_i$ typing rules are very similar to the LF typing rules, the only differences being the ability to deal with

Objects :

$$\frac{\Sigma(c) = A}{\Gamma \overset{i}{\vdash} c : A} \quad \frac{\Gamma(x) = A}{\Gamma \overset{i}{\vdash} x : A} \quad \frac{\Gamma \overset{i}{\vdash} M : A \quad A \equiv_\beta B \quad PF(A)}{\Gamma \overset{i}{\vdash} M : B}$$

$$\frac{\Gamma, x : A \overset{i}{\vdash} M : B}{\Gamma \overset{i}{\vdash} \lambda x : *.M : \Pi x : A.B} \quad \frac{\Gamma, x : A \overset{i}{\vdash} M : B}{\Gamma \overset{i}{\vdash} \lambda x : A.M : \Pi x : A.B}$$

$$\frac{\Gamma \overset{i}{\vdash} M : \Pi x : A.B \quad \Gamma \overset{i}{\vdash} N : A \quad PF(A)}{\Gamma \overset{i}{\vdash} M \ N : [N/x]B}$$

$$\frac{\Gamma \overset{i}{\vdash} M : \Pi x : A.B \quad \Gamma \overset{i}{\vdash} N : A \quad PF(A)}{\Gamma \overset{i}{\vdash} M * : [N/x]B}$$

Equivalence :

$$\overline{(\lambda x : A.M)N \equiv_\beta [N/x]M} \quad \overline{(\lambda x : *.M)N \equiv_\beta [N/x]M}$$

Figure 9. The LF_i type system.

implicitly-typed abstractions and applications whose argument is implicit. The notation $PF(A)$ means that A is placeholder-free (i.e., it does not contain placeholders). Similarly, we write $PF(\Gamma)$ to denote that the the types contained in the type environment Γ do not contain placeholders. We have decided to restrict the LF_i typing to situations where the types involved do not contain placeholders. This does not seem to affect the representation of proofs, but it simplifies considerably the various proofs of correctness of LF_i.

The LF_i typing rules cannot be used as the basis for an implementation of reconstruction, because the instantiation of placeholders is not completely determined by the context. Nevertheless, the LF_i typing system is an adequate basis for arguing the soundness of a reconstruction algorithm. This property is established through Theorem 1 that guarantees the existence of a well-typed placeholder-free object M' if the implicit object M can be typed in the LF_i typing discipline.

(SOUNDNESS OF LF_i TYPING.) 1. *If* $\Gamma \overset{i}{\vdash} M : A$ *and* $PF(\Gamma)$, $PF(A)$, *then there exists* M' *such that* $\Gamma \overset{LF}{\vdash} M' : A$.

The proof of Theorem 1 is by induction on the structure of the LF_i typing derivation. The fully-explicit term M' is constructed from M by replacing the placeholders with actual LF terms as specified by the typing derivation. This theorem is the keystone in the proof of adequacy of proof checking by using reconstruction algorithms that are sound with respect to LF_i.

3.3. THE RECONSTRUCTION ALGORITHM

The reconstruction algorithm performs the same functions as the LF type-checking algorithm, except that it also finds well-typed instantiations for the placeholders that occur in objects. Since a precise understanding of the reconstruction algorithm is not necessary for the purposes of this document, we only show here an informal sketch of the operation of the algorithm on the implicit proof object of Figure 8. The interested reader can find more details in [11, 9].

The reconstruction goal in this case is to verify that the implicit proof representation of Figure 8 has type

$$\mathtt{pf} \; (\mathtt{imp} \; \ulcorner P \urcorner \; (\mathtt{and} \; \ulcorner P \urcorner \; \ulcorner P \urcorner))$$

Based on this goal type and on the declared type of the constant $\mathtt{impi}$ (the head of the top-level application), the algorithm collects the following constraints:

$$*_1 \equiv \ulcorner P \urcorner$$
$$*_2 \equiv \mathtt{and} \; \ulcorner P \urcorner \; \ulcorner P \urcorner$$
$$\vdash (\lambda x : *_3.\mathtt{andi} \; *_4 \; *_5 \; x \; x)$$
$$: \; \mathtt{pf} \; \ulcorner P \urcorner \to \mathtt{pf} \; (\mathtt{and} \; \ulcorner P \urcorner \; \ulcorner P \urcorner)$$

The first two constraints lead to substitutions for $*_1$ and $*_2$, which are guaranteed by construction to be well-typed representation of predicates, since they originate in the trusted predicate to be proved. From the last constraint, using the rule for abstraction we obtain the following system of constraints:

$$*_3 \equiv \mathtt{pf} \; \ulcorner P \urcorner$$
$$x : \mathtt{pf} \; \ulcorner P \urcorner \vdash \mathtt{andi} \; *_4 \; *_5 \; x \; x \; : \; \mathtt{pf} \; (\mathtt{and} \; \ulcorner P \urcorner \; \ulcorner P \urcorner)$$

The process continues until no more constraints are generated. Note that each system of constraints consists of typing constraints and unification constraints, which in turn are of the simple rigid-rigid or flex-rigid kinds that can be solved eagerly.

4. The Checker module. The oracle-based representation of proofs

One of the main impediments to scalability in the LF-based realization of PCC is that proofs can be very large. Our solution to this problem is motivated by the observation that LF-based PCC proof checking is not able to exploit domain-specific knowledge in order to reduce the size of the proofs. For example, the LF-based PCC proofs of type safety are an order of magnitude larger than the size of the typing annotations that the Typed Assembly Language (TAL) system [7] uses. The overhead in TAL is smaller because TAL is less general than PCC and targets a specific type-safety policy, for a specific language and type system. The TAL type checker can be viewed as a proof checker specialized and optimized for a specific logic.

What we need is a different PCC implementation strategy that allows the size of PCC proofs to adapt automatically to the complexity of the property being checked. As a result, we should not have to pay a proof-size price for the generality of PCC in those instances when we check relatively simple properties. There are several components to our new strategy. First is a slightly different view on how the proofs in PCC can assist the verification on the host side. We assume that the host uses a non-deterministic checker for the safety policy. Then, a proof can essentially be replaced by an oracle guiding the non-deterministic checker. Every time the checker must make a choice between N possible ways to proceed, it consults the next $\lceil \log_2 N \rceil$ bits from the oracle. There are several important points that this new view of PCC exposes:

- The possibility of using non-determinism simplifies the design of the checker and enables the code receiver to use a simple checker even for checking a complex property.

- This view of verification exposes a three-way tradeoff between the complexity of the safety policy, the complexity and "smartness" of the checker, and the oracle size. If the verification problem is highly directed, as is the case with typical type-checking problems, the number of non-deterministic choices is usually small, and thus the required oracles are small. If the checker is "smart" and can narrow down further the choices, the oracle becomes even smaller. At an extreme, the checker might explore by itself small portions of the search space and require guidance from the oracle only in those situations when the search would be either too costly or not guaranteed to terminate.

- In the particular case of type-checking, even for assembly language, the checking problem is so directed that in many situations there is only one applicable choice, meaning that no hand-holding from the oracle is needed. This explains the large difference between the size of the typing derivation (i.e. the size of the actual proof) and the size of the oracles in our experiments.

- This view of PCC makes direct use of the defining property of the complexity class NP. This suggests that one of the benefits of PCC is that it allows the checker to check the solutions of problems in NP in polynomial time (with help from the oracle). But PCC can also help with checking even of solutions for semi-decidable problems, provided the checker and the oracle negotiate before hand a limit on the number of inference steps to be carried during verification.

– Since oracles are just streams of bits and no lookahead is necessary, they can be used in an online fashion, which is harder to do with syntactic representations of proofs. This leads to a smaller memory footprint for the checker, which is important in certain applications of PCC for embedded devices.

The Checker module in the proposed architecture is simply a non-deterministic logic interpreter whose logic program consists of the safety policy rules of inference formulated as Horn clauses and whose goals are the formulas produced by VCGen. For example, the rd inference rule of Figure 4 is expressed as the following clause written in Prolog notation:

$$\texttt{saferd(add}(A,\ I)) \ \texttt{:-} \ \ \texttt{of}(A,\ \texttt{array}(T,\ L),\ I \geq 0,\ L \geq I.$$

There are two major differences between a traditional logic interpreter and our Checker. One is that in PCC the logic program is dynamic since assumptions (represented as logic clauses) are added and retracted from the system as the symbolic execution follows different paths through the agent. However, this happens only in between two separate invocations of the Checker. The second and more important difference is that while a traditional interpreter selects clauses by trying them in order and backtracking on failure, the Checker is a non-deterministic logic interpreter meaning that it "guesses" the right clause to use at each step. This means that the Checker can avoid backtracking and it is thus significantly simpler and generally faster than a traditional interpreter. In essence the Checker contains a first-order unification engine and a simple control-flow mechanism that records and processes all the subgoals in a depth-first manner. The last element of the picture is that the "guesses" that the Checker makes are actually specified as a sequence of clause names as part of the "explanation" of safety that accompanies the code. In this sense the proof is replaced by an oracle that guides the non-deterministic interpreter to success.

As an example of how this works, consider the invocation of the Checker on the goal "$\texttt{saferd(add}(d_0,\ i_1))$". For this invocation the active assumptions are:

$$A_0 \;=\; \text{of}(d_0, \, \text{array}(\text{bool}, l_0))$$
$$A_1 \;=\; l_0 \geq i_1$$
$$A_2 \;=\; \text{ge}(i_1, 0)$$

The fragment of oracle for this checking goal is the following sequence of clause names "rd, A_0, geq, A_2, A_1". To verify the current goal the Checker obtains the name of the next clause to be used (rd) from the oracle and unifies its head with the current goal. This leads to the following subgoals, where T and L are not-yet-instantiated logic variables:

$$\text{of}(d_0, \, \text{array}(\text{T, L}))$$
$$i_1 \geq 0$$
$$\text{L} \geq i_1$$

To solve the first subgoal the Checker extracts the next clause name (A_0) from the oracle and unifies the subgoal with its head. The unification succeeds and it instantiates the logic variables T and L to bool and l_0 respectively. Then the oracle guides the interpreter to use the rule geq followed by assumption A_2 to validate the subgoal "$i_1 \geq 0$" and assumption A_1 to validate the subgoal "$l_0 \geq \text{i1}$". The oracle for checking all of the goals pertaining to our example is:

geqid, bool1, rd, A_0, geq, A_2, A_1, bool0, dec, A_1, A_2, eqid, eqid

4.1. AN OPTIMIZED CHECKER

So far we have not yet achieved a major reduction in the size of the "explanation". Since the oracle contains mentions of every single proof rule its size is comparable with that of a proof represented as an LF_i term. What we need now is to make the Checker "smarter" and ask it to narrow down the number of clauses that could possibly match the current goal, before consulting the oracle. If this number turns out to be zero then the Checker can reject the goal thus terminating the verification. In the fortunate case when the number of usable clauses is exactly one then the Checker can proceed with that clause without needing assistance from the oracle. And in the case when the number of usable clauses is larger than one the Checker needs only enough bits from the oracle to address among these usable clauses.

Such an optimization is also useful for a traditional logic interpreter because it reduces the need for backtracking. By having phrased the checking problem as a logic interpretation problem we can simply use off-the-shelf logic program optimizations to reduce the amount of non-determinism and thus to reduce the size of the necessary oracle.

Among the available logic program optimizations we selected automata-driven term indexing(ATI) [13] because it has a relatively small complexity and good clause selectivity. The basic idea behind ATI is that the conclusions of all the active clauses are scanned and a decision tree is build from them. Intuitively, the leaves of the decision tree are labeled with sets of clauses whose conclusions could match a goal whose structure is encoded by the path corresponding to the goal. This allows the quick computation of the set of clauses the could match a goal. For details on how the ATI technique is used in the implementation of the Checker the reader can consult [12].

In our example the ATI optimization works almost perfectly. For nearly all goals and subgoals it manages to infer that exactly one clause is usable. However, for three of the subgoals involving the predicate "$\geq$" the ATI technique confuses the correct clause (either A_1 or dec) with the geqid and the geq rules. This is because ATI is not able to encode exactly in the decision tree conclusions involving duplicate logical variables. Thus the conclusion of the geqid rule is encoded as "$E_1 \geq E_2$", which explains why ATI would think that this conclusion matches all subgoals involving the $\geq$ predicate.

Even with this minor lack of precision the oracle for our example is reduced to just 8 bits. For example, the proof of the first checking goal "$l_0 \geq l_0$" requires 1 bit of oracle because it is not clear whether to use the rule getid or geq. As another example, the proof of the goal "$l_0 \geq i_1$" (which occurs as a subgoal while proving saferd(add(d_0, i_1)) and $l_0 \geq$ sub($i_1, 1$)) requires two bits of oracle because it is not clear whether to use the rules geq or geqid or even the invariant assumption.

The original oracle consisted of 13 clause names and since we never had more than 16 active clauses in our example, we could use 52 bits to encode the entire oracle. In practice we observe savings of more than 30 times over the uncompressed oracles. This is because it is not uncommon to have even 1000 clauses active at one point. This would happen in deeply nested portions of the code where there are many assumptions active. And the ATI technique is so selective that even in these cases it is able to filter out most of the clauses.

5. Compiler support for scalability

So far we have described two techniques that proved essential for obtaining a scalable architecture for PCC: an oracle-based encoding of proofs and an interleaved execution of the VCGen and the Checker modules. Both of these techniques are implemented on the code-receiver end. But it would be incorrect to draw the conclusion that scalability can be achieved without cooperation from the code-producer end.

To see how the code producer can and should help, recall that VCGen considers both branches for each conditional and that symbolic execution stops when a return instruction is encountered or when an invariant is encountered for the second time. VCGen does verify that each cycle through the code has at least one invariant thus ensuring the termination of the symbolic evaluation.

But consider a program without loops and with a long sequence of conditionals, such as the following:

$$\text{if}(c_1)\ s_1\ \text{else}\ s_1';$$
$$\dots$$
$$\text{if}(c_n)\ s_n\ \text{else}\ s_n';$$
$$\text{return x}$$

VCGen considers each of the 2^n paths through this function because it might actually be the case that each such path has a different reason (and proof) to be safe. In most cases (and in all cases involving type safety) the verification can be modularized by placing an invariant annotation after each conditional. Then VCGen will have to verify only $2n$ path fragments from one invariant to another. Thus the code producer should place more than the minimum necessary number of invariants. If it fails to do so then the number of verification goals, and consequently the size of the oracle and the duration of the verification process can quickly become very large. We have verified in our experiments that this optimization is indeed very important for scalability.

6. Experimental results

The experiments discussed in this section are in the context of a PCC system that checks Intel x86 binaries for compliance with the Java type-safety policy. The binaries are produced using a bytecode-to-native optimizing compiler [1]. This compiler is responsible for generating the invariant annotations. The oracle is produced on the code producer side in a manner similar to oracle checking. The VCGen is used as usual to produce a sequence of goals and assumptions. Each of these goals is submitted to a modified Checker engine that first uses the ATI engine to compute a small set of usable clauses and then tries all such clauses in order, using

Program	Description	Source Size (LOC)	Code Size (bytes)
gnu-getopt	Command-line parser	1588	12644
linpack	Linear algebra routines	1050	17408
jal	SGI's Java Algorithm Library	3812	27080
nbody[†]	N-body simulation	3700	44064
lexgen	Lexical analyzer generator	7656	109196
ocaml[†]	Ocaml byte code interpreter	9400	112060
raja	Raytracer	8633	126364
kopi[†]	Java development tools	71200	760548
hotjava	Web browser	229853	2747548

Figure 10. Description of our test cases, along with the size of the Java source code and the machine-code size. † indicates that source code was not available and the Java source size is estimated based on the size of the bytecode.

backtracking. Then this modified Checker emits an oracle that encodes the sequence of clauses that led to success for each goal.

We carried out experiments on a set of nearly 300 Java packages of varying sizes. Some of the larger ones are described in Figure 10. The running times used in this paper are the median of five runs of the Checker on a 400MHz Pentium II machine with 128MB RAM.

Based on the data shown in Figure 10 we computed, for each example, how much smaller are the oracles compared to LF_i proofs (shown in Figure 12), what percentage of the code size are the oracles (shown in Figure 13) and how much slower is the logic interpreter compared to the LF_i checker (shown in Figure 14). These figures also show the geometric means for the corresponding ratios over these examples.

We observe that oracles are on average nearly 30 times smaller than LF_i proofs, and about 12% of the size of the machine code. While the binary representation of oracles is straightforward, for LF_i it is more complicated. In particular, one has to decide how to represent various syntactic entities. For the purpose of computing the size of LF_i proofs, we streamline LF_i terms as 16-bit tokens, each containing tag and data bits. The data can be the deBruijn index [2] for a variable, the index into the signature for constants, or the number of elements in an application or abstraction.

We also compared the performance of our technique with that obtained

Program	LF$_i$ Size (bytes)	LF$_i$ Time (ms)	Oracle Size (bytes)	Checking Time (ms)
gnu-getopt	49804	82	1936	223
linpack	65008	117	2360	319
jal	53328	84	1698	314
nbody	187026	373	7259	814
lexgen	413538	655	15726	1948
ocaml	415218	641	13607	1837
raja	371276	747	11854	2030
kopi	3380054	5321	96378	14693
hotjava	10813894	19757	354034	53491

Figure 11. For each of the test cases, the size and time to check the LF$_i$ representation of proofs, the size of the oracles and the logic interpretation time using oracles.

by using a popular off-the-shelf compression tool, namely gzip. We do more than 3 times better than gzip with maximum compression enabled, without incurring the decompression time or the addition of about 8000 lines of code to the server side of the PCC system. That is not to say that oracles could not benefit from further compression. There could be opportunities for Lempel-Ziv compression in oracles, in those situations when sequences of deduction rules are repeated. Instead, we are looking at the possibility of compressing these sequences at a semantic level, by discovering lemmas whose proof can be factored out.

It is also interesting to note that logic interpretation is about 3 times slower than LF$_i$ type checking. This is due to the overhead of constructing the decision trees used by ATI. There are some simple optimizations that one can do to reduce the checking time. For example, the results shown here are obtained by using an ATI truncated to depth 3. This saves time for the maintenance of the ATI but also loses precision thus leading to larger oracles. If we don't limit the size of the ATI we can save about 8% in the size of the oracles at the cost of increasing the checking time by 24%.

One interesting observation is that while LF$_i$ checking is faster than oracle checking, it also uses a lot more memory. While oracles can be consumed a few bits at a time, the LF$_i$ syntactic representation of a proof must be entirely brought in memory for checking. While we have not measured precisely the memory usage we encountered examples whose oracles

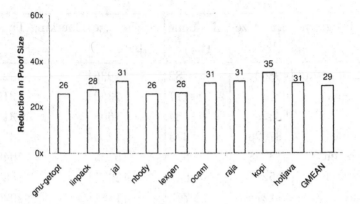

Figure 12. Ratio between LF$_i$ proof size and oracle size.

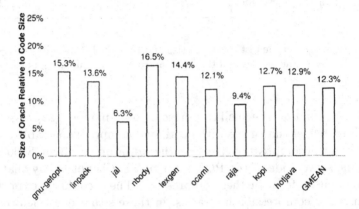

Figure 13. Ratio between oracle size and machine code.

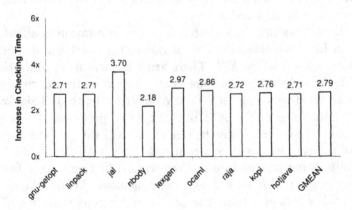

Figure 14. Ratio between logic-interpretation time and LF$_i$ type-checking time.

can be checked using less than 1Mbyte of memory while the checking of the corresponding LF$_i$ terms could not be performed even with 1Gbyte of virtual memory.

7. Conclusion

We presented in this paper an architecture for Proof-Carrying Code where proofs are replaced by an oracle guiding a non-deterministic checker for a given safety policy. The luxury of using non-determinism in the checker allows a simple checker to enforce even complex safety policies. Since many safety policies are relatively simple, the amount of non-determinism is low and this leads to small oracles that are required for checking compliance with such policies. In this sense the proposed PCC architecture is able to adapt the cost of verification to the complexity of the safety policy.

In designing this architecture we struggled to preserve a useful property of the previous implementation of PCC, namely that it can be easily configured to check different safety policies without changing the implementation. This has great software-engineering advantages and contributes to the trustworthiness of a PCC infrastructure since code that changes rarely is less likely to have bugs. To support this feature our choice for a non-deterministic checker is a non-deterministic logic interpreter that can be configured with the safety policy encoded as a logic program.

To achieve true scalability we had to solve several engineering problems, such as to design a low-cost interaction model between the various modules that compose the infrastructure. The code producer also must play an active role in ensuring that the verification process is quick. Through the combination of such techniques we have produced the first implementation of PCC that scales well to large programs at least in the context of a fairly simple type safety policy. What remains now to be seen is if Proof-Carrying Code can be practically applied to more complex safety policies.

Acknowledgments

I would like to thank Peter Lee for his guidance for developing the original Proof-Carrying Code system and Shree Rahul for the help with the collection of the experimental data presented in this paper. The certifying compiler for Java used in these experiments has been developed by Peter Lee, Mark Plesko, Chris Colby, John Gregorski, Guy Bialostocki and Andrew McCreight from Cedilla Systems Corporation.

288

References

1. Colby, C., P. Lee, G. C. Necula, F. Blau, M. Plesko, and K. Cline: 2000, 'A certifying compiler for Java'. *ACM SIGPLAN Notices* **35**(5), 95–107.
2. DeBruijn, N.: 1972, 'Lambda-calculus notation with nameless dummies, a tool for automatic formula manipulation'. *Indag. Mat.* **34**, 381–392.
3. Harper, R., F. Honsell, and G. Plotkin: 1993, 'A Framework for Defining Logics'. *Journal of the Association for Computing Machinery* **40**(1), 143–184.
4. Lindholm, T. and F. Yellin: 1997, *The Java Virtual Machine Specification*, The Java Series. Reading, MA, USA: Addison-Wesley.
5. McCanne, S. and V. Jacobson: 1993, 'The BSD Packet Filter: A New Architecture for User-level Packet Capture'. In: *The Winter 1993 USENIX Conference*. pp. 259–269.
6. Microsoft Corporation: 1996, 'Proposal for Authenticating Code Via the Internet'. http://www.microsoft.com/security/tech/authcode/authcode-f.htm.
7. Morrisett, G., D. Walker, K. Crary, and N. Glew: 1999, 'From system F to typed assembly language'. *ACM Transactions on Programming Languages and Systems* **21**(3), 527–568.
8. Necula, G. C.: 1997, 'Proof-Carrying Code'. In: *The 24th Annual ACM Symposium on Principles of Programming Languages*. pp. 106–119.
9. Necula, G. C.: 1998, 'Compiling with Proofs'. Ph.D. thesis, Carnegie Mellon University. Also available as CMU-CS-98-154.
10. Necula, G. C. and P. Lee: 1996, 'Safe Kernel Extensions without Run-Time Checking'. In: *Second Symposium on Operating Systems Design and Implementations*. pp. 229–243.
11. Necula, G. C. and P. Lee: 1998, 'Efficient Representation and Validation of Proofs'. In: *Thirteenth Annual Symposium on Logic in Computer Science*. Indianapolis, pp. 93–104.
12. Necula, G. C. and S. P. Rahul: 2001, 'Oracle-Based Checking of Untrusted Programs'. In: *The 28th Annual ACM Symposium on Principles of Programming Languages*. pp. 142–154.
13. Ramesh, R., I. V. Ramakrishnan, and D. S. Warren: 1995, 'Automata-Driven Indexing of Prolog Clauses'. *Journal of Logic Programming* **23**(2), 151–202.
14. Wahbe, R., S. Lucco, T. E. Anderson, and S. L. Graham: 1993, 'Efficient Software-Based Fault Isolation'. In: *14th ACM Symposium on Operating Systems Principles*. pp. 203–216.

ABSTRACTIONS AND REDUCTIONS IN MODEL CHECKING

ORNA GRUMBERG
Computer Science Department
The Technion
Haifa 32000
Israel
email: orna@cs.technion.ac.il

Abstract. We introduce the basic concepts of *temporal logic model checking* and its *state explosion problem*. We then focus on *abstraction*, which is one of the major methods for overcoming this problem. We distinguish between weak and strong preservations of properties by a given abstraction. We show how abstract models preserving ACTL* can be defined with human aid or automatically. When the abstraction is too coarse, we show how refinement can be applied to produce a more precise abstract model. Abstract interpretation is then introduced and applied in order to construct abstract models that are more precise and allow more ACTL properties to be proven. Finally, we show how to define abstract models that preserve ECTL* and full CTL*.

Keywords: model checking, temporal logic, abstraction, refinement, abstract interpretation, bisimulation, simulation preorder

1. Introduction

Temporal logic model checking is a procedure that gets as input a finite state model for a system and a property written in propositional temporal logic. It returns "yes" if the system has the desired property and returns "no" otherwise. In the latter case it also provides a counterexample that demonstrates how the system fails to satisfy the property. Model checking procedures can be quite efficient in time. However, they suffer from the state explosion problem: the number of states in the model grows exponen-

H. Schwichtenberg and R. Steinbrüggen (eds.), Proof and System-Reliability, 289–321.
© 2002 *Kluwer Academic Publishers. Printed in the Netherlands.*

tially with the number of system variables and components. This problem impedes the applicability of model checking to large systems, and much effort is invested in trying to avoid it.

Abstraction is a method for reducing the state space of the checked system. The reduction is achieved by hiding (abstracting away) some of the system details that might be irrelevant for the checked property. Abstract models are sometimes required to *strongly preserve* the checked properties. In this case, a property holds on the abstract model if and only if it holds on the original one. On the other hand, only *weak preservation*, may be required. In that case, if a property is true for the abstract model then it is also true for the original model. If a property is not true for the abstract model, then no conclusion can be reached with regards to the original model. The advantage of weak preservation is that it enables more significant reductions. However, it also increases the likelihood that we will be unable to determine the truth of a property in the system.

The decision as to which details are unnecessary for the verification task is made either manually (by the verification engineer) or automatically. In both cases, if the abstract model cannot determine the truth of the property in the system, then *refinement* is applied and additional details are introduced into the model.

The main goal of abstraction is to avoid the construction of the full system model. Thus, we need methods that derive an abstract model directly from some high-level description of the system (e.g. program text).

We will define conditions for strong and weak preservations for the temporal logic CTL* and its universal and existential fragments, ACTL* and ECTL*. We will show how to derive abstract models which preserve ACTL* from the program text, using nonautomatic and automatic abstractions. When the abstractions are too coarse, we will show how they can be refined.

The basic notions of abstract interpretation will be defined, and its use for deriving abstract models will be demonstrated. Abstract interpretation provides means for constructing more precise abstract models that allow us to prove more ACTL* properties. Within the framework of abstract interpretation we also show how to define abstract models that preserve ECTL* and full CTL*.

The rest of the paper is organized as follows. Section 2 defines temporal logics along with their semantics. It presents a model checking algorithm for CTL, and defines the notions of equivalence and preorder on models. Section 3 describes data abstraction. Approximated abstractions are also defined. It then shows how abstractions and approximations can be derived from a high level description of the program. Section 4 presents the ideas of counterexample-guided refinement in which both the initial abstraction and the refinement are constructed automatically. Section 5 develops abstract models within the abstract interpretation framework. Finally, Section 6 reviews the related work and Section 7 presents some concluding remarks.

2. Preliminaries

2.1. TEMPORAL LOGICS

We use finite state transition systems called *Kripke models* in order to model the verified systems.

Definition 2.1 (Kripke model) *Let AP be a set of atomic propositions. A* Kripke model M *over AP is a four-tuple* $M = (S, S_0, R, L)$, *where*

- *S is the set of states;*
- *$S_0 \subseteq S$ is the set of initial states;*
- *$R \subseteq S \times S$ is the transition relation, which must be* total, *i.e., for every state $s \in S$ there is a state $s' \in S$ such that $R(s, s')$;*
- *$L : S \to \mathcal{P}(AP)$ is a function that labels each state with the set of atomic propositions true in that state.*

A *path* in M starting from a state s is an infinite sequence of states $\pi = s_0 s_1 s_2 \ldots$ such that $s_0 = s$, and for every $i \geq 0$, $R(s_i, s_{i+1})$. The suffix of π from state s_i is denoted π^i.

We use propositional temporal logics as our specification languages. We present several subsets of the temporal logic CTL* [30] over a given finite set AP of atomic propositions. We will assume that formulas are expressed in *positive normal form*, in which negations are applied only to atomic propositions. This facilitates the definition of universal and existential subsets of CTL* [36]. Since negations are not allowed, both conjunction and disjunction are required. Negations applied to the *next-time* operator $\mathbf{X}$ can be "pushed inwards" using the logical equivalence $\neg(\mathbf{X} f) = \mathbf{X} \neg f$. The *unless* operator $\mathbf{R}$ (sometimes called the *release* operator), which is the dual of the *until* operator $\mathbf{U}$, is also added. Thus, $\neg(f \mathbf{U} g) = \neg f \mathbf{R} \neg g$.

Definition 2.2 (CTL*) *For a given set of atomic propositions AP, the logic CTL* is the set of state formulas, defined recursively by means of state formulas and path formulas as follows. State formulas are of the form:*

- *If $p \in AP$, then p and $\neg p$ are state formulas.*
- *If f and g are state formulas, then so are $f \wedge g$ and $f \vee g$.*
- *If f is a path formula, then $\mathbf{A} f$ and $\mathbf{E} f$ are state formulas.*

Path formulas are of the form:

- *If f is a state formula, then f is a path formula.*
- *If f and g are path formulas, then so are $f \wedge g$, and $f \vee g$.*

292

– *If f and g are path formulas, then so are $\mathbf{X}\,f$, $f\,\mathbf{U}\,g$, and $f\,\mathbf{R}\,g$.*

The abbreviations true, false and implication $\rightarrow$ are defined as usual. For path formula f, we also use the abbreviations $\mathbf{F}\,f \equiv true\,\mathbf{U}\,f$ and $\mathbf{G}\,f \equiv false\,\mathbf{R}\,f$. They express the properties that sometimes in the future f will hold and that f holds globally.

CTL [14] is a branching-time subset of CTL* in which every temporal operator is immediately preceded by a path quantifier, and no nesting of temporal operators is allowed. More precisely, CTL is the set of state formulas defined by:

– If $p \in AP$, then p and $\neg p$ are CTL formulas.
– If f and g are CTL formulas, then so are $f \wedge g$ and $f \vee g$.
– If f and g are CTL formulas, then so are $\mathbf{AX}\,f$, $\mathbf{A}(f\mathbf{U}g)$, $\mathbf{A}(f\mathbf{R}g)$ and $\mathbf{EX}\,f$, $\mathbf{E}(f\mathbf{U}g)$, $\mathbf{E}(f\mathbf{R}g)$.

ACTL* and ECTL* (*universal* and *existential* CTL*) are subsets of CTL* in which the only allowed path quantifiers are $\mathbf{A}$ and $\mathbf{E}$, respectively. ACTL and ECTL are the restriction of ACTL* and ECTL* to CTL.

LTL [55] can be defined as the subset of ACTL* consisting of formulas of the form $\mathbf{A}\,f$, where f is a path formula in which the only state subformulas permitted are Boolean combinations of atomic propositions. More precisely, f is defined (in positive normal form) by

1. If $p \in AP$ then p and $\neg p$ are path formulas.
2. If f and g are path formulas, then $f \wedge g$, $f \vee g$, $\mathbf{X}\,f$, $f\,\mathbf{U}\,g$, and $f\,\mathbf{R}\,g$ are path formulas.

We will refer to such f as an LTL *path formula.*

We now consider the semantics of the logic CTL* with respect to a Kripke model.

Definition 2.3 (Satisfaction of a formula) *Given a Kripke model M, satisfaction of a state formula f by a model M at a state s, denoted $M, s \models f$, and of a path formula g by a path π, denoted $M, \pi \models g$, is defined as follows (where M is omitted when clear from the context).*

1. $s \models p$ *if and only if* $p \in L(s)$; $s \models \neg p$ *if and only if* $p \notin L(s)$.
2. $s \models f \wedge g$ *if and only if* $s \models f$ *and* $s \models g$.
   $s \models f \vee g$ *if and only if* $s \models f$ *or* $s \models g$.
3. $s \models \mathbf{A}\,f$ *if and only if for every path π from s, $\pi \models f$.*
   $s \models \mathbf{E}\,f$ *if and only if there exists a path π from s such that $\pi \models f$.*
4. $\pi \models f$, *where f is a state formula, if and only if the first state of π satisfies the state formula.*
5. $\pi \models f \wedge g$ *if and only if* $\pi \models f$ *and* $\pi \models g$.
   $\pi \models f \vee g$ *if and only if* $\pi \models f$ *or* $\pi \models g$.

6.   (a) $\pi \models \mathbf{X} f$ if and only if $\pi^1 \models f$.

     (b) $\pi \models f \mathbf{U} g$ if and only if for some $n \geq 0$, $\pi^n \models g$ and for all $i < n$, $\pi^i \models f$.

     (c) $\pi \models f \mathbf{R} g$ if and only if for all $n \geq 0$, if for all $i < n$, $\pi^i \not\models f$ then $\pi^n \models g$.

$M \models f$ if and only if for every $s \in S_0$, $M, s \models f$.

In [30] it has been shown that CTL and LTL are incomparable in their expressive power, and that CTL* is more expressive than each.

Below we present several formulas together with their intended meaning and their syntactic association with some of the logics mentioned above.

- **Mutual exclusion:** $\mathbf{AG} \neg (cs_1 \wedge cs_2)$ where cs_i is an atomic proposition that is true in a state if and only if process i is in its critical section in that state. The formula means that it is invariantly true (in every state along every path) that processes 1 and 2 cannot be in their critical section at the same time.
  The formula is in CTL, LTL and CTL*.
- **Nonstarvation:** $\mathbf{AG}(request \longrightarrow \mathbf{AF}\, grant)$ means that every request will be granted along every execution.
  This formula is in CTL but not in LTL. However, it is equivalent to the LTL formula $\mathbf{AG}(request \longrightarrow \mathbf{F}\, grant)$.
- **"Sanity check":** The formula $\mathbf{EF}\, request$ is complementary to the nonstarvation formula. It excludes the case where the implication in the nonstarvation formula holds vacuously just because no request has been presented.
  This is a CTL formula that is not expressible in LTL.
- **Fairness:** $\mathbf{A}(\mathbf{G}\,\mathbf{F}\, enabled \longrightarrow \mathbf{G}\,\mathbf{F}\, executed)$ describes a fairness requirement that a transition which is infinitely often enabled ($\mathbf{G}\,\mathbf{F}\, enabled$) is also infinitely often executed ($\mathbf{G}\,\mathbf{F}\, executed$).
  This LTL formula is not expressible in CTL.
- **Reaching a reset state:** $\mathbf{AG}\,\mathbf{EF}\, reset$ describes a situation where in every state along every path there is a possible continuation that will eventually reach a reset state.
  The formula is in CTL and is not expressible in LTL.

2.2.   CTL MODEL CHECKING

In this section we briefly describe an algorithm for CTL model checking. CTL model checking [29] is widely used due to its efficient algorithm. LTL model checking [44] is also commonly used because many useful properties are easily expressed in LTL. CTL* model checking [31] can be built as a combination of the algorithms for LTL and CTL. For more details on these algorithms see [17].

The CTL model checking algorithm receives a Kripke model $M = (S, S_0, R, L)$ and a CTL formula f. It works iteratively on subformulas of f, from simpler subformulas to more complex ones. For each subformula g of f, it returns the set S_g of all states in M that satisfy g. That is, $S_g = \{s \mid M, s \models g\}$. An important property of the algorithm is that when it checks the formula g, all subformulas of g have already been checked. When the algorithm terminates, it returns $True$ if $S_0 \subseteq S_f$ and returns $False$ otherwise.

For CTL model checking, positive normal form is not necessary, in which case every formula can be expressed using the Boolean operators $\neg$ and $\wedge$ and the temporal operators **EX**, **EU**, and **EG**. The model checking algorithm consists of several procedures, each taking care of formulas in which the main operator is one of the above. Here, we present only the more complex procedures for formulas of the form $f = \mathbf{E}(g\mathbf{U}h)$ and $f = \mathbf{EG}\,g$.

The procedure CheckEU, presented in Figure 1, accepts as input the sets of states S_g and S_h and iteratively computes the set S_f of states that satisfy $f = \mathbf{E}(g\mathbf{U}h)$. The computation is based on the equivalence

$$\mathbf{E}(g\mathbf{U}h) = h \vee (g \wedge \mathbf{EX}(\mathbf{E}(g\mathbf{U}h))).$$

At each iteration, Q holds the set of states computed in the previous iteration while Q' holds the set of states computed in the current iteration. Initially, all states that satisfy h are introduced into Q'. At step i, all states in S_g that have a successor in Q are added. These are exactly the states that satisfy g and have a successor that satisfies $\mathbf{E}(g\mathbf{U}h)$. The computation stops when no more states can be added, i.e., a fixpoint is reached (in fact, this is a *least fixpoint*).

procedure CheckEU(S_g, S_h)
$Q := \emptyset;\ Q' := S_h;$
while $(Q \neq Q')$ **do**
$\quad Q := Q';$
$\quad Q' := Q \cup \{s \mid \exists s'[\ R(s, s') \wedge Q(s') \wedge S_g(s)\ ]\ \};$
end while;
$S_f := Q;$ **return** (S_f)

Figure 1. The procedure CheckEU for checking the formula $f = \mathbf{E}(g\mathbf{U}h)$

The procedure CheckEG, presented in Figure 2, accepts the set of states S_g and iteratively computes the set S_f of states that satisfy $f = \mathbf{EG}\,g$. The computation is based on the equivalence

$$\mathbf{EG}\,g = g \,\wedge\, \mathbf{EX}(\mathbf{EG}\,g).$$

Initially, all states that satisfy g are introduced into Q'. At any step, states that do not have a successor in Q are removed. When the computation terminates, each state in Q has a successor in Q. Since all states in Q satisfy g, they all satisfy $\mathbf{EG}\,g$. In this case, a *greatest fixpoint* is computed.

procedure CheckEG(S_g)
$Q := S;\ Q' := S_g;$
while $(Q \neq Q')$ **do**
    $Q := Q';$
    $Q' := Q \cap \{s \mid \exists s'[\, R(s, s') \wedge Q(s')\,]\ \};$
end while;
$S_f := Q;$ **return** (S_f)

Figure 2. The procedure CheckEG for checking the formula $f = \mathbf{EG}\,g$

Theorem 2.4 *[17] Given a model M and a CTL formula f, there is a model checking algorithm that works in time $O((|S| + |R|) \cdot |f|)$.*

2.3. EQUIVALENCES AND PREORDERS

In this section we define the bisimulation relation and the simulation pre-order over Kripke models. We will also state the relationships between these relations and logic preservation. Intuitively, two states are bisimilar if they are identically labeled and for every successor of one there is a bisimilar successor of the other. Similarly, one state is smaller than another by the simulation preorder if they are identically labeled and for every successor of the smaller state there is a corresponding successor of the greater one. The simulation preorder differs from bisimulation in that the greater state may have successors with no corresponding successors in the smaller state.

Let AP be a set of atomic propositions and let $M_1 = (S_1, S_{0_1}, R_1, L_1)$ and $M_2 = (S_2, S_{0_2}, R_2, L_2)$ be two models over AP.

Definition 2.5 *A relation $B \subseteq S_1 \times S_2$ is a bisimulation relation [54] over M_1 and M_2 if the following conditions hold:*

1. *For every $s_1 \in S_{0_1}$ there is $s_2 \in S_{0_2}$ such that $B(s_1, s_2)$. Moreover, for every $s_2 \in S_{0_2}$ there is $s_1 \in S_{0_1}$ such that $B(s_1, s_2)$.*
2. *For every $(s_1, s_2) \in B$,*

    − $L_1(s_1) = L_2(s_2)$ and

$$- \forall t_1[\ R_1(s_1, t_1) \ \longrightarrow \ \exists t_2[\ R_2(s_2, t_2) \ \wedge \ B(t_1, t_2) \ ]].$$
$$- \forall t_2[\ R_2(s_2, t_2) \ \longrightarrow \ \exists t_1[\ R_1(s_1, t_1) \ \wedge \ B(t_1, t_2) \ ]].$$

We write $s_1 \equiv s_2$ for $B(s_1, s_2)$. We say that M_1 and M_2 are *bisimilar* (denoted $M_1 \equiv M_2$) if there exists a bisimulation relation B over M_1 and M_2.

Definition 2.6 *A relation $H \subseteq S_1 \times S_2$ is a* simulation relation *[51] over M_1 and M_2 if the following conditions hold:*

1. *For every $s_1 \in S_{0_1}$ there is $s_2 \in S_{0_2}$ such that $H(s_1, s_2)$.*
2. *For every $(s_1, s_2) \in H$,*

   $$- \ L_1(s_1) = L_2(s_2) \ and$$
   $$- \ \forall t_1[\ R_1(s_1, t_1) \ \longrightarrow \ \exists t_2[\ R_2(s_2, t_2) \ \wedge \ H(t_1, t_2) \ ]].$$

We write $s_1 \preceq s_2$ for $H(s_1, s_2)$. M_2 *simulates* M_1 (denoted $M_1 \preceq M_2$) if there exists a simulation relation H over M_1 and M_2.

The relation $\equiv$ is an equivalence relation on the set of models, while the relation $\preceq$ is a preorder on this set. That is, $\equiv$ is reflexive, symmetric and transitive and $\preceq$ is reflexive and transitive. Note that if there is a bisimulation relation over M_1 and M_2, then there is a *unique* maximal bisimulation relation over M_1 and M_2, that includes any other bisimulation relation over M_1 and M_2. A similar property holds also for simulation.

The following theorem relates bisimulation and simulation to the logics they preserve[1].

Theorem 2.7

- *[9] Let $M_1 \equiv M_2$. Then for every CTL* formula f (with atomic propositions in AP), $M_1 \models f$ if and only if $M_2 \models f$.*
- *Let $M_1 \preceq M_2$.*

  - *[36] For every ACTL* formula f with atomic propositions in AP, $M_2 \models f$ implies $M_1 \models f$.*
  - *For every ECTL* formula f with atomic propositions in AP, $M_1 \models f$ implies $M_2 \models f$.*

The last part of Theorem 2.7 is a direct consequence of the previous part of this theorem. It is based on the observation that for every ECTL* formula f, there is an ACTL* formula which is equivalent to $\neg f$.

---

[1]Bisimulation and simulation also preserve the μ-calculus logic [40] and its universal [46] and existential subsets, respectively. The discussion of this is beyond the scope of this paper.

3. Data abstraction

The first abstraction that we present is *data abstraction* [15, 47]. In order to obtain a smaller model for the verified system, we abstract away some of the data information. At the same time, we make sure that each behavior of the system is represented in the reduced model. In fact, the reduced model may contain more behaviors than the concrete (full) model. Nevertheless, it is often smaller in size (number of states and transitions), which makes it easier to apply model checking.

Data abstraction is done by choosing, for every variable in the system, an abstract domain that is typically significantly smaller than the original domain. The abstract domain is chosen by the user. The user also supplies a mapping from the original domain onto the abstract one. An abstract model can then be defined in such a way that it is *greater by the simulation preorder* than the concrete model of the system. Theorem 2.7 can now be used to verify ACTL* properties of the system by checking them on the abstract model.

Clearly, a property verified for the abstract model can only refer to the abstract values of the program variables. In order for such a property to be meaningful in the concrete model we label the concrete states by atomic formulas of the form $\widehat{x_i} = a$. These atomic formulas indicate that the variable x_i has some value d that has been abstracted to a.

The definition of the abstract model is based on the definition of the concrete model. However, building it on top of the concrete model would defeat the purpose since the concrete model is often too large to fit into memory. Instead, we show how the abstract model can be derived directly from some high-level description of the program, e.g., from the program text. As we will see later, extracting a precise abstract model may not be an easy task. We therefore define an *approximated abstract model*. This model may have more behaviors than the abstract model, but it is easier to build from the program text.

Let P be a program with variables $x_1, \ldots, x_n$. For simplicity we assume that all variables are over the same domain D. Thus, the concrete (full) model of the system is defined over states s of the form $s = (d_1, \ldots, d_n)$ in $D \times \ldots \times D$, where d_i is the value of x_i in this state (denoted $s(x_i) = d_i$).

The first step in building an abstract model for P is choosing an abstract domain A and a surjection $h : D \to A$. The next step is to restrict the concrete model of P so that it reflects only the abstract values of its variables. This is done by defining a new set of atomic propositions

$$\widehat{AP} = \{ \ \widehat{x_i} = a \mid i = 1, \ldots, n \ \text{and} \ a \in A \ \}.$$

The notation $\hat{x}_i$ is used to emphasize that we refer to the abstract value of the variable x_i. The labeling of a state $s = (d_1, \ldots, d_n)$ in the concrete model will be defined by

$$L(s) = \{ \hat{x}_i = a \mid h(d_i) = a_i, \ i = 1, \ldots, n \}.$$

Example   3.1 *Let P be a program with a variable x over the integers. Let s, s' be two program states such that $s(x) = 2$ and $s'(x) = -7$. Following are two possible abstractions.*

Abstraction 1:

$$A_1 = \{a_-, a_0, a_+\} \ and$$

$$h_1(d) = \begin{cases} a_+ & if \ d > 0 \\ a_0 & if \ d = 0 \\ a_- & if \ d < 0 \end{cases}$$

The set of atomic propositions is $AP_1 = \{ \hat{x} = a_-, \ \hat{x} = a_0, \ \hat{x} = a_+ \}$.
The labeling of states in the concrete model induced by A_1 and h_1 is:
$L_1(s) = \{\hat{x} = a_+\}$ *and* $L_1(s') = \{\hat{x} = a_-\}$.
Abstraction 2:

$$A_2 = \{a_{even}, \ a_{odd}\} \ and$$

$$h_2(d) = \begin{cases} a_{even} & if \ even(|d|) \\ a_{odd} & if \ odd(|d|) \end{cases}$$

The set of atomic propositions is $AP_2 = \{ \hat{x} = a_{even}, \ \hat{x} = a_{odd} \}$.
The labeling of states in the concrete model induced by A_2 and h_2 is:
$L_2(s) = \{\hat{x} = a_{even}\}$ *and* $L_2(s') = \{\hat{x} = a_{odd}\}$.

By restricting the state labeling we lose the ability to refer to the actual values of the program variables. However, many of the states are now indistinguishable and can be collapsed into a single abstract state.

Given A and h as above, we can now define the most precise abstract model, M_r, called the *reduced model*. First we extend the mapping $h : D \to A$ to n-tuples in $D \times \ldots \times D$:

$$h((d_1, \ldots, d_n)) = (h(d_1), \ldots, h(d_n)).$$

An abstract state $(a_1, \ldots, a_n)$ of M_r will represent the set of all states $(d_1, \ldots, d_n)$ such that $h((d_1, \ldots, d_n)) = (a_1, \ldots, a_n)$. Concrete states s_1, s_2 are said to be *equivalent* ($s_1 \sim s_2$) if and only if $h(s_1) = h(s_2)$, that is, both states are mapped to the same abstract state. Thus, each abstract state represents an equivalence class of concrete states.

Definition 3.1 *Given a concrete model M, an abstract domain A, and an abstraction mapping $h : D \to A$, the reduced model $M_r = (S_r, S_{0_r}, R_r, L_r)$ is defined as follows:*

- $S_r = A \times \ldots \times A$.
- $S_{0_r}(s_r) \Leftrightarrow \exists s \in S_0 : h(s) = s_r$.
- $R_r(s_r, t_r) \Leftrightarrow \exists s, t[\; h(s) = s_r \wedge h(t) = t_r \wedge R(s, t)\;]$.
- *For $s_r = (a_1, \ldots, a_n)$, $L_r(s_r) = \{\; \widehat{x}_i = a_i \mid i = 1, \ldots, n \;\}$.*

This type of abstraction is called *existential abstraction*.

Lemma 3.2 *The reduced model M_r is greater by the simulation preorder than the concrete model M; that is, $M \preceq M_r$.*

To see why the lemma is true, note that the relation $H = \{\; (s, s_r) \mid h(s) = s_r \;\}$ is a simulation preorder between M and M_r. The following corollary is a direct consequence of this lemma and Theorem 2.7.

Corollary 3.3 *For every ACTL* formula φ, if $M_r \models \varphi$ then $M \models \varphi$.*

3.1. DERIVING MODELS FROM THE PROGRAM TEXT

In the next section we explain how the reduced and approximated model for the system can be derived directly from a high-level description of the system. In order to avoid having to choose a specific programming language, we argue that the program can be described by means of first-order formulas. In this section we demonstrate how this can be done.

Let P be a program, and let $\overline{x} = (x_1, \ldots, x_n)$ and $\overline{x}' = (x'_1, \ldots, x'_n)$ be two copies of the program variables, representing the current and next state, respectively. The program will be given by two first-order formulas, $\mathcal{S}_0(\overline{x})$ and $\mathcal{R}(\overline{x}, \overline{x}')$, describing the set of initial states and the set of transitions. Let $\overline{d} = (d_1, \ldots, d_n)$ be a vector of values. The notation $\mathcal{S}_0[\overline{x} \leftarrow \overline{d}]$ indicates that for every $i = 1, \ldots, n$, the value d_i is substituted for the variable x_i in the formula $\mathcal{S}_0$. A similar notation is used for substitution in the formula $\mathcal{R}$.

Definition 3.4 *Let $S = D \times \ldots \times D$ be the set of states in a model M. The formulas $\mathcal{S}_0(\overline{x})$ and $\mathcal{R}(\overline{x}, \overline{x}')$ define the set of initial states S_0 and the set of transitions R in M as follows. Let $s = (d_1, \ldots, d_n)$ and $s' = (d'_1, \ldots, d'_n)$ be two states in S.*

- $S_0(s) \Leftrightarrow \mathcal{S}_0(\overline{x})[\overline{x} \leftarrow \overline{d}]$ *is true.*
- $R(s, s') \Leftrightarrow \mathcal{R}(\overline{x}, \overline{x}')[\overline{x} \leftarrow \overline{d}, \overline{x}' \leftarrow \overline{d}']$ *is true.*

The following example demonstrates how a program can be described by means of first-order formulas. A more elaborate explanation can be found in [17]. We assume that each statement in the program starts and ends with labels that uniquely define the corresponding locations in the program. The program locations will be represented in the formula by the variable pc (the *program counter*), which ranges over the set of program labels.

Example 3.2 *Given a program with one variable x that starts at label l_0, in any state in which x is even, the set of its initial states is described by the formula:*

$$S_0(pc, x) = pc = l_0 \land even(x).$$

The statement $l: x := e \; l'$ is described by the formula:

$$\mathcal{R}(pc, x, pc', x') = pc = l \land x' = e \land pc' = l'.$$

The statement $l: \text{if } x = 0 \text{ then } l_1: x := 1 \text{ else } l_2: x := x + 1 \; l'$ is described by the formula:

$$
\begin{aligned}
\mathcal{R}(pc, x, pc', x') = \; & ((\; pc = l \; \land \; x = 0 \; \land x' = x \land pc' = l_1) \lor \\
& (\; pc = l \; \land \; x \neq 0 \; \land x' = x \land pc' = l_2) \lor \\
& (pc = l_1 \land x' = 1 \land pc' = l') \lor \\
& (pc = l_2 \land x' = x + 1 \land pc' = l')).
\end{aligned}
$$

Note that checking the condition of the *if* statement takes one transition, along which the value of the program variable is checked but not changed. If the program contains an additional variable y, then $y' = y$ will be added to the description of each of the transitions above. This captures the fact that variables that are not assigned a new value keep their previous value.

3.2. DERIVING ABSTRACT MODELS

Given S_0 and $\mathcal{R}$ that describe a concrete model M, we would like to define formulas $\widehat{S_0}$ and $\widehat{\mathcal{R}}$ that describe the reduced model M_r. The new formulas will be defined over variables $\widehat{x_i}$, which range over the abstract domain. The formulas will determine for abstract states whether they are initial and whether there is a transition connecting them. For this purpose, we first define a derivation of a formula over variables $\widehat{x_1}, \ldots, \widehat{x_k}$ from a formula over $x_1, \ldots, x_k$.

Definition 3.5 *Let ϕ be a first-order formula over variables $x_1, \ldots, x_k$. The formula $[\phi]$ over $\widehat{x_1}, \ldots, \widehat{x_k}$ is defined as follows:*

$$[\phi](\widehat{x_1}, \ldots, \widehat{x_k}) = \exists x_1 \ldots x_k \left(\bigwedge_{i=1}^{k} h(x_i) = \widehat{x_i} \land \phi(x_1, \ldots, x_k) \right).$$

Lemma   3.6 *Let S_0 and $\mathcal{R}$ be the formulas describing a model M over states in $D \times \ldots \times D$. Then the formulas $\widehat{S}_0 = [S_0]$ and $\widehat{\mathcal{R}} = [\mathcal{R}]$ describe the reduced model M_r over $A \times \ldots \times A$.*

The lemma holds since M_r is defined by existential abstraction (see Definition 3.1). This is directly reflected in $[S_0]$ and $[\mathcal{R}]$.

Using $\widehat{S}_0$ and $\widehat{\mathcal{R}}$ allows us to build the reduced model M_r without first building the concrete model M. However, the formulas S_0 and $\mathcal{R}$ might be quite large. Thus, applying existential quantification to them might be computationally expensive. We therefore define a transformation $\mathcal{T}$ on first-order formulas. The idea of $\mathcal{T}$ is to push the existential quantification inwards, so that it is applied to simpler formulas.

Definition   3.7 *Let ϕ be a first-order formula in positive normal form. Then the following holds:*

1. *If p is a primitive relation, then $\mathcal{T}(p(x_1, \ldots, x_k)) = [p](\widehat{x}_1, \ldots, \widehat{x}_k)$ and $\mathcal{T}(\neg p(x_1, \ldots, x_k)) = [\neg p](\widehat{x}_1, \ldots, \widehat{x}_k)$.*
2. *$\mathcal{T}(\phi_1 \wedge \phi_2) = \mathcal{T}(\phi_1) \wedge \mathcal{T}(\phi_2)$.*
3. *$\mathcal{T}(\phi_1 \vee \phi_2) = \mathcal{T}(\phi_1) \vee \mathcal{T}(\phi_2)$.*
4. *$\mathcal{T}(\forall x \phi) = \forall \widehat{x} \mathcal{T}(\phi)$.*
5. *$\mathcal{T}(\exists x \phi) = \exists \widehat{x} \mathcal{T}(\phi)$.*

We can now define an *approximated* abstract model M_a. It is defined over the same set of states as the reduced model, but its set of initial states and set of transitions are defined using the formulas $\mathcal{T}(S_0)$ and $\mathcal{T}(\mathcal{R})$. The following lemma ensures that every initial state of M_r is also an initial state of M_a. Moreover, every transition of M_r is also a transition of M_a.

Lemma   3.8 *For every first-order formula ϕ in positive normal form, $[\phi]$ implies $\mathcal{T}(\phi)$. In particular, $[S_0]$ implies $\mathcal{T}(S_0)$ and $[\mathcal{R}]$ implies $\mathcal{T}(\mathcal{R})$.*

Note that the other direction does not hold. Cases 2 and 4 of Definition 3.7 result in nonequivalent formulas.

Corollary   3.9 $M \preceq M_r \preceq M_a$.

By allowing M_a to have more behaviors than M_r, we increase the likelihood that it will falsify ACTL* formulas that are actually true in the concrete model and possibly true in M_r. This reflects the tradeoff between the precision of the model and its ease of computation.

In practice, there is no need to construct formulas in order to build the approximated model. The user should provide *abstract predicates* $[p]$ and $[\neg p]$ for every basic action p in the program (e.g. conditions, assignments of mathematical expressions). Based on these, the approximated model can be constructed automatically.

In [15, 47] several data abstractions have been suggested and used to verify meaningful properties of interesting programs.

4. Counterexample-guided abstraction refinement

In the previous section we showed how an abstract model can be constructed based on an abstract domain and a mapping, both provided by the user. Unfortunately, choosing a suitable abstraction is not trivial for large systems and requires considerable creativity.

In this section we describe a technique (presented in [16]) that determines the required domain and mapping automatically, based on an analysis of the program text. This technique differs from data abstraction in that the abstract domains and abstract mappings are defined for *clusters of variables* rather than single variables. The clusters are chosen according to dependencies that are found among the variables in the program. Abstracting clusters of variables results in an abstract model that reflects the system behavior more precisely.

As in the previous section, we use existential abstraction. Existential abstraction guarantees that ACTL* properties true of the abstract model are also true of the concrete model. However, if a property is false in the abstract model, then the counterexample produced by the model checking algorithm may be the result of some behavior that is not present in the concrete model. In this case, we refine the abstraction to eliminate the erroneous behavior from the abstract model. The refinement is determined by information obtained from the counterexample.

The suggested method has the following steps:

- **Generating an initial abstraction**.
  This involves the construction of variable clusters for variables which interfere with each other via conditions in the program.
- **Model-checking the abstract model**.
  If the formula is true, we conclude that the concrete model satisfies the formula and stop. If a counterexample $\widehat{T}$ is found, we check whether $\widehat{T}$ is a counterexample in the concrete model. If it is, we conclude that the concrete model does not satisfy the formula and stop. Otherwise, $\widehat{T}$ is a spurious counterexample, and we proceed to step 3.
- **Refining the abstraction**.
  This is done by splitting one abstract state so that $\widehat{T}$ is not included in the new abstract model. We then go back to step 2.

4.1. ASSUMPTIONS

There are several assumptions that are needed in order to make our method fully automatic and effective.

- The program is finite-state, i.e., each variable is over a finite domain.
- We use BDD-based algorithms. BDD [10] is a data structure for representing Boolean functions. BDDs are often very concise in their space requirements. Sets of states and sets of transitions of Kripke models can easily be represented by BDDs. Moreover, most operations applied in model checking algorithms can be implemented efficiently with BDDS. As a result, BDD-based model checking [13, 49] (called *symbolic model checking*) is very useful in practice.
- The full model is too large to fit into memory, even when represented by BDDs. However, subsets of the full state space can be held in memory. In particular, we will maintain as BDDs the sets of concrete states that are mapped to a specific abstract state.
- The transition relation of the full model is available. If it is too large, it is held *partitioned*. There are known techniques for handling partitioned transition relations in model checking [12].

4.2. GENERATING THE INITIAL ABSTRACTION

Let P be a program over variables $x_1, \ldots, x_n$. Suppose that each variable x_i is defined over domain D_{x_i}, which is finite. Finiteness of the domains is necessary in order for the method to be fully automatic. Let φ be the ACTL* formula to be checked on P.

Atomic formulas will be defined over program variables, constants and relation symbols. For instance, $x > y$ and $x = 1$ are atomic formulas. Boolean combinations of atomic formulas are used as conditions in the program. The logic ACTL* is also defined over atomic formulas of this type. We use *Atom* to denote the set of all atomic formulas appearing in P and φ.

Given a state $s = \bar{d} = (d_1, \ldots, d_n)$ and an atomic formula p over $x_1, \ldots, x_n$, we write $s \models p$ if $p[\bar{x} \leftarrow \bar{d}] = true$. That is, p is evaluated to *true* when each variable x_i is assigned the value d_i.

As before, the concrete model M of program P is defined over states $S = D_{x_1} \times \ldots \times D_{x_n}$, where each state in M is labeled with the set of atomic formulas from *Atom* that are true in that state.

We say that two atomic formulas *interfere* if the sets of variables appearing in them are not disjoint. Let $\equiv_I$ be that equivalence relation over *Atom* which is the reflexive, transitive closure of the interference relation. The equivalence class of an atomic formula $f \in Atom$ is called the *cluster of f* and is denoted by $[f]$. Note that if two atomic formulas f_1 and f_2 have nondisjoint sets of variables, then $[f_1] = [f_2]$. That is, a variable cannot occur in formulas that belong to different formula clusters.

Consequently, we can define an equivalence relation $\equiv_V$ on the program variables as follows:

$$x_i \equiv_V x_j \text{ if and only if } x_i \text{ and } x_j \text{ appear in atomic formulas}$$
$$\text{that belong to the same formula cluster.}$$

The equivalence classes of $\equiv_V$ are called variable clusters. Let $\{FC_1, \ldots, FC_m\}$ and $\{VC_1, \ldots, VC_m\}$ be the set of formula clusters and variable clusters, respectively. Each variable cluster VC_i is associated with a domain $D_{VC_i} = \prod_{x \in VC_i} D_x$, representing the value of all the variables in this cluster. Note that $S = D_{VC_1} \times \ldots \times D_{VC_m}$.

Example 4.1 *Let Atom* $= \{ x > y, \ x = 1, \ z = t \}$. *Then there are two formula clusters, $FC_1 = \{ x > y, \ x = 1 \}$ and $FC_2 = \{ z = t \}$, and two variable clusters, $VC_1 = \{x, y\}$ and $VC_2 = \{z, t\}$.*

The initial abstraction is defined by $h = (h_1, \ldots, h_m)$, where h_i is defined over D_{VC_i} as follows. For $(d_1, \ldots, d_k), (e_1, \ldots, e_k) \in D_{VC_i}$,

$$h_i((d_1, \ldots, d_k)) = h_i((e_1, \ldots, e_k)) \iff$$

$$\forall f \in FC_i : \ (d_1, \ldots, d_k) \models f \iff (e_1, \ldots, e_k) \models f.$$

Thus, two states are h_i-equivalent if and only if they satisfy the same formulas in FC_i. They are h-equivalent if and only if they satisfy the same formulas in *Atom*.

The reduced model for the abstraction h will be defined over abstract states that are the equivalence classes of the h-equivalence. Each equivalence class will be labeled by all formulas from *Atom* which are true in all states in the class. The initial states and transition relation are defined by the existential abstraction, as before.

Example 4.2 *Let P be a program with variables x, y over domain $\{0, 1\}$ and z, t over domain $\{true, false\}$. Let $FC_1 = \{ x > y, \ x = 1 \}$, $FC_2 = \{ z = t \}$, $VC_1 = \{x, y\}$ and $VC_2 = \{z, t\}$. Then, the h_1-equivalence classes are:*

$$E_{11} = \{(0, 0), (0, 1)\}, \ E_{12} = \{(1, 0)\}, \ E_{13} = \{(1, 1)\}.$$

The h_2-equivalence classes are:

$$E_{21} = \{(false, false), (true, true)\}, \ E_{22} = \{(false, true), (true, false)\}.$$

The reduced model contains six states labeled by:

$$
\begin{aligned}
L_r((E_{11}, E_{21})) &= \{z = t\} \\
L_r((E_{12}, E_{21})) &= \{x > y, x = 1, z = t\} \\
L_r((E_{13}, E_{21})) &= \{x = 1, z = t\} \\
L_r((E_{11}, E_{22})) &= \emptyset \\
L_r((E_{12}, E_{22})) &= \{x > y, x = 1\} \\
L_r((E_{13}, E_{22})) &= \{x = 1\}
\end{aligned}
$$

Example 4.2 shows that abstracting variable clusters rather than single variables allows smaller abstract domains that are more precise. Valuations of a whole variable cluster determine the truth of the conditions in the program, and thus influence its control flow. Abstracting a whole cluster allows us to identify all states that satisfy the same conditions and then abstract them together. Thus, the abstract model reflects more closely the control flow of the program.

4.3. IDENTIFYING SPURIOUS PATH COUNTEREXAMPLES

Once the reduced model is built, we can run the model checking algorithm on it. Suppose the model checking stopped with a path counterexample. Such a counterexample is produced, for instance, when the checked property is $\mathbf{AG}\,p$ for some atomic formula p. The path leads from an initial state to a state that falsifies p. Our goal is to find whether there is a corresponding path in the concrete model from an initial state to a state that falsifies p.

Let $\widehat{T} = \widehat{s}_1 \ldots \widehat{s}_n$ be the path counterexample in the reduced model. For an abstract state $\widehat{s}$, $h^{-1}(\widehat{s})$ denotes the set of concrete states that are mapped to $\widehat{s}$, i.e., $h^{-1}(\widehat{s}) = \{\, s \mid h(s) = \widehat{s}\,\}$. $h^{-1}(\widehat{T})$ denotes the set of all concrete paths from an initial state that correspond to $\widehat{T}$, i.e.,

$$
h^{-1}(\widehat{T}) = \{\, s_1 \ldots s_n \mid \bigwedge_{i=1}^{n} h(s_i) = \widehat{s}_i \,\wedge\, S_0(s_1) \,\wedge\, \bigwedge_{i=1}^{n-1} R(s_i, s_{i+1}) \,\}.
$$

We say that a path $\widehat{T} = \widehat{s}_1 \ldots \widehat{s}_n$ *corresponds to a real counterexample* if there is a path $\pi = s_1 \ldots s_n$ such that for all $1 \leq i \leq n$, $h(s_i) = \widehat{s}_i$. Moreover, π starts at an initial state and its final state falsifies p. Note that if $h(s_i) = \widehat{s}_i$, then s_i and $\widehat{s}_i$ satisfy the same atomic formulas. Thus, it immediately follows that $\widehat{T}$ corresponds to a real counterexample if and only if $h^{-1}(\widehat{T})$ is not empty.

The algorithm SplitPATH in Figure 3 checks whether $h^{-1}(\widehat{T})$ is empty. In fact, it computes a sequence $S_1, \ldots, S_n$ of sets of states. For each i, S_i

contains states that correspond to $\widehat{s_i}$ and are also successors of states in S_{i-1}. It starts with $S_1 = h^{-1}(\widehat{s_1}) \cap S_0$, which includes all initial states that correspond to $\widehat{s_1}$. If some S_i turns out to be empty, then SplitPATH returns both the place i where the failure occurred and the last nonempty set S_{i-1}.

SplitPATH uses the transition relation R of the concrete model M in order to compute $Img(S_{i-1}, R)$, which is the set of all successors of states in S_{i-1}. All operations in SplitPATH, including Img, are effectively implemented with BDDs (symbolic implementation). The following lemma proves the correctness of SplitPATH.

Lemma 4.1 $\widehat{T}$ *corresponds to a real counterexample if and only if for all* $1 \le i \le n$, $S_i \ne \emptyset$.

Algorithm SplitPATH($\widehat{T}$)

$S := h^{-1}(\widehat{s_1}) \cap S_0$;
$j := 1$;
while $(S \ne \emptyset$ and $j < n)$ **do**
    $j := j + 1$;
    $S_{\text{prev}} := S$;
    $S := Img(S, R) \cap h^{-1}(\widehat{s_j})$;
end while;
if $S \ne \emptyset$ **then** output "counterexample exists"
    else output j, S_{prev}

Figure 3. SplitPATH checks if an abstract path is spurious.

4.4. IDENTIFYING SPURIOUS LOOP COUNTEREXAMPLES

Suppose that the model checking returns a loop counterexample. This may occur, for instance, when the property **AF** p is checked. The counterexample exhibits an infinite path along which p never holds. A loop counterexample will be of the form

$$\widehat{T} = \widehat{s_1} \ldots \widehat{s_i} \langle \widehat{s_{i+1} \ldots s_n} \rangle^{\omega},$$

where the sequence $\widehat{s_{i+1}} \ldots \widehat{s_n}$ repeats forever.

Some difficulties arise when trying to determine whether a loop counterexample in the abstract model corresponds to a real counterexample in the concrete model. First, an abstract loop may correspond to different loops of different sizes in the concrete model. Furthermore, the loops may start at different stages of the unwinding. Clearly, the unwinding must eventually result in a periodic path. However, a careful analysis is needed in order to see that a polynomial number of unwindings is sufficient. More precisely, let

$$min = \min\{ |h^{-1}(\widehat{s_{i+1}})|, \ldots, |h^{-1}(\widehat{s_n})| \}.$$

That is, min is the size of the smallest set of concrete states that corresponds to one of the abstract states on the loop. Then $min + 1$ unwindings are sufficient. This is formalized in the following lemma. Let $\widehat{T} = \widehat{s_1} \ldots \widehat{s_i} \langle \widehat{s_{i+1}} \ldots \widehat{s_n} \rangle^\omega$ and $\widehat{T}_{unwind} = \widehat{s_1} \ldots \widehat{s_i} \langle \widehat{s_{i+1}} \ldots \widehat{s_n} \rangle^{min+1}$.

Lemma 4.2 $\widehat{T}$ *corresponds to a concrete counterexample if and only if* $h^{-1}(\widehat{T}_{unwind})$ *is not empty.*

Following is an intuitive explanation for the correctness of the "if" clause of the lemma. Assume $h^{-1}(\widehat{T}_{unwind}) \neq \emptyset$. For any concrete path π in $h^{-1}(\widehat{T}_{unwind})$, the suffix π^{i+1} of π goes $min + 1$ times through each of the sets $h^{-1}(\widehat{s_j})$, for $j = i + 1, \ldots, n$. Suppose $min = |h^{-1}(\widehat{s_M})|$. Then at least one state in $h^{-1}(\widehat{s_M})$ repeats twice along π^{i+1}, thus forming a concrete loop. Replacing π^{i+1} with this loop in π results in a loop counterexample in the concrete model.

From this lemma we conclude that the loop counterexample can be reduced to a path counterexample. Figure 4 presents the algorithm Split-LOOP, which applies SplitPATH to $\widehat{T}_{unwind}$ in order to check if an abstract loop is spurious. **LoopIndex**(j) computes the index in the unwound $\widehat{T}_{unwind}$ of the abstract state at position j. That is,

LoopIndex$(j) =$ if $j \leq n$ then j else$((j - (i + 1)mod(n - i)) + (i + 1))$.

Thus, SplitLOOP returns two indices k and p, which are consecutive on the loop, and the set $S_{prev} \subseteq h^{-1}(\widehat{s_k})$. In the refinement step, the loop counterexample can now be treated similarly to the path counterexample.

4.5. REFINING THE ABSTRACTION

Once we have realized that a path or a loop counterexample is spurious, we would like to eliminate it by refining our abstraction. We will only describe the refinement process for path counterexamples. The treatment of loop counterexamples is similar. Let j, S_{prev} be the output of SplitPATH, where $S_{prev} \subseteq h^{-1}(\widehat{s_{j-1}})$. We observe that the states in $h^{-1}(\widehat{s_{j-1}})$ can be partitioned into three subsets:

- S_D: **dead-end states**, which are reachable from an initial state in $h^{-1}(\widehat{s_1})$ but have no outgoing transition to states in $h^{-1}(\widehat{s_j})$. Note that $S_{prev} = S_D$.

Algorithm SplitLOOP$(\widehat{T})$

$min = \min\{\ |h^{-1}(\widehat{s_{i+1}})|, \ldots, |h^{-1}(\widehat{s_n})|\ \}$;
$\widehat{T}_{\text{unwind}} = \mathbf{unwind}(\widehat{T}, min + 1)$;
Compute j and S_{prev} as in **SplitPATH**$(\widehat{T}_{\text{unwind}})$;
$k := \mathbf{LoopIndex}(j)$;
$p := \mathbf{LoopIndex}(j + 1)$;
output S_{prev}, k, p

Figure 4. SplitLOOP checks if an abstract loop is spurious

- S_B: **bad states**, which are not reachable but have outgoing transitions to states in $h^{-1}(\widehat{s_j})$.
- S_I: **irrelevant states**, which are neither reachable nor dead.

Since existential abstraction is used, the dead-end states induce a path to $\widehat{s_{j-1}}$ in the abstract model. The bad states induce a transition from $\widehat{s_{j-1}}$ to $\widehat{s_j}$. Thus, a spurious path leading to $\widehat{s_j}$ is obtained.

In order to eliminate this path we need to refine the abstraction mapping h so that the dead-end states and bad states do not belong to the same abstract state.

Recall that each abstract state corresponds to an h-equivalence class of concrete states. The goal is to find a refinement that keeps the number of new equivalence classes as small as possible. It turns out that when irrelevant states are present, the problem of finding the optimal refinement is NP-complete. However, if the set of irrelevant states is empty, the problem can be solved in polynomial time. A possible heuristic associates S_I with S_B and then applies refinement, which separates them from S_D. The resulting refinement is not optimal, but gives good results in practice.

We now show how the model is refined in case $S_I = \emptyset$. Recall that $S = D_{VC_1} \times \ldots \times D_{VC_m}$ and $h = (h_1, \ldots, h_m)$, where h_i is defined over D_{VC_i}. The equivalence relations $\equiv (\equiv_i)$ are the sets of pairs of h-equivalent (h_i-equivalent) elements. For $S_D \subseteq S$, $i \in \{1, \ldots, m\}$, and $a \in D_{VC_i}$, we define the projection set $proj(S_D, i, a)$ by:

$$Proj(S_D, i, a) = \{\ (d_1, \ldots, d_{i-1}, d_{i+1}, \ldots, d_m)\ |$$

$$(d_1, \ldots, d_{i-1}, a, d_{i+1}, \ldots, d_m) \in S_D\ \}.$$

The *refinement procedure* checks, for each i and for each pair (a, b) in $\equiv_i$, whether $proj(S_D, i, a) = proj(S_D, i, b)$. If not, then (a, b) is eliminated from $\equiv_i$. By eliminating (a, b) from $\equiv_i$, we partition the equivalence class $h^{-1}(\widehat{s_{j-1}})$. Since $proj(S_D, i, a) \neq proj(S_D, i, b)$, there are states

$$s_a = (d_1, \ldots, d_{i-1}, a, d_{i+1}, \ldots, d_m) \text{ and } s_b = (d_1, \ldots, d_{i-1}, b, d_{i+1}, \ldots, d_m)$$

such that $s_a \in S_D$ and $s_b \notin S_D$ (or vice versa). This implies that $s_b \in S_B$ (since S_I is empty), and therefore s_a and s_b should not be in the same equivalence class. Removing (a, b) from $\equiv_i$ also removes (s_a, s_b) from $\equiv$, as required.

The refinement procedure continues to refine the abstraction mapping by partitioning equivalence classes until a real counterexample is found or until the ACTL* formula holds. Since each equivalence class is finite and nonempty, the process always terminates.

Theorem 4.3 *Given a model M and an ACTL* formula φ whose counterexample is either path or loop, the refinement algorithm finds a model $\widehat{M}$ such that $\widehat{M} \models \varphi \Leftrightarrow M \models \varphi$.*

In [16], several practical improvements are presented. The method is also experimented on large examples.

5. Abstract interpretation

In this section we show how abstractions preserving full CTL* and its universal and existential subsets can be defined within the framework of *abstract interpretation*. We will see that abstract interpretation can be used to obtain abstract models which are more precise and therefore preserve more properties than the existential abstraction presented previously [23, 22]. The abstraction suggested in this section abstracts the state space rather than the data domain and can be applied to infinite as well as finite sets. It actually preserves the full μ-calculus (see [22] for more details).

Given an abstract domain, abstract interpretation provides a general framework for automatically "interpreting" systems on the abstract domain. The classical abstract interpretation framework [19] was used to prove safety properties, and does not consider temporal logic or model checking. Hence, it usually abstracts sets of states. Here, on the other hand, we are interested in properties of computations. We therefore abstract Kripke models, including their sets of states and transitions.

The models we use in this section are slightly different from the Kripke models presented in Definition 2.1. Instead of using the set AP of atomic propositions to label states, we use *literals* from the set

$$Lit = AP \cup \{\neg p \mid p \in AP\}.$$

The Kripke model $M = (S, S_0, R, L)$ is defined as before for S, S_0, and R. The labeling function $L : S \to \mathcal{P}(Lit)$ is required to satisfy

$$p \in L(s) \Rightarrow \neg p \notin L(s) \text{ and } \neg p \in L(s) \Rightarrow p \notin L(s).$$

Recall that labeling a state with a formula means that the formula is true in that state. Thus, by the above, p and $\neg p$ cannot be true together in a state. It is not required, however, that $p \in L(s) \Leftrightarrow \neg p \notin L(s)$. Hence, it is possible that neither p nor $\neg p$ will be true in s.

It is straightforward to extend the semantics of CTL* formulas for these models. Only for literals should it be changed. The remaining semantics is identical to Definition 2.3. For literals, we can define the semantics by

1. If $p \in AP$ then
   $s \models p$ if and only if $p \in L(s)$; $s \models \neg p$ if and only if $\neg p \in L(s)$.

As a result of this change, however, the semantics of any CTL* formula may now be such that neither $s \models \varphi$ nor $s \models \neg\varphi$.

We will now present some basic notions required for the development of an abstract interpretation framework in the context of Kripke models and temporal logic specifications. For more details see [22, 24].

Abstract interpretation assumes two partially ordered sets, $(C, \sqsubseteq)$ and $(A, \leq)$, of the concrete and abstract domains. In addition, two mappings are used: The abstraction mapping $\alpha : C \to A$ and the concretization mapping $\gamma : A \to C$.

Definition 5.1 $(\alpha : C \to A, \gamma : A \to C)$ *is a* Galois connection *from* $(C, \sqsubseteq)$ *to* $(A, \leq)$ *if and only if*

- α *and* γ *are total and monotonic.*
- *for all* $c \in C$, $\gamma(\alpha(c)) \sqsupseteq c$.
- *for all* $a \in A$, $\alpha(\gamma(a)) \leq a$.

(α, γ) *is a* Galois insertion *if, in addition, the order* $\leq$ *on* A *is defined by the order* $\sqsupseteq$ *on* C *as follows:*

$$a \leq a' \Leftrightarrow \gamma(a) \sqsubseteq \gamma(a').$$

Note that the requirement for Galois insertion is stronger than the requirement for monotonicity of γ. Assume that $\leq$ and $\sqsupseteq$ are partial orders in which two elements are *equal* if and only if each is smaller than the other. Then, $\gamma(a_1) = \gamma(a_2)$ implies $a_1 = a_2$. Therefore, for Galois insertion

$$\alpha(\gamma(a)) = a.$$

For $a \leq a'$ we say that a is *more precise* than a', or a' *approximates* a. Similarly, for $c \sqsubseteq c'$.

We now present our framework in which the choice of concrete and abstract domains is motivated by the goal of model abstraction. Given a model M, we choose $(\mathcal{P}(S), \subseteq)$ as the concrete domain. We choose a set of abstract states $\widehat{S}$ and use the Galois insertion so that the partial order for $\widehat{S}$ is determined by

$$a \leq a' \Leftrightarrow \gamma(a) \subseteq \gamma(a').$$

Since γ and α are total and monotonic, $\widehat{S}$ must include a greatest element *top*, denoted $\top$, so that $\alpha(S) = \top$ and $\gamma(\top) = S$. Moreover, for every $S' \subseteq S$, there must be $a \in \widehat{S}$ such that $\alpha(S') = a$. However, associating each subset of concrete states with an abstract state does not imply that there must be a *different* abstract state for each subset.

Example 5.1 *A correct (though uninteresting) abstraction chooses $\widehat{S} = \{\top\}$ with $\gamma(\top) = S$ and for all $S' \subseteq S$, $\alpha(S') = \top$.*

The following abstraction is more interesting.

Example 5.2 *Let S be the set of all states with one variable x over the natural numbers. Let $\widehat{S} = \{ \text{grt\_5, leq\_5, } \top \}$ where $\gamma(\text{grt\_5}) = S_{>5} = \{ s \in S | s(x) > 5 \}$ and $\gamma(\text{leq\_5}) = S_{\leq 5} = \{ s \in S | s(x) \leq 5 \}$. (Also, $\gamma(\top) = S$, as is always the case).*

In order to guarantee that $\alpha(\gamma(a)) = a$, we must define $\alpha(S_{>5}) = \text{grt\_5}$ and $\alpha(S_{\leq 5}) = \text{leq\_5}$. This also guarantees that $\gamma(\alpha(S_{>5})) \supseteq S_{>5}$ and $\gamma(\alpha(S_{\leq 5})) \supseteq S_{\leq 5}$.

Consider now the set $S_{>6} = \{ s \in S | s(x) > 6 \}$. Both $\alpha(S_{>6}) = \text{grt\_5}$ and $\alpha(S_{>6}) = \top$ will satisfy $\gamma(\alpha(S_{>6})) \supseteq S_{>6}$. However, more precise abstraction is desired. Thus, since $\text{grt\_5} \leq \top$, we choose $\alpha(S_{>6}) = \text{grt\_5}$.

On the other hand, for $S_{>2} = \{ s \in S | s(x) > 2 \}$ the only correct choice is $\alpha(S_{>2}) = \top$.

Remark: The Galois insertion is less restrictive than existential abstraction in the sense that it allows nondisjoint subsets of states to be mapped to different abstract states. Furthermore, concrete states mapped to the same abstract state do not necessarily satisfy the same atomic formulas. In contrast, existential abstraction partitions the concrete state space into disjoint equivalence classes, so that all states in the same class satisfy the same set of atomic formulas.

5.1. THE ABSTRACT MODEL

The abstraction and concretization mappings defined so far determine the set of abstract states and their relationship with the set of concrete states. In order to define the abstract model we still need to define the state labeling, the set of initial states, and the transition relation of the abstract model.

We start with a definition that will be used when we present the abstract transition relation.

Definition 5.2 *Let A and B be sets and $R \subseteq A \times B$. The relations $R^{\exists\exists}, R^{\forall\exists} \subseteq \mathcal{P}(A) \times \mathcal{P}(B)$ are defined as follows:*

- $R^{\exists\exists} = \{ (X, Y) \mid \exists x \in X \ \exists y \in Y : R(x, y) \}$
- $R^{\forall\exists} = \{ (X, Y) \mid \forall x \in X \ \exists y \in Y : R(x, y) \}$

If R is a transition relation and X and Y are subsets of states, then $R^{\exists\exists}(X, Y)$ if and only if some state in X can make a transition to some state in Y. $R^{\forall\exists}(X, Y)$ if and only if every state in X can make a transition to some state in Y. Note that the transition relation defined by existential abstraction can be viewed as an $R^{\exists\exists}$ relation over the equivalence classes represented by the abstract states.

Given a set $\widehat{S}$ of abstract states, our goal is to define a precise abstract model $\widehat{M} = (\widehat{S}, \widehat{S_0}, \widehat{R}, \widehat{L})$ such that for every ACTL* formula φ over atomic formulas in Lit, and for every abstract state $a \in \widehat{S}$, the following requirement holds:

$$\widehat{M}, a \models \varphi \implies M, \gamma(a) \models \varphi. \tag{1}$$

5.1.1. The abstract labeling function

The abstract labeling function $\widehat{L}$ is defined so that Requirement (1) holds for the literals in Lit. For every $p \in Lit$,

$$p \in \widehat{L}(a) \iff \forall s \in \gamma(a) : p \in L(s).$$

Thus, an abstract state is labeled by literal p if and only if all states in its concretization are labeled by p. However, since our abstraction mapping does not require that all these states be identically labeled, it is possible that neither $p \in \widehat{L}(a)$ nor $\neg p \in \widehat{L}(a)$.

Explicitly labeling the negation of atomic formulas allows us to distinguish between the case in which "all concrete states do not satisfy p" and the one in which "not all concrete states satisfy p."

The following lemma states that less precise states satisfy fewer literals. Consequently, it is desirable to map subsets of states to their most precise abstraction (see Example 5.2).

Lemma 5.3 *For $a, a' \in \widehat{S}$, if $a' \geq a$ then $\forall p \in Lit, \ a' \models p \Rightarrow a \models p$.*

5.1.2. The abstract initial states

The set of initial abstract states is defined by

$$\widehat{S_0} = \{\alpha(\{s\}) \mid s \in S_0 \}.$$

This guarantees Requirement (1) on the level of models. That is, $\widehat{M} \models \varphi \Rightarrow M \models \varphi$. To see why this is true, note that

$$\widehat{M} \models \varphi \implies \forall a \in \widehat{S_0} : \widehat{M}, a \models \varphi \implies$$

$$\forall a \in \widehat{S_0} : M, \gamma(a) \models \varphi \implies \forall s \in S_0 : M, s \models \varphi \implies M \models \varphi.$$

As alternative definition might be $\widehat{S_0} = \alpha(S_0)$. It also satisfies Requirement (1). However, for each $s \in S_0$, $\alpha(\{s\}) \leq \alpha(S_0)$. Thus, the alternative definition suggests a single initial state which is less precise and therefore enables verification of fewer properties.

5.1.3. The abstract transition relation

A definition that is similar to existential abstraction could work in the case where ACTL* must be preserved. Using the notation of Definition 5.2, the abstract transition relation can be defined by:

$$\widehat{R}(a, b) \iff R^{\exists\exists}(\gamma(a), \gamma(b)).$$

However, as for the other components of the abstract model, the abstract interpretation framework provides the means for a more precise definition. Next we present an abstract transition relation that is more precise. It is denoted by $\widehat{R^A}$ in order to emphasize that it preserves ACTL*.

$$\widehat{R^A}(a, b)) \iff b \in \{\alpha(Y) \mid Y \in min\{Y' \mid R^{\exists\exists}(\gamma(a), Y')\}\}.$$

The difference between $\widehat{R}$ and $\widehat{R^A}$ can be explained as follows. Given an abstract state a, consider all $Y' \subseteq S$ such that there is a transition from some state in $\gamma(a)$ to some state in Y' (i.e., $R^{\exists\exists}(\gamma(a), Y')$). Then, $\widehat{R}$ connects a to $\alpha(Y')$ for each of these Y'. On the other hand, $\widehat{R^A}$ connects a to all $\alpha(Y)$ which are minimal (by the inclusion order) among these Y'. Clearly, $\widehat{R^A}$ connects a to fewer states, which are more precise. Note also that minimal Y's are always singletons.

Example 5.3 *The following example shows the difference between $\widehat{R^A}$ and $\widehat{R}$. It also demonstrates the ability of $\widehat{R^A}$ to verify more properties than $\widehat{R}$.*
Let $M = (S, S_0, R, L)$ where

314

- $S = \{s_1, s_2, s_3\}$,
- $S_0 = \{s_1\}$,
- $R = \{(s_1, s_2), (s_1, s_3), (s_2, s_2), (s_3, s_2)\}$;
- $L(s_1) = \{p\}$, $L(s_2) = \{p, q\}$, $L(s_3) = \{\neg p, q\}$.

$\widehat{M}$ is defined by

- $\widehat{S} = \{a_1, a_{23}, \top\}$ where

  • $\gamma(a_1) = \{s_1\}$, $\gamma(a_{23}) = \{s_2, s_3\}$ and $\gamma(\top) = S$.

  • $\alpha(\{s_1\}) = a_1$, $\alpha(\{s_2\}) = \alpha(\{s_3\}) = \alpha(\{s_2, s_3\}) = a_{23}$,
    $\alpha(\{s_1, s_2\}) = \alpha(\{s_1, s_3\}) = \alpha(S) = \top$.

- $\widehat{S_0} = \{a_1\}$,
- $\widehat{L}(a_1) = \{p\}$, $\widehat{L}(a_{23}) = \{q\}$ and $\widehat{L}(\top) = \emptyset$.
- $\widehat{R^A} = \{(a_1, a_{23}), (a_{23}, a_{23}), (\top, a_{23})\}$ and
  $\widehat{R} = \widehat{R^A} \cup \{(a_1, \top), (a_{23}, \top)(\top, a_{23}), (\top, \top)\}$.

Suppose we would like to verify the property $\mathbf{AX\,AG}\,q$ for M. When we check the property on the model defined by $\widehat{R}$, we find that it is false, and therefore we do not know whether it holds for M. However, if we check it on the model defined by $\widehat{R^A}$, since it is true for this model we can conclude that it is true for M as well.

Lemma 5.4 Let M be a model and $\widehat{M} = (\widehat{S}, \widehat{S_0}, \widehat{R^A}, \widehat{L})$ be an abstract model for M. Then $M \preceq \widehat{M}$. Thus, for every $ACTL^*$ formula φ and for every abstract state $a \in \widehat{S}$, $\widehat{M}, a \models \varphi \Rightarrow M, \gamma(a) \models \varphi$.

To show that $M \preceq \widehat{M}$, we define a simulation relation $H \subseteq S \times \widehat{S}$. In order to enable the simulation to relate abstract and concrete states we need to change the requirement on the state labeling: If $H(s, a)$ then $\widehat{L}(a) \subseteq L(s)$. The relation $H(s, a) \Leftrightarrow s \in \gamma(a)$ can now be shown to be a simulation preorder. For $ACTL^*$, Theorem 2.7 still holds with the new definition of $\preceq$ and thus implies the lemma.

5.2. ABSTRACT MODEL PRESERVING ECTL*

Until now we have only been concerned with abstractions preserving $ACTL^*$. In this section we show how to define an abstraction which preserves $ECTL^*$. The abstract model is defined in such a way that if an $ECTL^*$ formula is true for that model then it is also true for the concrete model. In the next section we show how to combine the abstractions for $ACTL^*$ and $ECTL^*$ into one abstraction that weakly preserves all of CTL^*. Recall that bisimulation also preserves full CTL^*. However, bisimulation provides strong preservation and therefore usually allows less reduction in the abstract model.

The ECTL*-preserving abstract model is identical to the ACTL*-preserving model $\widehat{M}$ defined above, except that the transition relation $\widehat{R^A}$ is replaced by a different transition relation, $\widehat{R^E}$.

The following observation explains the difference between $\widehat{R^A}$ and $\widehat{R^E}$. In order to preserve ACTL*, the set of abstract transitions should represent each of the concrete transitions. Additional transitions are also allowed. The abstract model then includes every behavior of the concrete model. Hence, every ACTL* property true for the abstract model is also true for the concrete model.

On the other hand, in order to preserve ECTL*, the set of abstract transitions must include only representatives of concrete transitions and nothing else. Thus, any behavior of the abstract model appears also in the concrete model. As a result, every ECTL* property true for the abstract model is true for the concrete model as well.

We will therefore have an abstract transition from a to b only if for *every* state in $\gamma(a)$ there is a transition to some state in $\gamma(b)$ (i.e., $R^{\forall\exists}(\gamma(a), \gamma(b))$). However, as for $\widehat{R^A}$, we suggest a better definition that connects a to fewer abstract states, which are more precise:

$$\widehat{R^E}(a,b)) \;\Leftrightarrow\; b \in \{\alpha(Y)|Y \in min\{Y'|R^{\forall\exists}(\gamma(a), Y')\}\}.$$

As in the case of $\widehat{R^A}$ and $R^{\exists\exists}$, $\widehat{R^E}$ contains less transitions than $R^{\forall\exists}$. Still, it does not allow to prove more ECTL* properties.

Lemma 5.5 *Let M be a model and $\widehat{M} = (\widehat{S}, \widehat{S_0}, \widehat{R^E}, \widehat{L})$ be an abstract model. Then $\widehat{M} \preceq M$. Thus, for every ECTL* formula φ, and for every abstract state $a \in \widehat{S}$, $\widehat{M}, a \models \varphi \Rightarrow M, \gamma(a) \models \varphi$.*

Here we define $H(a, s) \subseteq \widehat{S} \times S$. However, we relate exactly the same abstract and concrete states: $H(a, s) \;\Leftrightarrow\; s \in \gamma(a)$. As before we require that if s and a are related (here they are related by $H(a, s)$ rather than $H(s, a)$) then $\widehat{L}(a) \subseteq L(s)$. With these changes, Theorem 2.7 holds for ECTL* and thus the lemma holds.

Note that because of the minimality requirements in $\widehat{R^A}$, $\widehat{R^E}$ may not be included in $\widehat{R^A}$.

Example 5.4 *Consider the model of Example 5.3, in which the transition (s_3, s_2) is replaced by (s_3, s_1), resulting in a new transition relation R'. Then, $\widehat{R^A}' = \{(a_1, a_{23}), (a_{23}, a_{23}), (a_{23}, a_1), (\top, \top)\}$. On the other hand, $\widehat{R^E}' = \{(a_1, a_{23}), (a_{23}, \top), (\top, \top)\}$. Note that the transition $(a_{23}, \top) \in \widehat{R^E}'$ is not in $\widehat{R^A}'$ since $\top$ is not minimal in the set of states connected to a_{23}.*

316

Using the model with the $\widehat{R^E}'$ transition relation, we can verify for the concrete model the ECTL* property **EF EG** q.

5.3. ABSTRACT MODELS PRESERVING FULL CTL*

In order to (weakly) preserve full CTL*, we now define an abstract model with a *mixed* transition relation: $\widehat{M} = (\widehat{S}, \widehat{S_0}, \widehat{R^A}, \widehat{R^E}, \widehat{L})$. This model has two types of paths: *A-paths*, defined along $\widehat{R^A}$ transitions, and *E-path*, defined along $\widehat{R^E}$ transitions. The semantics of CTL* with respect to this model differs from the semantics in Definition 2.3 only in item 3.

- $s \models \mathbf{A} f$ if and only if for every A-path π from s, $\pi \models f$.
- $s \models \mathbf{E} f$ if and only if there exists an E-path π from s such that $\pi \models f$.

Theorem 5.6 *Let M be a model and $\widehat{M} = (\widehat{S}, \widehat{S_0}, \widehat{R^A}, \widehat{R^E}, \widehat{L})$ be a mixed abstract model. Then for every CTL* formula φ and for every abstract state $a \in \widehat{S}$, $\widehat{M}, a \models \varphi \Rightarrow M, \gamma(a) \models \varphi$.*

This theorem can be proved by induction of the formula structure. It can also be proved based on a *mixed simulation* over M and $\widehat{M}$. For details see [22].

In [22] approximations were defined in the context of abstract interpretation. Similarly to the framework of data abstraction, approximations here allow different levels of precision. It has also been shown how an approximation can be extracted from the program text.

6. Related work

Several works have applied data abstraction in order to reduce the state space. Wolper and Lovinfosse [60] characterize a class of *data-independent* systems in which the data values never affect the control flow of the computation. Therefore, the datapath can be abstracted away entirely. Van Aelten et al. [2] have discussed a method for simplifying the verification of synchronous processors by abstracting away the datapath. Abstracting the datapath using uninterpreted function symbols is very useful for verifying pipeline systems [7, 11, 39].

In this paper we present a methodology for automatic construction of an initial abstract model, based on atomic formulas extracted from the program text. The atomic formulas are similar to the *predicates* used for abstraction by Graf and Saidi [35]. However, predicates are used to generate an abstract model, while atomic formulas are used to construct an *abstraction mapping*.

The use of counterexamples to refine abstract models has been investigated by a number of researchers. The localization reduction by Kurshan [41] is an iterative technique in which both the initial abstraction and the counterexample-guided refinements are based on the *variable dependency graph*. The localization reduction either leaves a variable unchanged or replaces it by a nondeterministic assignment. A similar approach has been described by Balarin et al. in [3] and by Lind-Nielson and Andersen [45]. The method presented here, on the other hand, applies abstraction mapping that makes it possible to distinguish many degrees of abstraction for each variable.

Lind-Nielson and Andersen [45] also suggest a model checker that uses upper and lower approximations in order to handle all of CTL. Their approximation techniques avoid the need to recheck the entire model after each refinement, yet still guarantee completeness.

A number of other papers [42, 52, 53] have proposed abstraction refinement techniques for CTL model checking. However, these papers do not use counterexamples to refine the abstraction. The methods described in these papers are orthogonal to the techniques presented here and may be combined with them in order to achieve better performance. The technique proposed by Govindaraju and Dill [33] is a first step in this direction. The paper only handles safety properties and path counterexamples; it uses random choice to extend the counterexample it constructs.

Many abstraction techniques can be viewed as applications of the abstract interpretation framework [19, 59, 20]. Bjorner, Browne and Manna use abstract interpretation to automatically generate invariants for general infinite state systems [8]. Abstraction techniques for the μ-calculus have been suggested in [21, 22, 46].

Abstraction techniques for *infinite state systems* have been proposed in [1, 5, 43, 48]. The *predicate abstraction* technique, suggested by Graf and Saidi [35], is also aimed at abstracting an infinite state system into a finite state system. Later, a number of optimization techniques were developed in [6, 26, 25]. Saidi and Shankar have integrated predicate abstraction into the PVS system, which could easily determine when to abstract and when to model check [58]. Variants of predicate abstraction have been used in the Bandera Project [28] and the SLAM project [4].

Colón and Uribe [18] have presented a way to generate finite state abstractions using a decision procedure. As in predicate abstraction, their abstraction is generated using abstract Boolean variables.

A number of researchers have modeled or verified industrial hardware systems using abstraction techniques [32, 34, 37, 38]. In many cases, their abstractions are generated manually and combined with theorem-proving techniques [56, 57]. Dingel and Filkorn have used data abstraction and assume-guarantee reasoning, combined with theorem-proving techniques, in order to verify infinite state systems [27]. Recently, McMillan has incorporated a new type of data abstraction, along with assume-guarantee reasoning and theorem-proving techniques, into his Cadence SMV system [50].

7. Conclusion

In this work, three notions of abstraction have been introduced. They are all based on the idea that in order to check a specific property, some of the system states are in fact indistinguishable and can be collapsed into an abstract state that represents them.

Several concepts are common to all of these abstractions. Even though they were introduced within the framework of one of the abstractions, they are applicable with some changes to the other notions:

- The abstractions are derived from a high-level description of the program.
- Since deriving precise abstraction is usually difficult, the notion of approximations that are easier to compute is introduced.
- Refinement is required in case the abstraction is too coarse to enable the verification or the falsification of a given property.
- The abstractions provide weak preservation. Most of the discussion in this paper has been devoted to the preservation of ACTL*. However, preservation of ECTL* and full CTL* can be defined for each of the abstractions.

References

1. P. A. Abdulla, A. Bouajjani, B. Jonsson, and M. Nilsson. Verification of infinite-state systems by combining abstraction and reachability analysis. In *Computer-Aided Verification*, July 1999.
2. F. Van Aelten, S. Liao, J. Allen, and S. Devadas. Automatic generation and verification of sufficient correctness properties for synchronous processors. In *International Conference of Computer-Aided Design*, 1992.
3. F. Balarin and A. L. Sangiovanni-Vincentelli. An iterative approach to language containment. In *Computer-Aided Verification*, volume 697 of *LNCS*, pages 29–40, 1993.
4. T. Ball, R. Majumdar, T. Millstein, and S. K. Rajamani. Automatic predicate abstraction of C programs. In *ACM SIGPLAN 2001 Conference on Programming Language Design and Implementation (PLDI)*, 2001.
5. S. Bensalem, A. Bouajjani, C. Loiseaux, and J. Sifakis. Property preserving simulations. In *Computer-Aided Verification*, July 1992.
6. S. Bensalem, Y. Lakhnech, and S. Owre. Computing abstractions of infinite state systems compositionally and automatically. In *Computer-Aided Verification*, June 1998.

7. S. Berezin, A. Biere, E. Clarke, and Y. Zhu. Combining symbolic model checking with uninterpreted functions for out-of-order processor verification. In *Formal Methods in Computer-Aided Design*, pages 369–386, 1998.

8. N. S. Bjorner, A. Browne, and Z. Manna. Automatic generation of invariants and intermediate assertions. *Theoretical Computer Science*, 173(1):49–87, 1997.

9. M. C. Browne, E. M. Clarke, and O. Grumberg. Characterizing finite kripke structures in propositional temporal logic. *Theor.Comp.Science*, 59(1–2), July 1988.

10. R. E. Bryant. Graph-based algorithms for boolean function manipulation. *IEEE Transactions on Computers*, C-35(8):677–691, August 1986.

11. J. Burch and D. Dill. Automatic verification of pipelined microprocessor control. In *Computer-Aided Verification*, volume 818 of *LNCS*, pages 68–80, 1994.

12. J. R. Burch, E. M. Clarke, and D. E. Long. Symbolic model checking with partitioned transition relations. In A. Halaas and P. B. Denyer, editors, *Proceedings of the 1991 International Conference on Very Large Scale Integration*, August 1991.

13. J. R. Burch, E. M. Clarke, K. L. McMillan, D. L. Dill, and L. J. Hwang. Symbolic model checking: 10^{20} states and beyond. *Information and Computation*, 98(2):142–170, June 1992.

14. E. M. Clarke and E. A. Emerson. Synthesis of synchronization skeletons for branching time temporal logic. In D. Kozen, editor, *Logic of Programs: Workshop, Yorktown Heights, NY, May 1981*, volume 131 of *Lecture Notes in Computer Science*. Springer-Verlag, 1981.

15. E. M. Clarke, O. Grumberg, and D. E. Long. Model checking and abstraction. In *Proceedings of the Nineteenth Annual ACM Symposium on Principles of Programming Languages*. Association for Computing Machinery, January 1992.

16. E.M. Clarke, O. Grumberg, S. Jha, Y. Lu, and H. Veith. Counterexample-guided abstraction refinement. In *12th International Conference on Computer Aided Verification (CAV '00)*, LNCS, Chicago, USA, July 2000.

17. E.M. Clarke, O. Grumberg, and D.A. Peled. *Model Checking*. MIT press, December 1999.

18. M. A. Colón and T. E. Uribe. Generating finite-state abstraction of reactive systems using decision procedures. In *Computer-Aided Verification*, pages 293–304, 1998.

19. P. Cousot and R. Cousot. Abstract interpretation : A unified lattice model for static analysis of programs by construction or approximation of fixpoints. *ACM Symposium of Programming Language*, pages 238–252, 1977.

20. P. Cousot and R. Cousot. Refining model checking by abstract interpretation. *Automated Software Engineering*, 6:69–95, 1999.

21. D. Dams. *Abstract Interpretation and Partition Refinement for Model Checking*. PhD thesis, Technical University of Eindhoven, Eindhoven, The Netherlands, 1995.

22. D. Dams, R. Gerth, and O. Grumberg. Abstract interpretation of reactive systems. *ACM Transactions on Programming Languages and System (TOPLAS)*, 19(2), 1997.

23. D. Dams, O. Grumberg, and R. Gerth. Abstraction interpretation of reactive systems: The preservation of CTL*. In *Programming Concepts, Methods and Calculi (ProCoMet)*. North Holland, 1994.

24. Dennis Dams. *Abstract Interpretation and Partition Refinement for Model Checking*. PhD thesis, Eindhoven university, Holland, July 1996.

25. S. Das and D. L. Dill. Successive approximation of abstract transition relations. In *Proc. of the Sixteenth Annual IEEE Symposium on Logic in Computer Science (LICS)*, 2001.

26. S. Das, D. L. Dill, and S. Park. Experience with predicate abstraction. In *Computer-Aided Verification*, volume 1633 of *LNCS*, pages 160–171. Springer Verlag, July 1999.

27. J. Dingel and T. Filkorn. Model checking for infinite state systems using data abstraction, assumption-commitment style reasoning and theorem proving. In P. Wolper, editor, *Proceedings of the 7th International Conference On Computer Aided*

Verification, volume 939 of *Lecture Notes in Computer Science*, pages 54–69, Liege, Belgium, July 1995. Springer Verlag.

28. M. B. Dwyer, J. Hatcliff, R. Joehanes, S. Laubach, C. S. Pasareanu, Robby, W. Visser, and H. Zheng. Tool-supported program abstraction for finite-state verification. In *Proceedings of the 23rd International Conference on Software Engineering (ICSE)*, 2001.

29. E. A. Emerson and E. M. Clarke. Characterizing correctness properties of parallel programs using fixpoints. In *Lecture Notes in Computer Science 85*, pages 169–181. Automata, Languages and Programming, July 1980.

30. E. A. Emerson and J. Y. Halpern. "Sometimes" and "Not Never" revisited: On branching time versus linear time. *J. ACM*, 33(1):151–178, 1986.

31. E.A. Emerson and Chin Laung Lei. Modalities for model checking: Branching time strikes back. *Twelfth Symposium on Principles of Programming Languages, New Orleans, La.*, January 1985.

32. D.A. Fura, P.J. Windley, and A.K. Somani. Abstraction techniques for modeling real-world interface chips. In J.J. Joyce and C.-J.H. Seger, editors, *International Workshop on Higher Order Logic Theorem Proving and its Applications*, volume 780 of *Lecture Notes in Computer Science*, pages 267–281, Vancouver, Canada, August 1993. University of British Columbia, Springer Verlag, published 1994.

33. S. G. Govindaraju and D. L. Dill. Verification by approximate forward and backward reachability. In *Proceedings of International Conference on Computer-Aided Design*, November 1998.

34. S. Graf. Verification of distributed cache memory by using abstractions. In David L. Dill, editor, *Proceedings of the sixth International Conference on Computer-Aided Verification CAV*, volume 818 of *Lecture Notes in Computer Science*, pages 207–219, Standford, California, USA, June 1994. Springer Verlag.

35. S. Graf and H. Saidi. Construction of abstract state graphs with PVS. In *Computer-Aided Verification*, volume 1254 of *LNCS*, pages 72–83, June 1997.

36. O. Grumberg and D.E. Long. Model checking and modular verification. *ACM Trans. on Programming Languages and Systems*, 16(3):843–871, 1994.

37. P.-H. Ho, A. J. Isles, and T. Kam. Formal verification of pipeline control using controlled token nets and abstract interpretation. In *International Conference of Computer-Aided Design*, pages 529–536, 1998.

38. R. Hojati and R. K. Brayton. Automatic datapath abstraction in hardware systems. In P. Wolper, editor, *Proceedings of the 7th International Conference On Computer Aided Verification*, volume 939 of *Lecture Notes in Computer Science*, pages 98–113, Liege, Belgium, July 1995. Springer Verlag.

39. R. B. Jones, J. U. Skakkebak, and D. L. Dill. Reducing manual abstraction in formal verification of out-of-order execution. In *Formal Methods in Computer-Aided Design*, pages 2–17, 1998.

40. D. Kozen. Results on the propositional μ-calculus. *TCS*, 27, 1983.

41. R. P. Kurshan. *Computer-Aided Verification of Coordinating Processes*. Princeton University Press, 1994.

42. W. Lee, A. Pardo, J. Jang, G. Hachtel, and F. Somenzi. Tearing based abstraction for CTL model checking. In *International Conference of Computer-Aided Design*, pages 76–81, 1996.

43. D. Lesens and H. Sadi. Automatic verification of parameterized networks of processes by abstraction. In *International Workshop on Verification of Infinite State Systems (INFINITY)*, Bologna, July 1997.

44. O. Lichtenstein and A. Pnueli. Checking that finite state concurrent programs satisfy their linear specification. In *Proceedings of the Twelfth Annual ACM Symposium on Principles of Programming Languages*, pages 97–107. Association for Computing Machinery, January 1985.

45. J. Lind-Nielsen and H. R. Andersen. Stepwise CTL model checking of state/event

systems. In *Computer-Aided Verification*, volume 1633 of *LNCS*, pages 316–327. Springer Verlag, 1999.

46. C. Loiseaux, S. Graf, J. Sifakis, A. Bouajjani, and S. Bensalem. Property preserving abstractions for the verification of concurrent systems. *Formal Methods in System Design*, 6:11–45, 1995.

47. D. E. Long. *Model Checking, Abstraction, and Compositional Reasoning*. PhD thesis, Carnegie Mellon University, 1993.

48. Z. Manna, M. Colon, B. Finkbeiner, H. Sipma, and T. Uribe. Abstraction and modular verification of infinit-state reactive systems. In *Requirements Targeting Software and Systems Engineering (RTSE)*, 1998.

49. K. L. McMillan. *Symbolic Model Checking: An Approach to the State Explosion Problem*. PhD thesis, Carnegie Mellon University, 1992.

50. K. L. McMillan. Verification of infinite state systems by compositional model checking. In *Correct Hardware Design and Verification Methods*, September 1999.

51. R. Milner. An algebraic definition of simulation between programs. In *Proceedings of the Second Internation Joint Conference on Artificial Intelligence*, pages 481–489, September 1971.

52. A. Pardo. *Automatic Abstraction Techniques for Formal Verification of Digital Systems*. PhD thesis, University of Colorado at Boulder, Dept. of Computer Science, August 1997.

53. A. Pardo and G.D. Hachtel. Incremental CTL model checking using BDD subsetting. In *Design Automation Conference*, pages 457–462, 1998.

54. D. Park. Concurrency and automata on infinite sequences. In *5th GI-Conference on Theoretical Computer Science*, pages 167–183. Springer-Verlag, 1981. LNCS 104.

55. A. Pnueli. The Temporal Semantics of Concurrent Programs. *Theor.Comp.Science*, 13:45–60, 1981.

56. J. Rushby. Integrated formal verification: using model checking with automated abstraction, invariant generation, and theorem proving. In *Theoretical and practical aspects of SPIN model checking: 5th and 6th international SPIN workshops*, pages 1–11, 1999.

57. V. Rusu and E. Singerman. On proving safety properties by integrating static analysis, theorem proving and abstraction. In *Intl. Conference on Tools and Algorithms for the Construction and Analysis of Systems*, pages 178–192, 1999.

58. H. Saidi and N. Shankar. Abstract and model checking while you prove. In *Computer-Aided Verification*, number 1633 in LNCS, pages 443–454, July 1999.

59. J. Sifakis. Property preserving homomorphisms of transition systems. In *4th Workshop on Logics of Programs*, June 1983.

60. P. Wolper and V. Lovinfosse. Verifying properties of large sets of processes with network invariants. In *Proceedings of the 1989 International Workshop on Automatic Verification Methods for Finite State Systems*, volume 407 of *LNCS*, 1989.

HOARE LOGIC: FROM FIRST-ORDER TO PROPOSITIONAL FORMALISM *

JERZY TIURYN
Institute of Informatics
University of Warsaw
Tiuryn@mimuw.edu.pl

1. Introduction

Hoare Logic, intoduced by C.A.R. Hoare (Hoare, 1969), is a precursor of Dynamic Logic. It was one of the first formal verification systems designed for proving partial correctess assertions (PCAs) of deterministic **while** programs. It is related to the invariant assertion method of R. Floyd (Floyd, 1967). Both Hoare Logic and Dynamic Logic are examples of what is called the *exogenous* approach to the modal logic of programs. This means that programs are explicit and are part of well formed expressions, the logic is dealing with. It should be contrasted with another approach, called *endogenous*, which is exemplified by Temporal Logic (Pnueli, 1977). In the latter approach the program is fixed and it is viewed as part of the structure in which the logic is interpreted. The interested reader is referred to the surveys on Hoare Logic (Apt, 1981; Apt and Olderog, 1991) and on Temporal Logic (Emerson, 1990; Gabbay et al., 1994).

Most investigations in Hoare Logic are carried out in the context of first-order language. However, some properties of the formal system may be obscured by restricting attention to a very special form of atomic programs such as assignment statements. This raises the question: to which extend the metamathematical properties of Hoare Logic are a consequence of choosing assignment statements as the atomic programs? For this reason one can

---

* This work was supported by KBN Grant No. 7 T11C 028 20.

H. Schwichtenberg and R. Steinbrüggen (eds.), Proof and System-Reliability, 323–340.
© 2002 *Kluwer Academic Publishers. Printed in the Netherlands.*

consider to investigate Hoare Logic on the propositional level. Propositional Hoare Logic (PHL) was introduced by D. Kozen (Kozen, 1999). It is subsumed by other propositional program logics such as Propositional Dynamic Logic (PDL) (Fischer and Ladner, 1977), or Kleene Algebra with Tests (KAT) (Kozen, 1997). The Hoare PCA $\{b\}\, p\, \{c\}$ is expressed in PDL by the formula $b \rightarrow [p]c$ and in KAT by the equation $bp\bar{c} = 0$. The weakes liberal precondition of p with respect to c is expressed in PDL by $[p]c$.

As we will see, determinig the deductive strength of the original Hoare rules in a propositional context sheds some light on the boundary between Hoare logic proper and the expressiveness assumptions on the underlying domain. Instead of studying derivability in PHL of a single PCA $\{b\}\, p\, \{c\}$ we are concerned with derivability of rules of the form

$$\frac{\{b_1\}\, p_1\, \{c_1\} \dots \{b_n\}\, p_n\, \{c_n\}}{\{b\}\, p\, \{c\}}, \tag{1}$$

where $p_1, \dots, p_n$ are programs and $b_1, \dots, b_n, c_1, \dots, c_n, b, c$ are propositions.

In the present exposition we deal with regular expressions as programs, rather than **while** programs. This does not sacrifice the computational power and leads to simpler rules of inference.

The aim of this paper is to give an introduction to PHL. The paper is organized as follows. In Section 2 we introduce some notation including the class of deterministic **while** programs and their semantics. We also discuss briefly the notion of the weakest liberal precondition. Section 3 recalls some basic facts about the first-order Hoare Logic. In particular we discuss there the issues of incompleteness and relative completeness. Derivability and admissibility of rules is also briefly discussed there. In Section 4 we introduce Propositional Hoare Logic. We show that every valid rule (1) with atomic programs $p_1, \dots, p_n$ in the premises is derivable in PHL. Examples show that the atomicity assumption in this result is essential. We also show that if PHL is suitably extended by introducing a propositional version of weakes liberal preconditions, every valid rule (1) becomes derivable. Both these results are due to (Kozen and Tiuryn, 2000), We decided to include brief sketches of the proofs of these results since the publication of (Kozen and Tiuryn, 2000) is not easily available. Also the details of the presentation of the second result slightly differ from the original publication. In Section 4 we also mention the issue of complexity of deciding the validity of rules in PHL.

2. Preliminaries

A *signature* is a list of function and relation symbols together with a mapping which assigns an arity to each symbol. We assume that every signature always contains a binary relation symbol of equality, denoted by $=$. We also assume a countable set X of *individual variables*. *First-order formulas* are the expressions of the least set containing the atomic formulas, *falsehood* $\mathbf{0}$ and being closed under *implication* $\rightarrow$ and universal quantification $\forall x$ over individual variables x. A *structure* $\mathfrak{A}$ consists of a non-empty carrier A together with an interpretation of function and relation symbols by operations and relations of corresponding arity. A *valuation* is a mapping $v : X \rightarrow A$, which assigns a value $v(x)$ to each individual variable x. Valuations are sometimes called *states*. For a formula φ, $\mathrm{m}_{\mathfrak{A}}(\varphi)$ is the set of all valuations which satisfy φ in $\mathfrak{A}$. $v \in \mathrm{m}_{\mathfrak{A}}(\varphi)$ is sometimes denoted $\mathfrak{A}, v \models \varphi$. Let $S^{\mathfrak{A}}$ denote the set of all valuations in $\mathfrak{A}$.

Let us fix a signature Σ. **While** programs, denoted by $\alpha, \beta, \ldots$, are build from *tests* (denoted by σ, being quantifier-free first-order formulas) and *assignment statements* $x := t$ with help of the following programming constructs:

- *Composition*: $\alpha \,;\, \beta$.
- *Conditional*: **if** σ **then** α **else** β.
- *Iteration*: **while** σ **do** α.

Given a Σ-structure $\mathfrak{A}$. The *meaning* of a program α in $\mathfrak{A}$ is a binary *input-output* relation $\mathrm{m}_{\mathfrak{A}}(\alpha) \subseteq S^{\mathfrak{A}} \times S^{\mathfrak{A}}$.

- $\mathrm{m}_{\mathfrak{A}}(x := t) \overset{\text{def}}{=} \{(u, u[x/u(t)]) \mid u \in S^{\mathfrak{A}}\}$.
- $\mathrm{m}_{\mathfrak{A}}(\alpha \,;\, \beta) \overset{\text{def}}{=} \mathrm{m}_{\mathfrak{A}}(\alpha) \circ \mathrm{m}_{\mathfrak{A}}(\beta)$.
- $\mathrm{m}_{\mathfrak{A}}(\textbf{if } \sigma \textbf{ then } \alpha \textbf{ else } \beta) \overset{\text{def}}{=} \mathrm{m}_{\mathfrak{A}}(\sigma) \circ \mathrm{m}_{\mathfrak{A}}(\alpha) \cup \mathrm{m}_{\mathfrak{A}}(\neg\sigma) \circ \mathrm{m}_{\mathfrak{A}}(\beta)$.
- $\mathrm{m}_{\mathfrak{A}}(\textbf{while } \sigma \textbf{ do } \alpha) \overset{\text{def}}{=} (\mathrm{m}_{\mathfrak{A}}(\sigma) \circ \mathrm{m}_{\mathfrak{A}}(\alpha))^* \circ \mathrm{m}_{\mathfrak{A}}(\neg\sigma)$.

In the above equations $\circ$ denotes the operation of composition of relations and * denotes the transitive and reflexive closure of a binary relation. Moreover, we view $\mathrm{m}_{\mathfrak{A}}(\sigma)$ as partial identity: $\mathrm{m}_{\mathfrak{A}}(\sigma) = \{(v, v) \mid \mathfrak{A}, v \models \sigma\}$.

2.1. WEAKEST LIBERAL PRECONDITION

A *weakest liberal precondition* (Dijkstra, 1975) of a program α w.r.t. a formula ψ in a structure $\mathfrak{A}$ is the following predicate

$$WLP_{\mathfrak{A}}(\alpha, \psi) \overset{\text{def}}{=} \{u \in S^{\mathfrak{A}} \mid \forall v \ (u, v) \in \mathrm{m}_{\mathfrak{A}}(\alpha) \Longrightarrow v \in \mathrm{m}_{\mathfrak{A}}(\psi)\}.$$

This predicate is usually not first-order definable. A structure $\mathfrak{A}$ for which $WLP_{\mathfrak{A}}(\alpha, \psi)$ is first-order definable for all programs α and formulas ψ is called *expressive*.

Weakest liberal precondition for α w.r.t. ψ is expressed in Dynamic Logic by the *necessity statement* $[\alpha]\psi$, i.e. we have for every structure $\mathfrak{A}$,

$$WLP_{\mathfrak{A}}(\alpha, \psi) = m_{\mathfrak{A}}([\alpha]\psi).$$

The basic properties of weakest liberal preconditions are compiled in the following theorem. This provides an equivalent definition of the weakest liberal precondition.

THEOREM 2.1. $WLP(-, -)$ *is the least predicate satisfying the following equivalences.*

1. $\models WLP(x := t, \psi) \Longleftrightarrow \psi[x/t]$.
2. $\models WLP(\textbf{if } \sigma \textbf{ then } \alpha \textbf{ else } \beta, \psi) \Longleftrightarrow$
$$(\sigma \wedge WLP(\alpha, \psi)) \vee (\neg\sigma \wedge WLP(\beta, \psi)).$$
3. $\models WLP(\alpha\,;\,\beta, \psi) \Longleftrightarrow WLP(\alpha, WLP(\beta, \psi))$.
4. $\models WLP(\textbf{while } \sigma \textbf{ do } \alpha, \psi) \Longleftrightarrow$
$$(\neg\sigma \wedge \psi) \vee (\sigma \wedge WLP(\alpha, WLP(\textbf{while } \sigma \textbf{ do } \alpha, \psi))).$$

3. First-Order Hoare Logic

A *partial correctness assertion* (PCA) is a triple $\{\varphi\}\,\alpha\,\{\psi\}$, where α is a program and φ, ψ are first-order formulas. It expresses the property that if the program α is run in a state satisfying the *precondition* φ, then upon termination it will satisfy the *postcondition* ψ. Termination of α is not guaranteed, though. More formally, for a structure $\mathfrak{A}$,

$$\mathfrak{A} \models \{\varphi\}\,\alpha\,\{\psi\} \overset{\text{def}}{\Longleftrightarrow}$$

for all $u, v \in S^{\mathfrak{A}}$, if $u \in m_{\mathfrak{A}}(\varphi)$ and $(u, v) \in m_{\mathfrak{A}}(\alpha)$, then $v \in m_{\mathfrak{A}}(\psi)$.

A PCA $\{\varphi\}\,\alpha\,\{\psi\}$ is *valid*, denoted by $\models \{\varphi\}\,\alpha\,\{\psi\}$, if it holds in all structures. The next result relates partial correctness assertions to weakest liberal preconditions. It follows immediately from the definitions.

PROPOSITION 3.1. *For all formulas* φ, ψ *and every program* α,

$$\models \{\varphi\}\,\alpha\,\{\psi\} \Longleftrightarrow \varphi \to WLP(\alpha, \psi).$$

We cannot hope to have a sound proof system capable of deriving all valid PCAs. The reason is that the set of all valid PCAs is too complex. The next result is due to V. Pratt (Pratt, 1976).

THEOREM 3.2. (Complexity of PCA validity)

For sufficiently rich signatures Σ, the set of all valid PCAs is Π_2^0-complete.

Figure 1 defines Hoare Logic (HL). Note that **(Weakening)** refers to a given structure $\mathfrak{A}$. Thus, we are implicitly assuming the knowledge of the first-order theory of $\mathfrak{A}$. In this way we separate the complexity of the reasoning on the program level from the complexity of the structure in which the program is executed.

$$\textbf{(Assignment)} \quad \{\varphi[x/e]\}\, x := e \,\{\varphi\}$$

$$\textbf{(Composition)} \quad \frac{\{\varphi\}\,\alpha\,\{\xi\}, \quad \{\xi\}\,\beta\,\{\psi\}}{\{\varphi\}\,\alpha\,;\,\beta\,\{\psi\}}$$

$$\textbf{(Conditional)} \quad \frac{\{\varphi \wedge \sigma\}\,\alpha\,\{\psi\}, \quad \{\varphi \wedge \neg\sigma\}\,\beta\,\{\psi\}}{\{\varphi\}\,\textbf{if }\sigma\textbf{ then }\alpha\textbf{ else }\beta\,\{\psi\}}$$

$$\textbf{(While)} \quad \frac{\{\varphi \wedge \sigma\}\,\alpha\,\{\varphi\}}{\{\varphi\}\,\textbf{while }\sigma\textbf{ do }\alpha\,\{\varphi \wedge \neg\sigma\}}$$

$$\textbf{(Weakening)} \quad \frac{\mathfrak{A} \models \varphi' \to \varphi, \quad \{\varphi\}\,\alpha\,\{\psi\}, \quad \mathfrak{A} \models \psi \to \psi'}{\{\varphi'\}\,\alpha\,\{\psi'\}}$$

Figure 1. HL, Hoare Logic over a structure $\mathfrak{A}$.

We denote by $\vdash_{\mathfrak{A}} \{\varphi\}\,\alpha\,\{\psi\}$ the derivability of the PCA $\{\varphi\}\,\alpha\,\{\psi\}$ in HL over $\mathfrak{A}$.

THEOREM 3.3. (Soundness)

For every Σ-structure $\mathfrak{A}$, if $\vdash_{\mathfrak{A}} \{\varphi\}\,\alpha\,\{\psi\}$, then $\mathfrak{A} \models \{\varphi\}\,\alpha\,\{\psi\}$.

It follows from Theorem 3.2 that HL is incomplete. The particular reason for incompleteness of HL is that the *intermediate assertion* ξ in **(Composition)** rule and the *invariant assertion* φ in **(While)** rule need not be first-order definable. To illustrate this let us consider the following example (Wand, 1978). Let $\Sigma = \{f, r\}$, where f is a unary operation symbol and r is a unary relation symbol. Consider the structure $\mathfrak{A} = (A, f^{\mathfrak{A}}, r^{\mathfrak{A}})$,

where

$$A = \{a_i \mid i \in \mathbb{N}\} \cup \{b_i \mid i \in \mathbb{N}\},$$
$$r^{\mathfrak{A}} = \{a_{k^2} \mid k \in N\},$$

and $f^{\mathfrak{A}}$ is defined as follows (x stands here for a, or b)

$$f^{\mathfrak{A}}(x_i) = \begin{cases} x_0 & \text{if } i = 0, \\ x_{i-1} & \text{if } i > 0. \end{cases}$$

Hence f behaves in $\mathfrak{A}$ like predecessor (on two copies on natural numbers), and r defines an infinite subset of $\{a_i \mid i \in \mathbb{N}\}$ with an increasing distance between two consequtive elements in this subset.

Clearly we have $\mathfrak{A} \models \{r(x)\}$ **while** $x \neq f(x)$ **do** $x := f(x)$ $\{r(x)\}$, but

THEOREM 3.4. (Wand, 1978)
The PCA

$$\{r(x)\} \text{ \textbf{while} } x \neq f(x) \text{ \textbf{do} } x := f(x) \ \{r(x)\}$$

is not derivable in $\vdash_{\mathfrak{A}}$.

The reason why the PCA of Theorem 3.4 is not derivable is that derivability would imply that the set $\{a_i \mid i \in \mathbb{N}\}$ is first-order definable in $\mathfrak{A}$. However, it is not. The interested reader should try to prove both of these claims.

Let us recall that expressive structures are those for which the weakest precondition of every **while** program is first-order definable. The important examples of expressive structures are finite structures and the standard model of arithmetic, the latter is due to the encoding power of arithmetic.

The following well known result is due to S. Cook (Cook, 1978).

THEOREM 3.5. (Relative Completeness)
Hoare logic is relatively complete, i.e. for every expressive structure $\mathfrak{A}$, if $\mathfrak{A} \models \{\varphi\} \alpha \{\psi\}$, then $\vdash_{\mathfrak{A}} \{\varphi\} \alpha \{\psi\}$.

Outline of Proof. We will not give the full proof of this result since it is well documented in the literature (cf., eg. (Winskel, 1993)). One way of proving this result is to show that if we view the weakest liberal precondition as a first order formula, then in all structures $\mathfrak{A}$ we have

$$\vdash_{\mathfrak{A}} \{WLP(\alpha, \ \psi)\} \alpha \{\psi\}, \tag{2}$$

holds for all formulas ψ and programs α. The proof of (2) is by induction on α. To conclude the proof we observe that if $\mathfrak{A}$ is an expressive structure then (2) is obtainable by a legal derivation (i.e. all pre- and postconditions are first-order formulas) and if $\mathfrak{A} \models \{\varphi\} \, \alpha \, \{\psi\}$ holds, then by Proposition 3.1 and weakening applied to (2) we conclude that

$$\vdash_{\mathfrak{A}} \{\varphi\} \, \alpha \, \{\psi\}.$$

∎

Let us now briefly discuss the issue of admissibility vs. derivability of rules. Given a rule of the form

$$\textbf{(R)} \quad \frac{\{\varphi_1\} \, \alpha_1 \, \{\psi_1\} \quad \dots \quad \{\varphi_n\} \, \alpha_n \, \{\psi_n\}}{\{\varphi\} \, \alpha \, \{\psi\}}.$$

(R) is said to be *admissible* (in Hoare logic) if adding it to the rules of Hoare logic does not increase the set of theorems, i.e. for every structure $\mathfrak{A}$ and for every PCA $\{\varphi\} \, \alpha \, \{\psi\}$, this PCA is derivable in HL extended by **(R)** iff $\vdash_{\mathfrak{A}} \{\varphi\}\alpha\{\psi\}$. A stronger notion is that of derivability. **(R)** is said to *derivable* in Hoare logic if the conclusion $\{\varphi\} \, \alpha \, \{\psi\}$ can be derived in HL (uniformly for all structures $\mathfrak{A}$) from the premises $\{\varphi_1\} \, \alpha_1 \, \{\psi_1\} \dots \{\varphi_n\} \, \alpha_n \, \{\psi_n\}$. Another important notion is that of validity of a rule. **(R)** is said to be *valid* if for every structure $\mathfrak{A}$, if all premises of **(R)** are valid in $\mathfrak{A}$, then the conclusion is valid in $\mathfrak{A}$ as well.

Here is an example of an admissible rule. It will play an important role in the next Section.

PROPOSITION 3.6. *The following rule*

$$\textbf{(And/Or)} \quad \frac{\{\varphi_i\} \, \alpha \, \{\psi_j\} \quad i = 1, \dots, m; j = 1, \dots, n}{\{\bigvee_{i=1}^{m} \varphi_i\} \, \alpha \, \{\bigwedge_{j=1}^{n} \psi_j\}}$$

is admissible in Hoare logic.

Proof: The proof is by induction on α. We show that when the premises are derivable in HL, then the conclusion is derivable as well. When α is an assignment statement $x := t$, then we first observe that

$$\vdash_{\mathfrak{A}} \{\varphi\} \, x := t \, \{\psi\} \iff \mathfrak{A} \models \varphi \to \psi[x/t]. \tag{3}$$

We leave the proof of (3) to the reader. It follows from (3) that if $\{\varphi_i\}\alpha\{\psi_j\}$ is derivable, then $\mathfrak{A} \models \varphi_i \to \psi_j[x/t]$, for all $i = 1, \dots, m$ and $j = 1, \dots, n$. Thus

$$\mathfrak{A} \models \bigvee_{i=1}^{m} \varphi_i \to \bigwedge_{j=1}^{n} \psi_j[x/t]$$

and again by (3) we obtain the conclusion.

The case α being a conditional is immediate. If α is a composition $\beta \, ; \, \gamma$ and if $\xi_{i,j}$ is the intermediate assertion for $\vdash_{\mathfrak{A}} \{\varphi_i\} \, \beta \, ; \, \gamma \, \{\psi_j\}$, then $\bigwedge_{i,j} \xi_{i,j}$ is the intermediate assertion for $\vdash_{\mathfrak{A}} \{\bigvee_{i=1}^{m} \varphi_i\} \, \beta \, ; \, \gamma \, \{\bigwedge_{j=1}^{n} \psi_j\}$.

Finally let us consider the case of α being the iteration **while** σ **do** β. Assume that $\vdash_{\mathfrak{A}} \{\varphi_i\} \, \alpha \, \{\psi_j\}$. Hence there is an invariant $\xi_{i,j}$ such that

$$\varphi_i \rightarrow \xi_{i,j} \tag{4}$$
$$\xi_{i,j} \wedge \neg\sigma \rightarrow \psi_j \tag{5}$$

and

$$\vdash_{\mathfrak{A}} \{\xi_{i,j} \wedge \sigma\} \, \beta \, \{\xi_{i,j}\}.$$

Let

$$\xi \overset{\text{def}}{\Longleftrightarrow} \bigvee_i \bigwedge_j \xi i, j.$$

By the induction hypothesis we have

$$\vdash_{\mathfrak{A}} \{\sigma \wedge \bigvee_{i,j} \xi_{i,j}\} \, \beta \, \{\bigwedge_{i,j} \xi_{i,j}\}.$$

Since $\bigvee_i \bigwedge_j \xi_{i,j} \rightarrow \bigvee_{i,j} \xi_{i,j}$ and $\bigwedge_{i,j} \xi_{i,j} \rightarrow \bigvee_i \bigwedge_j \xi_{i,j}$, by weakening we obtain

$$\vdash_{\mathfrak{A}} \{\sigma \wedge \xi\} \, \beta \, \{\xi\}.$$

Thus $\vdash_{\mathfrak{A}} \{\xi\}$ **while** σ **do** $\beta \{\xi \wedge \neg\sigma\}$. By (4) $\bigvee_i \varphi_i \rightarrow \xi$ and by (5) $\xi \wedge \neg\sigma \rightarrow \bigwedge_j \psi_j$. Hence, by weakening we obtain $\vdash_{\mathfrak{A}} \{\bigvee_{i=1}^{m} \varphi_i\} \, \alpha \, \{\bigwedge_{j=1}^{n} \psi_j\}$. ∎

Consider now the following rule

$$\frac{\{\varphi\} \; \textbf{while} \; \sigma \; \textbf{do} \; \alpha \; \{\psi\}}{\{\varphi \wedge \sigma\} \, \alpha \, \{\neg\sigma \rightarrow \psi\}} \tag{6}$$

How can we argue that the above rule is valid? One way is to show the validity of (6) directly from the semantics of the **while** construct. But we can also show it by referring to the properties of the weakest liberal precondition listed in Theorem 2.1. We illustrate the latter method since it will be used in the next section. We will be little informal with our argument. Let us view the weakest liberal preconditions as first-order formulas. Let $\mathfrak{A}$ be any structure such that

$$\mathfrak{A} \models \{\varphi\} \; \textbf{while} \; \sigma \; \textbf{do} \; \alpha \; \{\psi\}. \tag{7}$$

Claim (2) in the proof of Theorem 3.5 states that $\vdash_{\mathfrak{A}} \{WLP(\alpha, \xi)\}\alpha\{\xi\}$ holds for all programs α and all formulas ξ. Let us choose for ξ the formula $WLP(\textbf{while } \sigma \textbf{ do } \alpha, \psi)$. Thus we have

$$\vdash_{\mathfrak{A}} \{WLP(\alpha, WLP(\textbf{while } \sigma \textbf{ do } \alpha, \psi))\}\alpha\{WLP(\textbf{while } \sigma \textbf{ do } \alpha, \psi)\}. \quad (8)$$

It follows from Theorem 2.1(4) that

$$\mathfrak{A} \models WLP(\textbf{while } \sigma \textbf{ do } \alpha, \psi) \to (\neg\sigma \to \psi) \quad (9)$$

and

$$\mathfrak{A} \models WLP(\textbf{while } \sigma \textbf{ do } \alpha, \psi) \wedge \sigma \to \\ WLP(\alpha, WLP(\textbf{while } \sigma \textbf{ do } \alpha, \psi)) \quad (10)$$

It follows from Proposition 3.1 applied to (7) that

$$\mathfrak{A} \models \varphi \to WLP(\textbf{while } \sigma \textbf{ do } \alpha, \psi).$$

Thus by (9), (10) and weakening applied to (8) we obtain $\vdash_{\mathfrak{A}} \{\varphi\wedge\sigma\}\alpha\{\neg\sigma \to \psi\}$. Hence

$$\mathfrak{A} \models \{\varphi \wedge \sigma\}\,\alpha\,\{\neg\sigma \to \psi\}.$$

4. Propositional Hoare Logic

On the propositional level of reasoning we start with two sorts of atoms: *atomic programs* and *atomic propositions*. *Propositions* are constructed from atomic propositions, and $\mathbf{0}$ (*falsehood*) with help of implication $\to$. We denote the negation $b \to \mathbf{0}$ by $\bar{b}$. The truth value $\mathbf{1}$ is defined as $\bar{\mathbf{0}}$. Disjunction and conjunction are defined in terms of $\to$ and $\mathbf{0}$ in the usual way. If C is a finite set of propositions, then $\bigvee C$ denotes the disjunction of its elements. In particular we set $\bigvee \emptyset = \mathbf{0}$. Similarly, $\bigwedge C$ denotes the conjunction of the elements of C and we take $\bigwedge \emptyset = \mathbf{1}$.

As in Propositional Dynamic Logic (see (Harel et al., 2000)), instead of a more traditional conditional and **while** constructs we base our programs on two more fundamental programming constructs: iteration * and nondeterministic binary choice $+$. *Programs* are inductively defined as follows:

- Every atomic program and every proposition is a program.
- If p, q are programs then the following expressions are programs as well.

  - $p\,;q$
  - $p + q$

- p^*

We add parenthesis when necessary. Because of the propositional level of reasoning we cannot reason about program variables and assignment statements — the atomic programs play the role of assignment statements, but we are not restricted to think of an atomic program as an assignment statement. The symbol · is overloaded in our approach. It denotes composition of programs or conjunction of formulas, depending on the context.

We interpret propositions and programs in Kripke frames. A *Kripke frame* $\mathfrak{K}$ consists of a set of states K and a map $\mathsf{m}_{\mathfrak{K}}$ associating a subset of K with each atomic proposition and a binary relation on K with each atomic program. The map $\mathsf{m}_{\mathfrak{K}}$ is extended inductively to compound programs and propositions (Harel et al., 2000). For the sake of completeness we give the definition. For propositions we have the following equations.

- $\mathsf{m}_{\mathfrak{K}}(\mathbf{0}) = \emptyset$.
- $\mathsf{m}_{\mathfrak{K}}(b \to c) = \{s \in K \mid s \notin \mathsf{m}_{\mathfrak{K}}(b) \text{ or } s \in \mathsf{m}_{\mathfrak{K}}(c)\}$.

We overload the operator $\mathsf{m}_{\mathfrak{K}}$ in the sense that the binary relation assigned to a proposition viewed as a program is a partial identity, rather than a set of states, as it is the case when the proposition is viewed in the usual way (i.e. as a proposition). It will be always clear from the context in which sense a given proposition is used in an expression. For programs the definitions are as follows.

- $\mathsf{m}_{\mathfrak{K}}(b) = \{(s,s) \in K \times K \mid s \in \mathsf{m}_{\mathfrak{K}}(b)\}$.
- $\mathsf{m}_{\mathfrak{K}}(p \,;\, q) = \mathsf{m}_{\mathfrak{K}}(p) \circ \mathsf{m}_{\mathfrak{K}}(q)$.
- $\mathsf{m}_{\mathfrak{K}}(p + q) = \mathsf{m}_{\mathfrak{K}}(p) \cup \mathsf{m}_{\mathfrak{K}}(q)$.
- $\mathsf{m}_{\mathfrak{K}}(p^*) = \bigcup_{n \geq 0} (\mathsf{m}_{\mathfrak{K}}(p))^n$.

We write $\mathfrak{K}, s \vDash b$ for $s \in \mathsf{m}_{\mathfrak{K}}(b)$ and $s \xrightarrow{p}_{\mathfrak{K}} t$ for $(s,t) \in \mathsf{m}_{\mathfrak{K}}(p)$.

It follows from the above semantics of programs that conditional and **while** constructs are definable in our formalism:

$$\textbf{if } b \textbf{ then } p \textbf{ else } q \overset{\text{def}}{=} bp + \bar{b}q \tag{11}$$

and

$$\textbf{while } b \textbf{ do } p \overset{\text{def}}{=} (bp)^* \bar{b}. \tag{12}$$

The PCA $\{b\}\, p\, \{c\}$ says intuitively that if b holds before executing p, then c must hold afterward. Formally, in a state s of a Kripke frame $\mathfrak{K}$,

$$\mathfrak{K}, s \vDash \{b\}\, p\, \{c\} \overset{\text{def}}{\Longleftrightarrow} (\mathfrak{K}, s \vDash b \implies \forall t\, (s \xrightarrow{p}_{\mathfrak{K}} t \implies \mathfrak{K}, t \vDash c)).$$

For a PCA φ and a set Φ of PCAs, we write

$$\mathfrak{K} \vDash \varphi \stackrel{\text{def}}{\Longleftrightarrow} \forall s \in \mathfrak{K} \quad \mathfrak{K}, s \vDash \varphi$$

$$\mathfrak{K} \vDash \Phi \stackrel{\text{def}}{\Longleftrightarrow} \forall \varphi \in \Phi \quad \mathfrak{K} \vDash \varphi$$

$$\Phi \vDash \varphi \stackrel{\text{def}}{\Longleftrightarrow} \forall \mathfrak{K} \quad \mathfrak{K} \vDash \Phi \Longrightarrow \mathfrak{K} \vDash \varphi.$$

Consider the general form of a rule of inference.

$$\textbf{(R)} \qquad \frac{\{b_1\}\, p_1\, \{c_1\}, \ldots, \{b_n\}\, p_n\, \{c_n\}}{\{b\}\, p\, \{c\}} \qquad (13)$$

The rule **(R)** is said to be *valid* if $\{\{b_1\}p_1\{c_1\}, \ldots, \{b_n\}p_n\{c_n\}\} \vDash \{b\}p\{c\}$. Call a PCA $\{b\}\, p\, \{c\}$ *atomic* if p is an atomic program. The rule **(R)** is said to be *atomic* if all its premises are atomic.

$$\textbf{(Test)} \quad \{b\}\, c\, \{bc\}$$

$$\textbf{(Composition)} \quad \frac{\{b\}\, p\, \{c\}, \quad \{c\}\, q\, \{d\}}{\{b\}\, p\,;\, q\, \{d\}}$$

$$\textbf{(Choice)} \quad \frac{\{b\}\, p\, \{c\}, \quad \{b\}\, q\, \{c\}}{\{b\}\, p + q\, \{c\}}$$

$$\textbf{(Iteration)} \quad \frac{\{b\}\, p\, \{b\}}{\{b\}\, p^*\, \{b\}}$$

$$\textbf{(Weakening)} \quad \frac{b' \to b, \quad \{b\}\, p\, \{c\}, \quad c \to c'}{\{b'\}\, p\, \{c'\}}$$

$$\textbf{(And/Or)} \quad \frac{\{b\}\, p\, \{c\} \quad b \in B,\ c \in C}{\{\bigvee B\}\, p\, \{\bigwedge C\}}$$

Figure 2. PHL, Propositional Hoare Logic. p, q are programs, b, c are propositions, B, C are finite sets of propositions.

We can rewrite the traditional Hoare rules for the propositional level — they would look exactly the same as in Figure 1, except that the assignment rule would be missing. The pre- and postconditons, as well as the tests are propositions on the propositional level . The conditional and while rules are replaced by simpler rules, as can be seen in Figure 2. We also need the and/or rule in PHL for reasons which will become clear a bit later.

It is immediately clear that all the rules of PHL are valid. Let us observe that without **(And/Or)** rule no atomic PCA is derivable. However, with help of this rule we can derive a few atomic PCAs: taking $B = \{b\}$ and $C = \emptyset$ we get $\vdash \{b\}\, p\, \{\mathbf{1}\}$; on the other hand, taking $B = \emptyset$ and $C = \{c\}$ we get $\vdash \{\mathbf{0}\}\, p\, \{c\}$. Thus **(And/Or)** rule is not admissible in PHL.

Even with the **(And/Or)** rule very few PCAs are derivable. The reason is that no specific axioms are assumed for atomic PCAs. Thus we investigate the derivability of valid atomic rules. Let us recall that derivability of the rule **(R)** in PHL means that the conclusion $\{b\}\, p\, \{c\}$ is derivable in PHL from the additional PCAs $\{b_1\}\, p_1\, \{c_1\}, \ldots, \{b_n\}\, p_n\, \{c_n\}$ treated as extra axioms. For example, translating the conditional rule under the definition (11) becomes

$$\textbf{(Cond)} \qquad \frac{\{bc\}\, p\, \{d\} \quad \{\bar{b}c\}\, q\, \{d\}}{\{c\}\, bp + \bar{b}q\, \{d\}}.$$

It is derivable in PHL. Indeed, assuming $\{bc\}\, p\, \{d\}$, by **(Test)** and **(Composition)** we obtain $\{c\}\, bp\, \{d\}$. In a similar way we obtain $\{c\}\, \bar{b}p\, \{d\}$. Thus, by **(Choice)** we get the conclusion $\{c\}\, bp + \bar{b}q\, \{d\}$. In a similar way one shows the derivability of the translation of **(While)** rule:

$$\textbf{(Wh)} \qquad \frac{\{bc\}\, p\, \{c\}}{\{c\}\, (bp)^*\bar{b}\, \{\bar{b}c\}}.$$

Atomic rules are potentially interesting for the reason that they express partial correctness of a compound program, subject to partiall correctness assumprions about its atomic components. The rules of PHL can be viewed in this way. For example, the **(Composition)** rule says that the composition of two programs is partially correct under the assumption of suitable partial correctness assertions for both of the programs. Hence one of the issues of completeness of PHL can be expressed as follows. Given an atomic rule which is valid. Is it derivable in PHL? An affirmative answer is given by the following result.

THEOREM 4.1. (Kozen and Tiuryn, 2000)
Every valid atomic rule of the form (13) is derivable in PHL.

Proof: We sketch the main steps of the proof of Theorem 4.1. For a finite set Φ of PCAs and a PCA φ we write $\Phi \vdash \varphi$ if the conclusion φ is derivable from the premises Φ in PHL. Suppose Φ is a set of atomic PCAs and φ a PCA such that $\Phi \nvdash \varphi$. A Kripke frame $\mathfrak{K}$ is constructed such that $\mathfrak{K} \vDash \Phi$ but $\mathfrak{K} \nvDash \varphi$.

Let us call an *atom* any maximal propositionally consistent conjunction of atomic propositions, which occur in Φ or φ, or their negations. Atoms are denoted $\alpha, \beta, \gamma, \ldots$. We identify an atom α with the conjunction of all formulas in α. The set K of states of our Kripke frame is the set of all atoms. For propositons b, c we write $b \leq c$ if $b \to c$ is a propositional tautology.

For atomic programs a and atomic propositions b, define

$$m_{\mathfrak{K}}(a) \stackrel{\text{def}}{=} \{(\alpha, \beta) \mid \Phi \not\vdash \{\alpha\}\, a\, \{\bar{\beta}\}\}$$

$$m_{\mathfrak{K}}(b) \stackrel{\text{def}}{=} \{\alpha \mid \alpha \leq b\}.$$

Since all premises in Φ are atomic, it follows that $\mathfrak{K} \models \Phi$.

To conclude the proof we show that for every program p and for all atoms α, β,

$$\Phi \not\vdash \{\alpha\}\, p\, \{\bar{\beta}\} \implies \alpha \xrightarrow[\mathfrak{K}]{p} \beta. \tag{14}$$

One shows the contrapositive of (14) by induction on p. Having proved (14) let us assume that $\Phi \not\vdash \{b\}\, p\, \{c\}$. By the **(And/Or)** rule, there must exist $\alpha \leq b$ and $\beta \leq \bar{c}$ such that $\Phi \not\vdash \{\alpha\}\, p\, \{\bar{\beta}\}$. Hence, by (14) we conclude that $\alpha \xrightarrow[\mathfrak{K}]{p} \beta$, i.e. $\mathfrak{K} \not\models \{b\}\, p\, \{c\}$. ∎

The reader may have started wondering whether thet reason for completeness in Theorem 4.1 is that the set of all valid atomic rules is relatively small. It follows from the next result that this is not the case: deciding validity of an atomic rule is as difficult as deciding validity of a quantified propositional formula.

THEOREM 4.2. (Kozen, 1999)
The set of all valid atomic rules of the form (13) is PSPACE-complete.

The original proof of the lower bound in the above theorem was by direct encoding of polynomial space-bounded deterministic Turing machines. A shorter proof can be found in (Cohen and Kozen, 2000), where the reduction is from the universality problem for nondeterministic finite automata.

Let us note that the assumption of atomicity of a rule in Theorem 4.1 is essential. Indeed, the following rule is valid

$$\textbf{(Iteration-Reverse)} \qquad \frac{\{b\}\, p^*\, \{c\}}{\{b\}\, p\, \{c\}} \tag{15}$$

but it is not derivable in PHL. The reason is that in the rules of PHL one can never get a conclusion PCA with a program which is simpler than any of the programs occuring in the premises. Hence, in order to obtain completeness for more general rules, we have to enrich the system. We will use the idea of weakest liberal preconditions.

4.1. WEAKEST PRECONDITIONS

We extend our assertion language by formulas of the form

$$b \rightarrow [p_1][p_2] \cdots [p_n]c,$$

where b, c are propositions and $p_1, \ldots, p_n$ are regular programs. We call such formulas *extended PCAs*. When $n = 0$, the above expression reduces to the ordinary proposition $b \rightarrow c$. In this sense extended PCAs contain propositions. We will abbreviate $1 \rightarrow [p_1][p_2] \cdots [p_n]c$ by $[p_1][p_2] \cdots [p_n]c$.

We assume that in every model a set of states is assigned to each extended PCA, i.e. that extended PCAs have got truth values in each state of the model. We assume that the interpretation is such that the following properties are satisfied.

$$[s \models [a]\psi] \leftrightarrow \forall t[(s,t) \in m_{\mathfrak{K}}(a) \rightarrow t \models \psi] \tag{16}$$

$$[p + q]\psi \leftrightarrow [p]\psi \wedge [q]\psi \tag{17}$$

$$[p\,;q]\psi \leftrightarrow [p][q]\psi \tag{18}$$

$$[p^*]\psi \leftrightarrow \psi \wedge [p][p^*]\psi \tag{19}$$

$$[b]\psi \leftrightarrow (b \rightarrow \psi). \tag{20}$$

In (16) a is an atomic program. This property establishes the expressiveness of the weakest precondition in the language of extended PCAs. Properties (17-20) are axioms of PDL (see (Harel et al., 2000)) and are related to the basic properties of the weakest liberal preconditions for the first-order case (cf. Theorem 2.1). We will use letters φ, ψ, γ to range over extended PCAs.

Now we enrich PHL in order to get the completeness for arbitrary valid rules. The extended system is obtained from PHL by adding an axiom (**Atom**) and a rule of inference (**Extended PCA Intro**). These are presented in Figure 3.

PROPOSITION 4.3. *For every Kripke frame $\mathfrak{K}$ satisfying (16-20), the axiom (**Atom**) holds in $\mathfrak{K}$ and the rule (**Extended PCA Intro**) is valid in $\mathfrak{K}$.*

Proof: Validity of (**Atom**) follows immediately from implication $\rightarrow$ in (16). Validity of (**Extended PCA Intro**) is proved by induction on p. For an atomic p it follows from the implication $\leftarrow$ in (16). Each of the induction steps is handled by one of the properties (17-20). ∎

$$\text{(Atom)} \quad \{[a]\psi\}\, a \,\{\psi\}$$

$$\text{(Extended PCA Intro)} \quad \frac{\{b\}\, p \,\{c\}}{b \to [p]c}$$

Figure 3. EPHL, Extended Propositional Hoare Logic: PHL, as defined in Fig. 2, augmented with the above two rules. *a* in **(Atom)** is an atomic program.

The next result says that the axiom **(Atom)** can be extended to all programs. Let us denote by $\vdash_{EPHL} \{\varphi\}\, p \,\{\psi\}$ the derivability of the PCA $\{\varphi\}\, p \,\{\psi\}$ in EPHL.

PROPOSITION 4.4. *For every program p and for every extended PCA ψ, the PCA $\{[p]\psi\}\, p \,\{\psi\}$ is derivable in EPHL, i.e. $\vdash_{EPHL} \{[p]\psi\}\, p \,\{\psi\}$.*

Proof. The proof is by induction on p. The base step is just **(Atom)**. We just show the induction step for p being a composition $q_1 ; q_2$. By the induction hypothesis we have

$$\vdash_{EPHL} \{[q_1][q_2]\psi\}\, q_1 \,\{[q_2]\psi\}$$

and

$$\vdash_{EPHL} \{[q_2]\psi\}\, q_2 \,\{\psi\}.$$

By **(Composition)** we obtain

$$\vdash_{EPHL} \{[q_1][q_2]\psi\}\, q_1 ; q_2 \,\{\psi\}.$$

By (18) and weakening we obtain

$$\vdash_{EPHL} \{[q_1 ; q_2]\psi\}\, q_1 ; q_2 \,\{\psi\}.$$

∎

Let us show how to derive in EPHL the rule **(Iteration-Reverse)** (see (15)). Assume $\{b\}\, p^* \,\{c\}$, by **(Extended PCA Intro)** we get

$$b \to [p^*]c. \tag{21}$$

Since by (19)

$$[p^*]c \leftrightarrow c \wedge [p][p^*]c, \tag{22}$$

it follows by propositional reasoning that

$$[p^*]c \to c \tag{23}$$

and by (21) and (22)

$$b \to [p][p^*]c. \tag{24}$$

Now, we have an instance of (**Atom**):

$$\{[p][p^*]c\}\, p\, \{[p^*]c\}.$$

Thus by (23), (24) and weakening we obtain

$$\{b\}\, p\, \{c\}.$$

THEOREM 4.5. (Kozen and Tiuryn, 2000)
Every valid rule of the form (13) is derivable in EPHL.

Proof: For a given set X of extended PCAs we define the *Fisher–Ladner closure* $FL(X)$ in a similar way as in PDL (see (Harel et al., 2000)). A set X of extended PCAs is said to be *closed* if it satisfies the following closure properties:

- $b \to \psi \in X \;\Longrightarrow\; b \in X$ and $\psi \in X$
- $\mathbf{0} \in X$
- $[p+q]\psi \in X \;\Longrightarrow\; [p]\psi \in X$ and $[q]\psi \in X$
- $[p\,;q]\psi \in X \;\Longrightarrow\; [p][q]\psi \in X$ and $[q]\psi \in X$
- $[p^*]\psi \in X \;\Longrightarrow\; \psi \in X$ and $[p][p^*]\psi \in X$
- $[b]\psi \in X \;\Longrightarrow\; b \to \psi \in X$
- $[a]\psi \in X \;\Longrightarrow\; \psi \in X.$

$FL(X)$ is the least closed set containing X. The important property of this closure is that for a finite set X of extended PCAs, $FL(X)$ is again a finite set of extended PCAs.

Let $\Phi = \{b_1 \to [p_1]c_1, \ldots b_n \to [p_n]c_n\}$, where $\{b_1\}p_1\{c_1\}, \ldots, \{b_n\}p_n\{c_n\}$ are the premises of the rule (13).

An *atom* α is a set of formulas of $FL(\Phi)$ and their negations satisfying the following properties:

(i) for each $\psi \in FL(\Phi)$, exactly one of ψ, $\bar{\psi} \in \alpha$

(ii) for $b \to \psi \in FL(\Phi)$, $b \to \psi \in \alpha \iff (b \in \alpha \implies \psi \in \alpha)$

(iii) $\mathbf{0} \notin \alpha$

(iv) for $[p+q]\psi \in FL(\Phi)$, $[p+q]\psi \in \alpha \iff [p]\psi \in \alpha$ and $[q]\psi \in \alpha$

(v) for $[p;q]\psi \in FL(\Phi)$, $[p;q]\psi \in \alpha \iff [p][q]\psi \in \alpha$

(vi) for $[p^*]\psi \in FL(\Phi)$, $[p^*]\psi \in \alpha \iff \psi \in \alpha$ and $[p][p^*]\psi \in \alpha$

(vii) for $[b]\psi \in FL(\Phi)$, $[b]\psi \in \alpha \iff b \to \psi \in \alpha$

(viii) if $\{b\}\, p\, \{c\} \in \Phi$, then $b \to [p]c \in \alpha$.

Thus atoms represent consistency conditions not only implied by propositional logic, but also by the properties (17-20). It follows that since $FL(\Phi)$ is finite the set of all atoms is finite too. Let K be the set of all atoms.

Now we can construct a finite Kripke frame $\mathfrak{K}$ with states K. We define

$$m_{\mathfrak{K}}(a) \stackrel{\text{def}}{=} \{(\alpha,\beta) \mid \forall [a]\psi \in FL(\Phi)\ ([a]\psi \in \alpha \implies \psi \in \beta)\}$$

$$m_{\mathfrak{K}}(b) \stackrel{\text{def}}{=} \{\alpha \mid b \in \alpha\}$$

$$m_{\mathfrak{K}}([p]\psi) \stackrel{\text{def}}{=} \{\alpha \mid [p]\psi \in \alpha\}$$

for atomic programs a, atomic propositions b, and extended PCAs of the form $[p]\psi$. The meaning function $m_{\mathfrak{K}}$ is lifted to compound programs and propositions according to the usual inductive rules. It is easy to show that for every proposition b and a state α,

$$b \in \alpha \iff \alpha \models b.$$

The above equivalence can be strengthened to all extended PCAs which belong to $FL(\Phi)$. For a given extended PCA $\psi = b \to [p_1][p_2]\cdots[p_n]c$ let $\hat{\psi}$ denote the PCA $\{b\}\, p_1;\ldots;p_n\, \{c\}$. The following property can be proved in $\mathfrak{K}$: for every $\psi \in FL(\Phi)$ and for every state α,

$$\psi \in \alpha \implies \alpha \models \hat{\psi}. \tag{25}$$

To prove (25) we first show by induction on p that for an extended PCA $[p]\psi \in FL(\Phi)$ and atoms α, β,

$$[p]\psi \in \alpha \text{ and } \alpha \xrightarrow[\mathfrak{K}]{p} \beta \implies \psi \in \beta.$$

Then (25) follows by a simple induction on ψ.

Let $\Psi = \{\{b_1\}\, p_1\, \{c_1\}, \ldots, \{b_n\}\, p_n\, \{c_n\}\}$ be the set of premises of the rule (13). It follows from (25) that $\mathfrak{K} \models \Psi$.

Let us recall that, as in the proof of Theorem 4.1, we identify an atom α with the conjuction of all formulas in α. To complete the proof of Theorem 4.5 we prove that for every program p and all atoms α, β,

$$\Psi \nvdash_{EPHL} \{\alpha\}\, p\, \{\bar{\beta}\} \implies \alpha \xrightarrow[\mathfrak{K}]{p} \beta. \tag{26}$$

The proof of the contraposition of (26) is by induction on p. For the base case we use the axiom **(Atom)**, **(Weakening)** and the meaning function for atomic programs.

Now, if $\Psi \not\vdash_{EPHL} \{b\}p\{c\}$, then it is easy to show by the **(And/Or)** rule that there exist states α, β such that $\alpha \models b$, $\beta \models \bar{c}$ and $\Psi \not\vdash_{EPHL} \{\alpha\}p\{\bar{\beta}\}$. Thus by (26) we obtain $\alpha \xrightarrow{p}_{\mathcal{R}} \beta$ and therefore $\alpha \not\models \{b\}p\{c\}$. This completes the proof. ∎

References

Apt, K. R.: 1981, 'Ten years of Hoare's logic: a survey—part I'. *ACM Trans. Programming Languages and Systems* **3**, 431–483.

Apt, K. R. and E.-R. Olderog: 1991, *Verification of Sequential and Concurrent Programs*. Springer-Verlag.

Cohen, E. and D. Kozen: 2000, 'A Note on the complexity of propositional Hoare logic'. *Trans. Computational Logic* **1**(1), 171–174.

Cook, S. A.: 1978, 'Soundness and completeness of an axiom system for program verification'. *SIAM J. Comput.* **7**, 70–80.

Dijkstra, E.: 1975, 'Guarded Commands, Nondeterminacy and Formal Derivation of Programs'. *CACM* **18**(8), 453–457.

Emerson, E. A.: 1990, 'Temporal and modal logic'. In: J. van Leeuwen (ed.): *Handbook of theoretical computer science*, Vol. B: formal models and semantics. Elsevier, pp. 995–1072.

Fischer, M. J. and R. E. Ladner: 1977, 'Propositional modal logic of programs'. In: *Proc. 9th Symp. Theory of Comput.* pp. 286–294.

Floyd, R.: 1967, 'Assigning Meaning to Programs'. In: J. Schwartz (ed.): *Proc. Symp. in Applied Mathematics.* pp. 19–32.

Gabbay, D., I. Hodkinson, and M. Reynolds: 1994, *Temporal Logic: Mathematical Foundations and Computational Aspects.* Oxford University Press.

Harel, D., D. Kozen, and J. Tiuryn: 2000, *Dynamic Logic.* Cambridge, MA: MIT Press.

Hoare, C.: 1969, 'An Axiomatic Basis for Computer Programming'. *CACM* **12**(10), 576–580.

Kozen, D.: 1997, 'Kleene algebra with tests'. *Transactions on Programming Languages and Systems* **19**(3), 427–443.

Kozen, D.: 1999, 'On Hoare logic and Kleene algebra with tests'. In: *Proc. Conf. Logic in Computer Science (LICS'99).* pp. 167–172.

Kozen, D. and J. Tiuryn: 2000, 'On the completeness of propositional Hoare logic'. In: J. Desharnais (ed.): *Proc. 5th Int. Seminar Relational Methods in Computer Science (RelMiCS 2000).* pp. 195–202.

Pnueli, A.: 1977, 'The temporal logic of programs'. In: *Proc. 18th Symp. Found. Comput. Sci.* pp. 46–57.

Pratt, V. R.: 1976, 'Semantical considerations on Floyd-Hoare logic'. In: *Proc. 17th Symp. Found. Comput. Sci.* pp. 109–121.

Wand, M.: 1978, 'A new incompleteness result for Hoare's system'. *J. Assoc. Comput. Mach.* **25**, 168–175.

Winskel, G.: 1993, *The formal semantics of programming languages.* MIT Press.

HOARE LOGICS IN ISABELLE/HOL

TOBIAS NIPKOW

Technische Universität München, Institut für Informatik

nipkow@in.tum.de

Abstract. This paper describes Hoare logics for a number of imperative language constructs, from while-loops via exceptions to mutually recursive procedures. Both partial and total correctness are treated. In particular a proof system for total correctness of recursive procedures in the presence of unbounded nondeterminism is presented. All systems are formalized and shown to be sound and complete in the theorem prover Isabelle/HOL.

1. Introduction

Hoare logic is a well developed area with a long history (by computer science standards). The purpose of this report is

- to present, in a unified notation, Hoare logics for a number of different programming language constructs such as loops, exceptions, expressions with side effects, and procedures, together with clear presentations of their soundness and especially completeness proofs, and
- to show that this can and argue that this should be done in a theorem prover, in our case Isabelle/HOL [20].

Thus one can view this report as a relative of Apt's survey papers [2, 3], but with new foundations. Instead of on paper, all formalizations and proofs have been carried out with the help of a theorem prover. A first attempt in this direction [16, 17] merely formalized and debugged an existing completeness proof. Kleymann [27] went one step further and formalized a new and slick Hoare logic for total correctness of recursive procedures. This problem has an interesting history. The logic presented by Apt [2] was later found to be unsound by America and de Boer [1], who modified the system and proved its soundness and completeness. Their proof system, however, suffers from three additional rules with syntactic side conditions. The first really simple system is the one by Kleymann [27] — and it was embedded in a theorem prover.

H. Schwichtenberg and R. Steinbrüggen (eds.), Proof and System-Reliability, 341–367.
© 2002 *Kluwer Academic Publishers. Printed in the Netherlands.*

It may be argued that Kleymann's proof system has nothing to do with the use of a theorem prover. Although the two things are indeed independent of each other, theorem provers tend to act like Occam's razor: the enormous amount of detail one has to deal with when building a model of anything inside a theorem prover often forces one to discover unexpected simplifications. Thus programming logics are an ideal application area for theorem provers: both correctness and simplicity of such logics is of vital importance, and, as Apt himself says [2], "various proofs given in the literature are awkward, incomplete, or even incorrect."

In modelling the assertion language, we follow the *extensional* approach [15] where assertions are identified with functions from states to propositions in the logic of the theorem prover. That is, we model only the semantics but not the syntax of assertions. This is common practice (with the exception of [7], but they do not consider completeness) because the alternative, embedding an assertion language with quantifiers in a theorem prover, is not just hard work but also orthogonal to the problem of embedding the computational part of the Hoare logic. As a consequence we have solved the question of *expressiveness*, i.e. whether the assertion language is strong enough to express all intermediate predicates that may arise in a proof, by going to a higher order logic. Thus our completeness results do not automatically carry over to other logical systems, say first order arithmetic. The advantage of the extensional approach is that it separates reasoning about programs from expressiveness considerations — the latter can then be conducted in isolation for each assertion language.

Much of this report is inspired by the work of Kleymann [10]. He used the theorem prover LEGO [26] rather than Isabelle/HOL, which makes very little difference except in one place (Sect. 2.4). Although in the end, none of our logics are identical to any of his, they are closely related. The main differences are that we consider many more language constructs, we generalize from deterministic to nondeterministic languages, and we consider partial as well as total correctness.

The whole paper is generated directly from the Isabelle input files (which include the text as comments). That is, if you see a lemma or theorem, you can be sure its proof has been checked by Isabelle. But as the report does not go into the details of the proofs, no previous exposure to theorem provers is a prerequisite for reading it.

Isabelle/HOL is an interactive theorem prover for HOL, higher order logic. Most of the syntax of HOL will be familiar to anybody with some background in functional programming and logic. We just highlight some of Isabelle's nonstandard notation.

The syntax $[P;\ Q] \implies R$ should be read as an inference rule with the two premises P and Q and the conclusion R. Logically it is just a shorthand for $P \implies Q \implies R$. *Note that semicolon will also denote sequential composition of programs!* There are actually two implications $\longrightarrow$ and $\implies$. The two mean the same thing, except that $\longrightarrow$ is HOL's "real" implication, whereas $\implies$ comes from Isabelle's meta-logic and expresses inference rules. Thus $\implies$ cannot appear inside a HOL formula. For the purpose of this paper the two may be identified. However, beware that $\longrightarrow$ binds more tightly than $\implies$: in $\forall x.\ P \longrightarrow Q$ the $\forall x$ covers $P \longrightarrow Q$, whereas in $\forall x.\ P \implies Q$ it covers only P.

Set comprehension is written $\{x.\ P\}$ rather than $\{x \mid P\}$ and is also available for tuples, e.g. $\{(x,\ y,\ z).\ P\}$.

1.1. STRUCTURE OF THE PAPER

In Sect. 2 we discuss a simple while-language and modular extensions to nondeterminism and local variables. In Sect. 3 we add exceptions and in Sect. 4 side effects in expression evaluation. In Sect. 5 we add a single parameterless but potentially recursive procedure, later extending it with nondeterminism and local variables. In Sect. 6 the single procedure is replaced by multiple mutually recursive procedures. In Sect. 7 we treat a single function with a single value parameter.

In each case we present syntax, operational semantics and Hoare logic for partial correctness, together with soundness and completeness theorems. For loops and procedures, we also cover total correctness.

The guiding principle is to cover each language feature in isolation and in the simplest possible version. Combinations are only discussed if there is some interference. We do not expect the remaining combinations to present any problems not encountered before.

The choice of language features covered was inspired by the work of von Oheimb [22, 23] who presents an Isabelle/HOL embedding of a Hoare logic for a subset of Java. Based on the proof systems in this report, we have meanwhile designed a simpler Hoare logic for a smaller subset of Java [24].

2. A simple while-language

2.1. SYNTAX AND OPERATIONAL SEMANTICS

We start by declaring the two types *var* and *val* of variables and values:

```
typedecl var
typedecl val
```

They need not be refined any further. Building on them, we define states as functions from variables to values and boolean expressions (*bexp*) as functions from states to the booleans:

types *state = var ⇒ val*
      bexp = state ⇒ bool

Most of the time the type of states could have been left completely unspecified, just like the types *var* and *val*.

Our model of boolean expressions requires a few words of explanation. Type *bool* is HOL's predefined type of propositions. Thus all the usual logical connectives like ∧ and ∨ are available. Instead of modelling the syntax of boolean expressions, we model their semantics. For example, the programming language expression x != y becomes $\lambda s.\ s\ x \neq s\ y$, where $s\ x$ expresses the lookup of the value of variable x in state s.

Now it is time to describe the (abstract and concrete) syntax of our programming language, which is easily done with a recursive datatype such as found in most functional programming languages:

datatype *com = Do (state ⇒ state)*
      | *Semi com com*   (-; - [60, 60] 10)
      | *Cond bexp com com*   (IF - THEN - ELSE - 60)
      | *While bexp com*   (WHILE - DO - 60)

Statements in this language are called *commands*. They are modelled as terms of type *com*. *Do f* represents an atomic command that changes the state from s to $f\ s$ in one step. Thus the command that does nothing, often called **skip**, can be represented by *Do* ($\lambda s.\ s$). More interestingly, an assignment x := e, where e is some expression, can be modelled as follows: represent e by a function e from *state* to *val*, and the assignment by *Do* ($\lambda s.\ s(x := e\ s)$), where $f(a := v)$ is a predefined construct in HOL for updating function f at argument a with value v. Again we have chosen to model the semantics rather than the syntax, which simplifies matters enormously. Of course it means that we can no longer talk about certain syntactic matters, but that is just fine.

The constructors *Semi*, *Cond* and *While* represent sequential composition, conditional and while-loop. The annotations allow us to write

    c1; *c2*   *IF b THEN c1 ELSE c2*   *WHILE b DO c*

instead of *Semi c1 c2*, *Cond b c1 c2* and *While b c*.

Now it is time to define the semantics of the language, which we do operationally, by the simplest possible scheme, a so-called *evaluation* or *big step* semantics. Execution of commands is defined via triples of the form $s\ -c\rightarrow\ t$ which should be read as "execution of c starting in state

s may terminate in state t". This allows for two kinds of nondeterminism: there may be other executions $s\ -c\rightarrow\ u$ with $t \neq u$, and there may be nonterminating computations as well. For the time being we do not model nontermination explicitly. Only if for some s and c there is no triple $s\ -c\rightarrow\ t$ does this signal what we intuitively view as nontermination. We start with a simple deterministic language and assertions about terminating computations. Nondeterminism and nontermination are treated later. The semantics of our language is defined inductively. Beware that semicolon is used both as a separator of premises and for sequential composition.

$s\ -Do\ f\rightarrow\ f\ s$

$[\![\ s0\ -c1\rightarrow\ s1;\ s1\ -c2\rightarrow\ s2\ ]\!] \Longrightarrow s0\ -c1;c2\rightarrow\ s2$

$[\![\ b\ s;\ s\ -c1\rightarrow\ t\ ]\!] \Longrightarrow s\ -IF\ b\ THEN\ c1\ ELSE\ c2\rightarrow\ t$
$[\![\ \neg b\ s;\ s\ -c2\rightarrow\ t\ ]\!] \Longrightarrow s\ -IF\ b\ THEN\ c1\ ELSE\ c2\rightarrow\ t$

$\neg b\ s \Longrightarrow s\ -WHILE\ b\ DO\ c\rightarrow\ s$
$[\![\ b\ s;\ s\ -c\rightarrow\ t;\ t\ -WHILE\ b\ DO\ c\rightarrow\ u\ ]\!] \Longrightarrow s\ -WHILE\ b\ DO\ c\rightarrow\ u$

2.2. HOARE LOGIC FOR PARTIAL CORRECTNESS

We continue our semantic approach by modelling assertions just like boolean expressions, i.e. as functions:

types $assn\ =\ state \Rightarrow bool$

Hoare triples are triples of the form $\{P\}\ c\ \{Q\}$, where the assertions P and Q are the so-called pre and postconditions. Such a triple is *valid* (denoted by $\models$) iff every (terminating) execution starting in a state satisfying P ends up in a state satisfying Q:

$$\models \{P\}c\{Q\} \equiv \forall s\ t.\ s\ -c\rightarrow\ t \longrightarrow P\ s \longrightarrow Q\ t$$

The $\equiv$ sign denotes definitional equality.

This notion of validity is called *partial correctness* because it does not require termination of c.

Finally we come to the core of this paper, Hoare logic, i.e. inference rules for deriving (hopefully valid) Hoare triples. As usual, derivability is indicated by $\vdash$, and defined inductively:

$\vdash \{\lambda s.\ P(f\ s)\}\ Do\ f\ \{P\}$

$[\![\ \vdash \{P\}c1\{Q\};\ \vdash \{Q\}c2\{R\}\ ]\!] \Longrightarrow \vdash \{P\}\ c1;c2\ \{R\}$

$$[\![\vdash \{\lambda s.\ P\ s \wedge b\ s\}\ c1\ \{Q\};\ \vdash \{\lambda s.\ P\ s \wedge \neg b\ s\}\ c2\ \{Q\}]\!]$$
$$\Longrightarrow \vdash \{P\}\ IF\ b\ THEN\ c1\ ELSE\ c2\ \{Q\}$$

$$\vdash \{\lambda s.\ P\ s \wedge b\ s\}\ c\ \{P\} \Longrightarrow \vdash \{P\}\ WHILE\ b\ DO\ c\ \{\lambda s.\ P\ s \wedge \neg b\ s\}$$

$$[\![\forall s.\ P'\ s \longrightarrow P\ s;\ \vdash \{P\}c\{Q\};\ \forall s.\ Q\ s \longrightarrow Q'\ s]\!] \Longrightarrow \vdash \{P'\}c\{Q'\}$$

The final rule is called the *consequence rule*.

Soundness is proved by induction on the derivation of $\vdash \{P\}\ c\ \{Q\}$:

theorem $\vdash \{P\}c\{Q\} \Longrightarrow \models \{P\}c\{Q\}$

Only the *While*-case requires additional help in the form of a lemma:

$$[\![s\ -WHILE\ b\ DO\ c \rightarrow t;\ P\ s;\ \forall s\ s'.\ P\ s \wedge b\ s \wedge s\ -c \rightarrow s' \longrightarrow P\ s']\!]$$
$$\Longrightarrow P\ t \wedge \neg\ b\ t$$

This lemma is the operational counterpart of the *While*-rule of Hoare logic. It is proved by induction on the derivation of $s\ -WHILE\ b\ DO\ c \rightarrow t$.

Completeness is not quite as straightforward, but still easy. The proof is best explained in terms of the *weakest precondition*:

$$wp :: com \Rightarrow assn \Rightarrow assn$$
$$wp\ c\ Q \equiv \lambda s.\ \forall t.\ s\ -c \rightarrow t \longrightarrow Q\ t$$

Loosely speaking, $wp\ c\ Q$ is the set of all start states such that all (terminating) executions of c end up in Q. This is appropriate in the context of partial correctness. Dijkstra calls this the weakest *liberal* precondition to emphasize that it corresponds to partial correctness. We use "weakest precondition" all the time and let the context determine if we talk about partial or total correctness — the latter is introduced further below.

The following lemmas about wp are easily derived:

lemma $wp\ (Do\ f)\ Q = (\lambda s.\ Q(f\ s))$
lemma $wp\ (c1;c2)\ R = wp\ c1\ (wp\ c2\ R)$
lemma $wp\ (IF\ b\ THEN\ c1\ ELSE\ c2)\ Q = (\lambda s.\ wp\ (if\ b\ s\ then\ c1\ else\ c2)\ Q\ s)$
lemma $wp\ (WHILE\ b\ DO\ c)\ Q =$
$\qquad (\lambda s.\ if\ b\ s\ then\ wp\ (c;WHILE\ b\ DO\ c)\ Q\ s\ else\ Q\ s)$

Note that $if-then-else$ is HOL's predefined conditional expression.

By induction on c one can easily prove

lemma $\forall Q. \vdash \{wp\ c\ Q\}\ c\ \{Q\}$

from which completeness follows more or less directly via the rule of consequence:

theorem $\models \{P\}c\{Q\} \Longrightarrow \vdash \{P\}c\{Q\}$

2.3. MODULAR EXTENSIONS OF PURE WHILE

We discuss two modular extensions of our simple while-language: nondeterminism and local variables. By modularity we mean that we can add these features without disturbing the existing setup. In fact, even the proofs can be extended modularly: the soundness proofs acquire two new easy cases, and for the completeness proofs we merely have to provide suitable lemmas about how wp behaves on the new constructs.

2.3.1. *Nondeterminism*

We add a choice construct at the level of commands: $c1 \mid c2$ is the command that can nondeterministically choose to execute either $c1$ or $c2$:
$$s -c1 \rightarrow t \implies s -c1 \mid c2 \rightarrow t \qquad s -c2 \rightarrow t \implies s -c1 \mid c2 \rightarrow t$$

The proof rule is analogous. If we want to make sure that all executions of $c1 \mid c2$ fulfill their specification, both $c1$ and $c2$ must do so:
$$[\vdash \{P\}\ c1\ \{Q\}; \vdash \{P\}\ c2\ \{Q\}] \implies \vdash \{P\}\ c1 \mid c2\ \{Q\}$$

The behaviour of wp (required for the completness proof) is obvious:
$$wp\ (c1 \mid c2)\ Q = (\lambda s.\ wp\ c1\ Q\ s \wedge wp\ c2\ Q\ s)$$

2.3.2. *Local variables*

We add a new command $VAR\ x = e;\ c$ that assigns x the value of e, executes c, and then restores the old value of x:
$$s(x := e\ s) -c \rightarrow t \implies s -VAR\ x = e;\ c \rightarrow t(x := s\ x)$$

The corresponding proof rule
$$\forall v. \vdash \{\lambda s.\ P\ (s(x := v)) \wedge s\ x = e\ (s(x := v))\}\ c\ \{\lambda s.\ Q\ (s(x := v))\} \implies$$
$$\vdash \{P\}\ VAR\ x = e;\ c\ \{Q\}$$

needs a few words of explanation. The main problem is how to refer to the initial value of x in the postcondition. In some related calculi like VDM [9], this is part of the logic, but in plain Hoare logic we have to remember the old value of x explicitly by equating it to something we can refer to in the postcondition. That is the *raison d'être* for v. Of course this should work for every value of x. Hence the $\forall v$. If you are used to more syntactic presentations of Hoare logic you may prefer a side condition that v does not occur free in P, e, c or Q. However, since we embed Hoare logic in a language with quantifiers, why not use them to good effect?

The behaviour of wp mirrors the execution of VAR:

$$wp \ (VAR \ x = e; \ c) \ Q = (\lambda s. \ wp \ c \ (\lambda t. \ Q \ (t(x := s \ x))) \ (s(x := e \ s)))$$

2.4. HOARE LOGIC FOR TOTAL CORRECTNESS

2.4.1. *Termination*

Although partial correctness appeals because of its simplicity, in many cases one would like the additional assurance that the command is guaranteed to termiate if started in a state that satisfies the precondition. Even to express this we need to define when a command is guaranteed to terminate. We can do this without modifying our existing semantics by merely adding a second inductively defined judgement $c \downarrow s$ that expresses guaranteed termination of c started in state s:

$$Do \ f \downarrow s$$

$$[\![\ c1 \downarrow s0; \ \forall s1. \ s0 \ -c1 \to \ s1 \ \longrightarrow \ c2 \downarrow s1 \ ]\!] \Longrightarrow (c1;c2) \downarrow s0$$

$$[\![\ b \ s; \ c1 \downarrow s \ ]\!] \Longrightarrow IF \ b \ THEN \ c1 \ ELSE \ c2 \downarrow s$$
$$[\![\ \neg b \ s; \ c2 \downarrow s \ ]\!] \Longrightarrow IF \ b \ THEN \ c1 \ ELSE \ c2 \downarrow s$$

$$\neg b \ s \Longrightarrow WHILE \ b \ DO \ c \downarrow s$$
$$[\![\ b \ s; \ c \downarrow s; \ \forall t. \ s \ -c \to \ t \ \longrightarrow \ WHILE \ b \ DO \ c \downarrow t \ ]\!] \Longrightarrow WHILE \ b \ DO \ c \downarrow s$$

The rules should be self-explanatory.

Now that we have termination, we can define total validity, $\models_t$, as partial validity and guaranteed termination:

$$\models_t \{P\}c\{Q\} \ \equiv \ \models \{P\}c\{Q\} \wedge (\forall s. \ P \ s \ \longrightarrow \ c{\downarrow}s)$$

2.4.2. *Hoare logic*

Derivability of Hoare triples in the proof system for total correctness is written $\vdash_t \{P\} \ c \ \{Q\}$ and defined inductively. The rules for $\vdash_t$ differ from those for $\vdash$ only in the one place where nontermination can arise: the *While*-rule. Hence we only show that one rule:

$$[\![wf \ r; \ \forall s'. \ \vdash_t \{\lambda s. \ P \ s \wedge b \ s \wedge s' = s\} \ c \ \{\lambda s. \ P \ s \wedge (s,s') \in r\}]\!]$$
$$\Longrightarrow \vdash_t \{P\} \ WHILE \ b \ DO \ c \ \{\lambda s. \ P \ s \wedge \neg b \ s\}$$

The rule is like the one for partial correctness but it requires additionally that with every execution of the loop body a wellfounded relation $(wf \ r)$ on the state space decreases: wellfoundedness of r means there is no infinite descending chain $\ldots$, $(s2, s1) \in r$, $(s1, s0) \in r$. To compare the value of the state before and after the execution of the loop body we again use the trick discussed in connection with local variables above: a locally $\forall$-quantified variable.

This is almost the rule by Kleymann [10], except that we do not have a wellfounded relation on some arbitrary set *together* with a measure function on states, but have collapsed this into a wellfounded relation on states. This does not just shorten the rule but it also simplifies it logically: now we know that wellfounded relations on the state space suffice and we do not need to drag in other types. I should mention that this simplification was forced on me by Isabelle: since Isabelle does not allow local quantification over types, I could not even express Kleymann's rule, which requires just that.

The soundness theorem

theorem $\vdash_t \{P\}c\{Q\} \implies \models_t \{P\}c\{Q\}$

is again proved by induction over c. But in the *While*-case we do not appeal to the same lemma as in the proof for $\vdash \{P\} \ c \ \{Q\} \implies \models \{P\} \ c \ \{Q\}$. Instead we perform a local proof by wellfounded induction over the given relation r.

The completeness proof proceeds along the same lines as the one for partial correctness. First we have to strengthen our notion of weakest precondition to take termination into account:

$$wp_t \ c \ Q \equiv \lambda s. \ wp \ c \ Q \ s \wedge c{\downarrow}s$$

The lemmas proved about wp_t are the same as those for wp, except for the *While*-case, which we deal with locally below. The key lemma

lemma $\forall Q. \vdash_t \{wp_t \ c \ Q\} \ c \ \{Q\}$

is again proved by induction on c. The *While*-case is interesting because we now have to furnish a suitable wellfounded relation. Of course the execution of the loop body directly yields the required relation, as the following lemma shows. Remember that set comprehension in Isabelle/HOL uses "." rather than "|".

$$wf \ \{(t, s). \ WHILE \ b \ DO \ c \downarrow s \wedge b \ s \wedge s \ -c{\rightarrow} \ t\}$$

This lemma follows easily from the lemma that if $WHILE \ b \ DO \ c \downarrow s$ then there is no infinite sequence of executions of the body, which is proved by induction on $WHILE \ b \ DO \ c \downarrow s$.

The actual completeness theorem follows directly, in the same manner as for partial correctness.

theorem $\models_t \{P\}c\{Q\} \implies \vdash_t \{P\}c\{Q\}$

2.4.3. *Modular extensions of pure while*

Nondeterministic choice and local variables can be added without disturbing anything. Their proof rules remain exactly the same as in the case of partial correctness. We have carried this out as well; the details are straightforward.

3. Exceptions

We extend our pure while-language with exceptions, a modification that is decidedly non-modular as it changes the state space and the semantics.

3.1. SYNTAX AND SEMANTICS

Our exceptions are very simple: there is only one exception, which we call *error*, and it can be raised and handled. Semantically we treat errors by extending our state space with a new boolean component that indicates if the error has been raised. This new state space *estate* is defined as a record with two components:

record *estate* = *st* :: *state*
$\qquad\qquad$ *err* :: *bool*

Record selectors are simply projection functions. Records are constructed as in $(\!| st = s,\ err = False |\!)$ and updated selectively as in $es(\!| err := True |\!)$. We also introduce *ok s* as a shorthand for $\neg err\ s$.
$\qquad$ Boolean expressions are now defined as

types *bexp* = *estate* $\Rightarrow$ *bool*

The syntax of the language with errors is the same as the simple while language, but extended with a construct for handling errors:

datatype *com* = *Do* (*estate* $\Rightarrow$ *estate*)
$\qquad\qquad$ | *Semi com com* $\qquad\qquad$ (-; - [60, 60] 10)
$\qquad\qquad$ | *Cond bexp com com* $\qquad$ (*IF - THEN - ELSE -* 60)
$\qquad\qquad$ | *While bexp com* $\qquad\qquad$ (*WHILE - DO -* 60)
$\qquad\qquad$ | *Handle com com* $\qquad\qquad$ (- *HANDLE -* 60)

How is an error raised? Simply by $Do(\lambda s.\ s(\!| err := True |\!))$. And how is it handled? Command *c1 HANDLE c2* executes *c1*, and, if this raises an error, resets the error and continues with executing *c2*.
$\qquad$ Having the error flag as part of the state allows us to execute commands only if the state is *ok* and to skip execution otherwise [19]. It leads to the following set of rules for command execution:

$ok\ s \implies s\ -Do\ f \rightarrow f\ s$

$[\![\ s0\ -c1\rightarrow s1;\ s1\ -c2\rightarrow s2\ ]\!] \implies s0\ -c1;c2\rightarrow s2$

$[\![\ ok\ s;\ \ b\ s;\ s\ -c1\rightarrow t\ ]\!] \implies s\ -IF\ b\ THEN\ c1\ ELSE\ c2\rightarrow t$
$[\![\ ok\ s;\ \neg b\ s;\ s\ -c2\rightarrow t\ ]\!] \implies s\ -IF\ b\ THEN\ c1\ ELSE\ c2\rightarrow t$

$[\![\ ok\ s;\ \neg b\ s\ ]\!] \implies s\ -WHILE\ b\ DO\ c\rightarrow s$
$[\![\ ok\ s;\ b\ s;\ s\ -c\rightarrow t;\ t\ -WHILE\ b\ DO\ c\rightarrow u\ ]\!] \implies s\ -WHILE\ b\ DO\ c\rightarrow u$

$[\![\ ok\ s0;\ s0\ -c1\rightarrow s1;\ ok\ s1\ ]\!] \implies s0\ -c1\ HANDLE\ c2\rightarrow s1$
$[\![\ ok\ s0;\ s0\ -c1\rightarrow s1;\ err\ s1;\ s1(\!|err:=False|\!)\ -c2\rightarrow s2\ ]\!]$
$\implies s0\ -c1\ HANDLE\ c2\rightarrow s2$

$err\ s \implies s\ -c\rightarrow s$

3.2. HOARE LOGIC

The Hoare logic follows the same lines as the one for the language without exceptions. The main change is that assertions are now functions of *estate*:

types $assn = estate \Rightarrow bool$

Hence pre and postconditions can talk about whether an error has been raised or not. Validity $\models \{P\}\ c\ \{Q\}$ is again partial correctness as defined in Sect. 2.2. The proof rules for the logic are the following:

$\vdash \{\lambda s.\ if\ err\ s\ then\ P\ s\ else\ P(f\ s)\}\ Do\ f\ \{P\}$

$[\![\ \vdash \{P\}c1\{Q\}; \vdash \{Q\}c2\{R\}\ ]\!] \implies \vdash \{P\}\ c1;c2\ \{R\}$

$[\![\ \vdash \{\lambda s.\ ok\ s \wedge P\ s \wedge b\ s\}\ c1\ \{Q\}; \vdash \{\lambda s.\ ok\ s \wedge P\ s \wedge \neg b\ s\}\ c2\ \{Q\}\ ]\!]$
$\implies \vdash \{\lambda s.\ if\ err\ s\ then\ Q\ s\ else\ P\ s\}\ IF\ b\ THEN\ c1\ ELSE\ c2\ \{Q\}$

$\vdash \{\lambda s.\ ok\ s \wedge P\ s \wedge b\ s\}\ c\ \{P\}$
$\implies \vdash \{P\}\ WHILE\ b\ DO\ c\ \{\lambda s.\ P\ s \wedge (ok\ s \longrightarrow \neg b\ s)\}$

$[\![\ \vdash \{\lambda s.\ ok\ s \wedge P\ s\}\ c1\ \{\lambda s.\ if\ err\ s\ then\ Q(s(\!|err:=False|\!))\ else\ R\ s\};$
$\vdash \{Q\}\ c2\ \{R\}\ ]\!]$
$\implies \vdash \{\lambda s.\ if\ err\ s\ then\ R\ s\ else\ P\ s\}\ c1\ HANDLE\ c2\ \{R\}$

$[\![\ \forall s.\ P'\ s \longrightarrow P\ s; \vdash \{P\}c\{Q\}; \forall s.\ Q\ s \longrightarrow Q'\ s\ ]\!] \implies \vdash \{P'\}c\{Q'\}$

The conclusions of the rules for *Do*, *Cond* and *Handle* follow the pattern that the precondition has been augmented with a conditional that reduces to the "normal" precondition P if no error is present but collapses to the postcondition Q otherwise, because in that case the command does nothing. In the *While* rule we had to ensure that invariance of P only needs to be proved for *ok* states, and that after the loop we can only infer the negation of the loop test if we are in an *ok* state — otherwise we may have left the loop by the error exit instead of normally. The most puzzling rule may be the one for *Handle*: why does it always require $\vdash \{Q\}\ c2\ \{R\}$, even if $c1$ cannot raise an error? The answer is that we can simply set Q to *False*, in which case $\vdash \{Q\}\ c2\ \{R\}$ is always provable.

Soundness is proved as usual by induction on c, almost exactly as for the basic while-language in Sect. 2.2, just with a few more case distinctions.

theorem $\vdash \{P\}c\{Q\} \implies \models \{P\}c\{Q\}$

The weakest precondition *wp* is also defined as in Sect. 2.2, but we obtain a different set of derived laws:

lemma $wp\ (Do\ f)\ Q = (\lambda s.\ if\ err\ s\ then\ Q\ s\ else\ Q(f\ s))$
lemma $wp\ (c1;c2)\ R = wp\ c1\ (wp\ c2\ R)$
lemma $wp\ (IF\ b\ THEN\ c1\ ELSE\ c2)\ Q =$
$\qquad (\lambda s.\ if\ err\ s\ then\ Q\ s\ else\ wp\ (if\ b\ s\ then\ c1\ else\ c2)\ Q\ s)$
lemma $wp\ (WHILE\ b\ DO\ c)\ Q =$
$\qquad (\lambda s.\ if\ err\ s\ then\ Q\ s\ else\ if\ b\ s\ then\ wp\ (c;WHILE\ b\ DO\ c)\ Q\ s\ else\ Q\ s)$
lemma $wp\ (c1\ HANDLE\ c2)\ R =$
$\qquad (\lambda s.\ if\ err\ s\ then\ R\ s$
$\qquad\qquad else\ wp\ c1\ (\lambda t.\ if\ err\ t\ then\ (wp\ c2\ R)(t(\!|err{:=}False|\!))\ else\ R\ t)\ s)$

As in Sect. 2.2, the key lemma is now proved without much fuss

lemma $\forall Q.\ \vdash \{wp\ c\ Q\}\ c\ \{Q\}$

and completeness follows directly:

theorem $\models \{P\}c\{Q\} \implies \vdash \{P\}c\{Q\}$

4. Side effects

We consider a language where the evaluation of expressions may have side effects. In practice this occurs because of side effecting operators like $++$ in C or because of user-defined functions with side effects. One trivial solution to this problem is require a program transformation step that eliminates compound expressions. For example, x := f(g(y)), where f and g may have side effects, is transformed into z := g(y); x := f(z). The resulting program is easy to deal with. Our aim is to show that one can reason about compound expressions directly and that the required Hoare logic is quite straightforward. The essential idea goes back to Kowaltowski [12]: specify the behaviour of expressions by Hoare triples where the postcondition can refer to the value of the expression. This was already formalized by von Oheimb [22]. We improve his rules a little by dropping the unnecessary dependence of the precondition on the expression value.

4.1. SYNTAX AND SEMANTICS

Types *var*, *val*, *state* and *bexp* are defined as in Sect. 2. But now we introduce a separate type of *expressions* because we want to study how expression evaluation interacts with side effecting function calls:

datatype $expr = Var\ var \mid Fun\ (state \Rightarrow val\ list \Rightarrow val \times state)\ (expr\ list)$

An expression can either be a variable or *Fun f es*, the application of a function *f* to a list of expressions *es*. Function *f* depends not just on a list of values but also on the state, and it returns not just a value but also a new state. Thus we now have a more syntactic representation of expressions (e.g. compared with *bexp*), but the notion of functions is still a semantic one.

Throughout this section, *e* always stands for an expression and *es* always for an expression list.

With the arrival of expressions, the syntax of commands changes: the generic *Do* is replaced with a proper assigment command, and *SKIP* is added as a separate command:

datatype $com = SKIP$
| *Assign var expr* $(- := - [60, 60]\ 10)$
| *Semi com com* $(-;\ - [60, 60]\ 10)$
| *Cond bexp com com* $(IF - THEN - ELSE -\ 60)$
| *While bexp com* $(WHILE - DO -\ 60)$

Now that expression evaluation can have side effects, the semantics of the language is defined by three transition relations:

$s -c\rightarrow t$ the familiar execution of commands,

$s -e\Rightarrow (v,t)$ the evaluation of an expression *e* which produces both a value *v* and a new state *t*, and

$s =es\Rightarrow (vs,t)$ the evaluation of an expression list *es* which produces both a list of values *vs* and a new state *t*.

Evaluation of expressions and expression lists is defined mutually inductively:

$s - Var\ x \Rightarrow (s\ x,s)$

$s = es \Rightarrow (vs,t) \implies s - Fun\ f\ es \Rightarrow f\ t\ vs$

$s = [] \Rightarrow ([],s)$

$[s - e \Rightarrow (v,t);\ t = es \Rightarrow (vs,u)] \implies s = e \# es \Rightarrow (v \# vs, u)$

Lists in Isabelle/HOL are built up from them empty list [] by the infix constructor #, where $x\ \#\ xs$ is the list with head x and tail xs.

Command execution is defined as usual. Hence we only show the rules for the two new commands:

$s - SKIP \rightarrow s$

$s - e \Rightarrow (v,t) \implies s - x := e \rightarrow t(x := v)$

4.2. HOARE LOGIC

Since expresssion evaluation may change the state, we need to reason about the individual expresssion evaluation steps as well. To reason about evaluation, we need to take the computed values into account, too. Thus there will be two new kinds of Hoare triples: $\{P\}e\{Q'\}$ and $\{P\}es\{Q''\}$, where P depends only on the state but where Q' depends also on the value of e and Q'' also on the value of es. Thus there are three types of assertions:

types $assn = state \Rightarrow bool$

$\quad vassn = val \Rightarrow state \Rightarrow bool$

$\quad vsassn = val\ list \Rightarrow state \Rightarrow bool$

Most of them time we use P, Q and R for all three kinds of assertions.

Validity of the three kinds of Hoare triples is denoted by $\models$, $\models_e$, $\models_{es}$. The definitions need no comments:

$\models \{P\}c\{Q\} \equiv \forall s\ t.\ s - c \rightarrow t \longrightarrow P\ s \longrightarrow Q\ t$

$\models_e \{P\}e\{Q\} \equiv \forall s\ t\ v.\ s - e \Rightarrow (v,t) \longrightarrow P\ s \longrightarrow Q\ v\ t$

$\models_{es} \{P\}es\{Q\} \equiv \forall s\ t\ vs.\ s = es \Rightarrow (vs,t) \longrightarrow P\ s \longrightarrow Q\ vs\ t$

Thus there are also three kinds of judgements: $\vdash \{P\}\ c\ \{Q\}$, $\vdash_e \{P\}\ e\ \{Q\}$ and $\vdash_{es} \{P\}\ es\ \{Q\}$. The latter two are defined by mutual induction:

$\vdash_e \{\lambda s.\ Q\ (s\ x)\ s\}\ Var\ x\ \{Q\}$

$\vdash_{es} \{P\}\ es\ \{\lambda vs\ s.\ Q\ (fst(f\ s\ vs))\ (snd(f\ s\ vs))\} \implies \vdash_e \{P\}\ Fun\ f\ es\ \{Q\}$

$\vdash_{es} \{P\ []\}\ []\ \{P\}$

$[\vdash_e \{P\}\ e\ \{Q\};\ \forall v.\ \vdash_{es} \{Q\ v\}\ es\ \{\lambda vs.\ R(v \# vs)\}\ ] \implies \vdash_{es} \{P\}\ e \# es\ \{R\}$

Functions fst and snd select the first and second component of a pair, i.e. the value and the state in the above rule.

If you wonder where the rules come from: they are derived from the proofs one would perform in ordinary Hoare logic on the program one obtains by removing nested expressions as indicated at the beginning of this section. You can still recognize the ordinary assigment axiom (first rule) and sequential composition (last rule).

As for the operational semantics, the rules for commands are the same as in the side effect free language, except of course for the two new commands, whose rules are straightforward:

$\vdash \{P\}\ SKIP\ \{P\}$

$\vdash_e \{P\}\ e\ \{\lambda v\ s.\ Q(s(x:=v))\} \implies \vdash \{P\}\ x:=e\ \{Q\}$

Soundness of $\vdash_e$ and $\vdash_{es}$ is easily proved by simultaneous induction on e and es:

theorem *ehoare-sound*:
$(\vdash_e \{P\}e\{Q'\} \longrightarrow \models_e \{P\}e\{Q'\}) \ \wedge\ (\vdash_{es} \{P\}es\{Q''\} \longrightarrow \models_{es} \{P\}es\{Q''\})$

Soundness of $\vdash$ is proved as usual, by induction on c. The *Assign*-case is solved by lemma *ehoare-sound* above.

theorem $\vdash \{P\}c\{Q\} \implies \models \{P\}c\{Q\}$

Completeness is also proved in the standard manner. But since we have three kinds of triples, we also need three weakest preconditions:

$wp\ ::\ com \Rightarrow assn \Rightarrow assn$
$wp\ c\ Q\quad \equiv\ (\lambda s.\ \forall t.\ s\ -c\rightarrow\ t\ \longrightarrow\ Q\ t)$
$wpe\ ::\ expr \Rightarrow vassn \Rightarrow assn$
$wpe\ e\ Q\quad \equiv\ (\lambda s.\ \forall v\ t.\ s\ -e\Rightarrow\ (v,t)\ \longrightarrow\ Q\ v\ t)$
$wpes\ ::\ expr\ list \Rightarrow vsassn \Rightarrow assn$
$wpes\ es\ Q \equiv\ (\lambda s.\ \forall vs\ t.\ s\ =es\Rightarrow\ (vs,t)\ \longrightarrow\ Q\ vs\ t)$

Of the laws proved about wp, wpe and $wpes$ we only show the "new" ones:

lemma $wp\ SKIP\ P = P$
lemma $wp\ (x:=e)\ Q = wpe\ e\ (\lambda v\ s.\ Q(s(x:=v)))$
lemma $wpe\ (Var\ x)\ Q = (\lambda s.\ Q\ (s\ x)\ s)$
lemma $wpe\ (Fun\ f\ es)\ Q = wpes\ es\ (\lambda vs\ s.\ Q\ (fst(f\ s\ vs))\ (snd(f\ s\ vs)))$
lemma $wpes\ []\ Q = Q\ []$
lemma $wpes\ (e\#es)\ Q = wpe\ e\ (\lambda v.\ wpes\ es\ (\lambda vs.\ Q(v\#vs)))$

Our standard lemma for the completeness theorem is first proved for expressions and expression lists by (an easy) simultaneous induction on e and es:

lemma $(\forall Q.\ \vdash_e \{wpe\ e\ Q\}\ e\ \{Q\}) \ \wedge\ (\forall Q.\ \vdash_{es} \{wpes\ es\ Q\}\ es\ \{Q\})$

With the help of this lemma in the *Assign*-case we can prove the key lemma by induction on c; the other cases go through as usual.

lemma $\forall Q. \vdash \{wp\ c\ Q\}\ c\ \{Q\}$

The completeness theorem follows directly:

theorem $\models \{P\}c\{Q\} \implies \vdash \{P\}c\{Q\}$

5. Procedures

So far, things were well-understood long before they were modelled in a theorem prover. Procedures, however, are different. In the introduction I have already sketched the history of Hoare logics for recursive procedures. As a motivation of the technical difficulties, consider the following parameterless recursive procedure:

```
proc = if i=0 then skip else i := i-1; CALL; i := i+1
```

A classic example of the subtle problems associated with reasoning about procedures is the proof that i is invariant: $\{$i=N$\}$ CALL $\{$i=N$\}$. This is done by induction: we assume $\{$i=N$\}$ CALL $\{$i=N$\}$ and have to prove $\{$i=N$\}$ **body** $\{$i=N$\}$, where **body** is the body of the procedure. The case i=0 is trivial. Otherwise we have to show $\{$i=N$\}$i:=i-1;CALL;i:=i+1$\{$i=N$\}$, which can be reduced to $\{$i=N-1$\}$ CALL $\{$i=N-1$\}$. But how can we deduce$\{$i=N-1$\}$ CALL $\{$i=N-1$\}$ from the induction hypothesis $\{$i=N$\}$ CALL $\{$i=N$\}$? Clearly, we have to instantiate N in the induction hypothesis — after all N is arbitrary as it does not occur in the program. The problems with procedures are largely due to unsound or incomplete adaption rules. We follow the solution of Morris and Kleymann and adjust the value of auxiliary variables like N with the help of the consequence rule. We also follow Kleymann in modelling auxiliary variables as a separate concept (as suggested in [4]). Our main contribution is a generalization from deterministic to nondeterministic languages.

5.1. SYNTAX AND OPERATIONAL SEMANTICS

Types *var*, *val*, *state* and *bexp* are defined as in Sect. 2. We start with a minimal set of commands:

datatype *com* = *Do* (*state* $\Rightarrow$ *state*)
         | *Semi com com*   (-; - [60, 60] 10)
         | *Cond bexp com com*   (*IF - THEN - ELSE - 60*)
         | *While bexp com*   (*WHILE - DO - 60*)
         | *CALL*

There is only one parameterless procedure in the program. Hence *CALL* does not even need to mention the procedure name. There is no separate syntax for procedure declarations. Instead we declare a HOL constant

consts *body* :: *com*

that represents the body of the one procedure in the program.

As before, command execution is described by transitions $s -c\rightarrow t$. The only new rule is the one for *CALL* — it requires no comment:

$$s -body\rightarrow t \implies s -CALL\rightarrow t$$

This semantics turns out not to be fine-grained enough. The soundness proof for the Hoare logic below proceeds by induction on the call depth during execution. To make this work we define a second semantics $s -c-n\rightarrow t$ which expresses that the execution uses at most n nested procedure invocations, where n is a natural number. The rules are straightforward: n is just passed around, except for procedure calls, where it is decremented (*Suc n* is $n + 1$):

$$s -Do\ f-n\rightarrow f\ s$$

$$[\![\ s0\ -c1-n\rightarrow s1;\ s1\ -c2-n\rightarrow s2\ ]\!] \implies s0\ -c1;c2-n\rightarrow s2$$

$$[\![\ b\ s;\ s\ -c1-n\rightarrow t\ ]\!] \implies s\ -IF\ b\ THEN\ c1\ ELSE\ c2-n\rightarrow t$$
$$[\![\ \neg b\ s;\ s\ -c2-n\rightarrow t\ ]\!] \implies s\ -IF\ b\ THEN\ c1\ ELSE\ c2-n\rightarrow t$$

$$\neg b\ s \implies s\ -WHILE\ b\ DO\ c-n\rightarrow s$$
$$[\![b\ s;\ s\ -c-n\rightarrow t;\ t\ -WHILE\ b\ DO\ c-n\rightarrow u]\!] \implies s\ -WHILE\ b\ DO\ c-n\rightarrow u$$

$$s\ -body-n\rightarrow t \implies s\ -CALL-Suc\ n\rightarrow t$$

By induction on $s -c-m\rightarrow t$ we show monotonicity w.r.t. the call depth:

lemma $s -c-m\rightarrow t \implies \forall n.\ m \leq n \longrightarrow s -c-n\rightarrow t$

With the help of this lemma we prove the expected relationship between the two semantics:

lemma *exec-iff-execn*: $(s -c\rightarrow t) = (\exists n.\ s -c-n\rightarrow t)$

Both directions are proved separately by induction on the operational semantics.

5.2. HOARE LOGIC FOR PARTIAL CORRECTNESS

Taking auxiliary variables seriously means that assertions must now depend on them as well as on the state. Initially we do not fix the type of auxiliary variables but parameterize the type of assertions with a type variable $'a$:

types $'a\ assn = 'a \Rightarrow state \Rightarrow bool$

Type constructors are written postfix.

The second major change is the need to reason about Hoare triples in a context: proofs about recursive procedures are conducted by induction where we assume that all *CALL*s satisfy the given pre/postconditions and have to show that the body does as well. The assumption is stored in a context, which is a set of Hoare triples:

types $'a\ cntxt = ('a\ assn \times com \times 'a\ assn)set$

In the presence of only a single procedure the context will always be empty or a singleton set. With multiple procedures, larger sets can arise.

Now that we have contexts, validity becomes more complicated. Ordinary validity (w.r.t. partial correctness) is still what it used to be, except that we have to take auxiliary variables into account as well:

$$\models \{P\}c\{Q\} \equiv \forall s\ t.\ s\ -c\rightarrow t \longrightarrow (\forall z.\ P\ z\ s \longrightarrow Q\ z\ t)$$

Auxiliary variables are always denoted by z.

Validity of a context and validity of a Hoare triple in a context are defined as follows:

$$\Vdash C \qquad \equiv \forall (P,c,Q) \in C.\ \models \{P\}c\{Q\}$$
$$C \models \{P\}c\{Q\} \equiv \Vdash C \longrightarrow \models \{P\}c\{Q\}$$

Note that $\{\} \models \{P\}\ c\ \{Q\}$ is equivalent to $\models \{P\}\ c\ \{Q\}$.

Unfortunately, this is not the end of it. As we have two semantics, $-c\rightarrow$ and $-c-n\rightarrow$, we also need a second notion of validity parameterized with the recursion depth n:

$$\models n\ \{P\}c\{Q\} \equiv \forall s\ t.\ s\ -c-n\rightarrow t \longrightarrow (\forall z.\ P\ z\ s \longrightarrow Q\ z\ t)$$
$$\Vdash\text{-}n\ C \equiv \forall (P,c,Q) \in C.\ \models n\ \{P\}c\{Q\}$$
$$C \models n\ \{P\}c\{Q\} \equiv \Vdash\text{-}n\ C \longrightarrow \models n\ \{P\}c\{Q\}$$

Finally we come to the proof system for deriving triples in a context:

$$C \vdash \{\lambda z\ s.\ P\ z\ (f\ s)\}\ Do\ f\ \{P\}$$

$$[\ C \vdash \{P\}c1\{Q\};\ C \vdash \{Q\}c2\{R\}\ ] \implies C \vdash \{P\}\ c1;c2\ \{R\}$$

$$[\ C \vdash \{\lambda z\ s.\ P\ z\ s \wedge b\ s\}c1\{Q\};\ C \vdash \{\lambda z\ s.\ P\ z\ s \wedge \neg b\ s\}c2\{Q\}\ ]$$

$\implies C \vdash \{P\}$ *IF b THEN c1 ELSE c2* $\{Q\}$

$C \vdash \{\lambda z\, s.\ P\, z\, s \wedge b\, s\}\ c\ \{P\}$
$\implies C \vdash \{P\}$ *WHILE b DO c* $\{\lambda z\, s.\ P\, z\, s \wedge \neg b\, s\}$

$[\![\ C \vdash \{P'\}c\{Q'\};\ \ \forall s\, t.\ (\forall z.\ P'\, z\, s \longrightarrow Q'\, z\, t) \longrightarrow (\forall z.\ P\, z\, s \longrightarrow Q\, z\, t)\ ]\!]$
$\implies C \vdash \{P\}c\{Q\}$

$\{(P, CALL, Q)\} \vdash \{P\}body\{Q\} \implies \{\} \vdash \{P\}\ CALL\ \{Q\}$

$\{(P, CALL, Q)\} \vdash \{P\}\ CALL\ \{Q\}$

The first four rules are familiar, except for their adaptation to auxiliary variables. The *CALL* rule embodies induction and has already been motivated above. Note that it is only applicable if the context is empty. This shows that we never need nested induction. For the same reason the assumption rule (the last rule) is stated with just a singleton context.

The only real surprise is the rule of consequence, which appears in print in this form for the first time as far as I am aware. Morris [14] and later Olderog [25] show that the consequence rule with side condition

$$\forall s.\ P\, z\, s \longrightarrow (\forall t.\ (\forall z'.\ P'\, z'\, s \longrightarrow Q'\, z'\, t) \longrightarrow Q\, z\, t)$$

(which is already close to our side condition) is "adaption complete" for partial correctness. Olderog also shows completeness of a proof system based on this rule. Hofmann [6] builds on [27] and shows soundness and completeness of a Hoare logic where the consequence rule has the following side condition:

$$\forall s\, t\, z.\ P\, z\, s \longrightarrow Q\, z\, t \vee (\exists z'.\ P'\, z'\, s \wedge (Q'\, z'\, t \longrightarrow Q\, z\, t))$$

The side conditions by Morris and Hofmann are logically equivalent to ours. But the symmetry of our new version appeals not just for aesthetic reasons but because one can actually remember it! Furthermore, its soundness proof is very direct: In order to show $C \models \{P\}\ c\ \{Q\}$ we assume the validity of C (which implies $\models \{P'\}\ c\ \{Q'\}$ because of $C \models \{P'\}\ c\ \{Q'\}$) and prove $\models \{P\}\ c\ \{Q\}$: assuming $s\ -c\rightarrow\ t$, $\models \{P'\}\ c\ \{Q'\}$ implies $\forall z.\ P'\, z\, s \longrightarrow Q'\, z\, t$, which, by the side condition of the consequence rule, implies $\forall z.\ P\, z\, s \longrightarrow Q\, z\, t$, which is exactly what we need for $\models \{P\}\ c\ \{Q\}$.

The proof of the soundness theorem

theorem $C \vdash \{P\}c\{Q\} \implies C \models \{P\}c\{Q\}$

requires a generalization: $\forall n.\ C \models_n \{P\}\ c\ \{Q\}$ is proved instead, from which the actual theorem follows directly via lemma *exec-iff-execn* in Sect. 5.1.

The generalization is proved by induction on c. The reason for the generalization is that soundness of the *CALL* rule is proved by induction on the maximal call depth, i.e. n.

The completeness proof is quite different from the ones we have seen so far. It employs the notion of a *most general triple* (or *most general formula*) due to Gorelick [5]:

$$MGT :: com \Rightarrow state\ assn \times com \times state\ assn$$
$$MGT\ c \equiv (\lambda z\ s.\ z = s,\ c,\ \lambda z\ t.\ z - c \rightarrow t)$$

Note that the type of z has been identified with *state*. This means that for every state variable there is an auxiliary variable, which is simply there to record the value of the program variables before execution of a command. This is exactly what, for example, VDM offers by allowing you to refer to the pre-value of a variable in a postcondition [11]. The intuition behind $MGT\ c$ is that it completely describes the operational behaviour of c. It is easy to see that, in the presence of the new consequence rule, $\{\} \vdash MGT\ c$ implies completeness:

lemma *MGT-implies-complete*:
$$\{\} \vdash MGT\ c \implies \{\} \models \{P\}c\{Q\} \implies \{\} \vdash \{P\}c\{Q::state\ assn\}$$

Note that the type constraint $Q::state\ assn$ is not inferred automatically: although both pre and postcondition of $MGT\ c$ are of type *state assn*, this does not force Q to have the same type.

In order to discharge $\{\} \vdash MGT\ c$ one proves

lemma *MGT-lemma*: $C \vdash MGT\ CALL \implies C \vdash MGT\ c$

The proof is by induction on c. In the *While*-case it is easy to show that $\lambda z\ t.\ (z,\ t) \in \{(s,\ t).\ b\ s \wedge s - c \rightarrow t\}^*$ is invariant. The precondition $\lambda z\ s.\ z = s$ establishes the invariant and a reflexive transitive closure induction shows that the invariant conjoined with $\neg\ b\ t$ implies the postcondition $\lambda z\ t.\ z - WHILE\ b\ DO\ c \rightarrow t$. The remaining cases are trivial.

Using the *MGT-lemma* (together with the *CALL* and the assumption rule) one can easily derive

lemma $\{\} \vdash MGT\ CALL$

Using the *MGT-lemma* once more we obtain $\{\} \vdash MGT\ c$ and thus by *MGT-implies-complete* completeness.

theorem $\{\} \models \{P\}c\{Q\} \implies \{\} \vdash \{P\}c\{Q::state\ assn\}$

5.3. MODULAR EXTENSIONS

Procedures can be extended with nondeterministic choice and local variables just as we extended the pure while-language in Sect. 2.2: the operational semantics is the same, and the Hoare rules just need to be extended with a context C. The proofs are identical as well. But in the case of local variables the resulting language may not be what one expects: it has the semantics of *dynamic scoping*: if the procedure body refers to some variable, say x, then the execution of that body during the execution of $VAR\ x = e;\ CALL$ will refer to the local x with initial value e. We have also studied a language with static scoping, but do not discuss the details in this paper.

5.4. HOARE LOGIC FOR TOTAL CORRECTNESS

This is the most complicated system in this paper. We only show the key elements and none of the proofs.

For termination we have just one new obvious rule:

$$body \downarrow s \implies CALL \downarrow s$$

Validity is defined as expected:

$$\models_t \{P\}c\{Q\} \equiv \models \{P\}c\{Q\} \wedge (\forall z\ s.\ P\ z\ s \longrightarrow c{\downarrow}s)$$
$$C \models_t \{P\}c\{Q\} \equiv (\forall (P',c',Q') \in C.\ \models_t \{P'\}c'\{Q'\}) \longrightarrow \models_t \{P\}c\{Q\}$$

Instead of the full set of proof rules we merely show those that differ from the system for partial correctness:

$$[\![wf\ r;\ \forall s'.\ C \vdash_t \{\lambda z\ s.\ P\ z\ s \wedge b\ s \wedge s' = s\}\ c\ \{\lambda z\ s.\ P\ z\ s \wedge (s,s') \in r\}]\!]$$
$$\implies C \vdash_t \{P\}\ WHILE\ b\ DO\ c\ \{\lambda z\ s.\ P\ z\ s \wedge \neg b\ s\}$$

$$[\![wf\ r;\ \forall s'.\ \{(\lambda z\ s.\ P\ z\ s \wedge (s,s') \in r,\ CALL,\ Q)\}$$
$$\vdash_t \{\lambda z\ s.\ P\ z\ s \wedge s = s'\}\ body\ \{Q\}]\!]$$
$$\implies \{\} \vdash_t \{P\}\ CALL\ \{Q\}$$

$$[\ C \vdash_t \{P'\}c\{Q'\};$$
$$(\forall s\ t.\ (\forall z.\ P'\ z\ s \longrightarrow Q'\ z\ t) \longrightarrow (\forall z.\ P\ z\ s \longrightarrow Q\ z\ t)) \wedge$$
$$(\forall s.\ (\exists z.\ P\ z\ s) \longrightarrow (\exists z.\ P'\ z\ s))]$$
$$\implies C \vdash_t \{P\}c\{Q\}$$

As in the case for the pure while-language, our rules for total correctness are very similar to those by Kleymann [11]. The side condition in our rule of consequence looks quite different from the one by Kleymann, but the two are in fact equivalent:

lemma $((\forall s\ t.\ (\forall z.\ P'\ z\ s \longrightarrow Q'\ z\ t) \longrightarrow (\forall z.\ P\ z\ s \longrightarrow Q\ z\ t)) \wedge$
$\qquad (\forall s.\ (\exists z.\ P\ z\ s) \longrightarrow (\exists z.\ P'\ z\ s)))$
$\quad = (\forall z\ s.\ P\ z\ s \longrightarrow (\forall t.\exists z'.\ P'\ z'\ s \wedge (Q'\ z'\ t \longrightarrow Q\ z\ t)))$

Kleymann's version is easier to use, whereas our new version clearly shows that it is a conjunction of the side condition for partial correctness with precondition strengthening.

The key difference to the work by Kleymann (and America and de Boer) is that soundness and completeness

theorem $C \vdash_t \{P\}c\{Q\} \implies C \models_t \{P\}c\{Q\}$
theorem $\{\} \models_t \{P\}c\{Q\} \implies \{\} \vdash_t \{P\}c\{Q::state\ assn\}$

are shown for arbitrary, i.e. unbounded nondeterminism. This is a significant extension and appears to have been an open problem. The details are found in a separate paper [18].

6. More procedures

We now generalize from a single procedure to a whole set of procedures following the ideas of von Oheimb [21]. The basic setup of Sect. 5.1 is modified only in a few places:

- We introduce a new basic type *pname* of procedure names.
- Constant *body* is now of type *pname* $\Rightarrow$ *com*.
- The *CALL* command now has an argument of type *pname*, the name of the procedure that is to be called.
- The call rule of the operational semantics now says

$$s - body\ p \rightarrow t \implies s - CALL\ p \rightarrow t$$

Note that this setup assumes that we have a procedure body for each procedure name. If you feel uncomfortable with the idea of an infinity of procedures, you may assume that *pname* is the finite subset of all procedure names that occur in some fixed program.

6.1. HOARE LOGIC

Types *assn* and and *cntxt* are defined as in Sect. 5.2, as are $\models \{P\}\ c\ \{Q\}$, $\models C$, $\models n\ \{P\}\ c\ \{Q\}$ and $\models$-$n\ C$. However, we now need an additional notion of validity $C \models D$ where D is a set as well. The reason is that we can now have mutually recursive procedures whose correctness needs to be established by simultaneous induction. Instead of sets of Hoare triples we may think of conjunctions. We define both $C \Vdash D$ and its relativized version:

$$C \Vdash D \quad \equiv \quad \Vdash C \longrightarrow \Vdash D$$
$$C \Vdash\text{-}n\ D \quad \equiv \quad \Vdash\text{-}n\ C \longrightarrow \Vdash\text{-}n\ D$$

Our Hoare logic now defines judgements of the form $C \Vdash D$ where both C and D are (potentially infinite) sets of Hoare triples; $C \vdash \{P\}\ c\ \{Q\}$ is simply an abbreviation for $C \Vdash \{(P,c,Q)\}$. With this abbreviation the rules for *Do*, *Semi*, *If*, *While* and consequence are exactly the same as in Sect. 5.2. The remaining rules are

$$[\![\ \forall (P,c,Q) \in C.\ \exists p.\ c = CALL\ p;$$
$$\quad C \Vdash \{(P,b,Q).\ \exists p.\ (P,CALL\ p,Q) \in C \wedge b = body\ p\}\ ]\!]$$
$$\implies \{\} \Vdash C$$

$$(P,CALL\ p,Q) \in C \implies C \vdash \{P\}\ CALL\ p\ \{Q\}$$

$$\forall (P,c,Q) \in D.\ C \vdash \{P\}c\{Q\} \implies C \Vdash D$$
$$[\![\ C \Vdash D;\ (P,c,Q) \in D\ ]\!] \implies C \vdash \{P\}c\{Q\}$$

The *CALL* and the assumption rule are straightforward generalizations of their counterparts in Sect. 5.2. The final two rules are structural rules and could be called conjunction introduction and elimination, because they put together and take apart sets of triples.

theorem $C \Vdash D \implies C \Vdash D$

As before, we prove a generalization of $C \Vdash D$, namely $\forall n.\ C \Vdash\text{-}n\ D$, by induction on $C \Vdash D$, with an induction on n in the *CALL* case.

The completeness proof resembles the one in Sect. 5.2 closely: the most general triple *MGT* is defined exactly as before, and the lemmas leading up to completness are simple generalizations:

lemma *MGT-implies-complete*:
$\{\} \Vdash \{MGT\ c\} \implies \vDash \{P\}c\{Q\} \implies \{\} \vdash \{P\}c\{Q::state\ assn\}$
lemma *MGT-lemma*: $\forall p.\ C \Vdash \{MGT(CALL\ p)\} \implies C \Vdash \{MGT\ c\}$
lemma $\{\} \Vdash \{mgt.\ \exists p.\ mgt = MGT(CALL\ p)\}$
theorem $\vDash \{P\}c\{Q\} \implies \{\} \vdash \{P\}c\{Q::state\ assn\}$

7. Functions

As our final variation on the theme of procedures we study a language with functions, i.e. procedures that take arguments and return results. Our functions have exactly one parameter which is passed by value.

364

7.1. SYNTAX AND OPERATIONAL SEMANTICS

The basic types *var*, *val*, *state* and *bexp* are defined as in Sect. 2. In addition we have

types *expr* = *state* ⇒ *val*

Commands are also defined as in Sect. 5.1, but extended with function calls and local variables:

| *Fun var expr*   (- := *FUN* - *60*)
| *Var var expr com*   (*VAR* - = -; -)

Command *Var* is familar. Command $x := FUN\ e$ is meant to pass the value of *e* as the (single) parameter to the (single) function in the program and to assign the result of the function to *x*. This syntax rules out nested expressions and thus all the complications treated in Sect. 4. The body of the function is again found in constant *body*. Furthermore there are two distinguished variables *arg*, the name of the formal parameter, and *res*, a variable for communicating the result back to the caller. Function calls are reduced to procedure calls as follows:

$$funcall\ x\ e \equiv (VAR\ arg = e;\ CALL);\ Do(\lambda s.\ s(x := s\ res))$$

The argument becomes a local variable and is thus not modified by further function calls in the body. Variable *res* is not protected like this and it is the programmer's responsibility to introduce a local variable if needed.

Note that *CALL* is still present to allow this two stage execution of function calls but it is not meant to be used on its own.

The semantics of *Fun*, the only new command, is obvious:

$$s\ -funcall\ x\ e \rightarrow t\ \implies\ s\ -x := FUN\ e \rightarrow t$$

The rule for local variables is the same as in Sect. 5.2, i.e. we have dynamic scoping, which is just what we need for the above implementation of parameter passing.

In essence, this is all there is to say about functions, because we have reduced function calls to the composition of previously studied language constructs.

7.2. HOARE LOGIC

Assertions and validity are defined as for procedures. The proof system is the one from Sect. 5.2 together with the rule for *Var* discussed in Sect. 5.2, extended with the obvious rule for function calls:

$$C \vdash \{P\}\mathit{funcall}\ x\ e\{Q\} \implies C \vdash \{P\}\ x := \mathit{FUN}\ e\ \{Q\}$$

Soundness and completeness are proved entirely as for procedures.

theorem $C \vdash \{P\}c\{Q\} \implies C \models \{P\}c\{Q\}$

theorem $\{\} \models \{P\}c\{Q\} \implies \{\} \vdash \{P\}c\{Q::\mathit{state\ assn}\}$

8. Conclusion

The preceding sections show that a variety of language constructs can be given a simple Hoare logic in the unified framework of HOL. In particular we have been able to settle the case of total correctness of recursive procedures combined with unbounded nondeterminism [18].

Although one would think that Hoare logics formalized in theorem provers should always come with a completeness proof, this is not so. The argument used by [8] to explain the absence of a completeness proof is the ability to fall back on a denotational semantics if the Hoare logic fails. The authors of [13] even claim that completeness is not a meaningful question in their setting. This paper has shown that completeness can be established by standard means even for complicated language constructs. Thus it should not be ignored.

Acknowledgments I am indebted to Thomas Kleymann and David von Oheimb for providing the logical foundations and to Markus Wenzel for the Isar extension of Isabelle: without it the production of the paper from the Isabelle theories, something I would no longer want to miss, would have been impossible. Gerwin Klein, Norbert Schirmer and Markus Wenzel commented on a draft version.

References

1. Pierre America and Frank de Boer. Proving total correctness of recursive procedures. *Information and Computation*, 84:129–162, 1990.
2. Krzysztof Apt. Ten Years of Hoare's Logic: A Survey — Part I. *ACM Trans. Programming Languages and Systems*, 3(4):431–483, 1981.
3. Krzysztof Apt. Ten Years of Hoare's Logic: A Survey — Part II: Nondeterminism. *Theoretical Computer Science*, 28:83–109, 1984.
4. Krzysztof Apt and Lambert Meertens. Completeness with finite systems of intermediate assertions for recursive program schemes. *SIAM Jornal on Computing*, 9(4):665–671, 1980.
5. Gerald Arthur Gorelick. A complete axiomatic system for proving assertions about recursive and non-recursive programs. Technical Report 75, Dept. of Computer Science, Univ. of Toronto, 1975.
6. Martin Hofmann. Semantik und Verifikation. Lecture notes, Universität Marburg. In German, 1997.

7. Peter V. Homeier and David F. Martin. Mechanical verification of mutually recursive procedures. In M.A. McRobbie and J.K. Slaney, editors, *Automated Deduction — CADE-13*, volume 1104 of *Lect. Notes in Comp. Sci.*, pages 201–215. Springer-Verlag, 1996.

8. Bart Jacobs and Erik Poll. A logic for the Java modeling language JML. In H. Hussmann, editor, *Fundamental Approaches to Software Engineering*, volume 2029 of *Lect. Notes in Comp. Sci.*, pages 284–299. Springer-Verlag, 2001.

9. Cliff B. Jones. *Systematic Software Development Using VDM*. Prentice-Hall, 2nd edition, 1990.

10. Thomas Kleymann. *Hoare Logic and VDM: Machine-Checked Soundness and Completeness Proofs*. PhD thesis, Department of Computer Science, University of Edinburgh, 1998. Report ECS-LFCS-98-392.

11. Thomas Kleymann. Hoare logic and auxiliary variables. *Formal Aspects of Computing*, 11:541–566, 1999.

12. Tomasz Kowaltowski. Axiomatic approach to side effects and general jumps. *Acta Informatica*, 7:357–360, 1977.

13. Linas Laibinis and Joakim von Wright. Functional procedures in higher-order logic. In M. Aagaard and J. Harrison, editors, *Theorem Proving in Higher Order Logics*, volume 1896 of *Lect. Notes in Comp. Sci.*, pages 372–387. Springer-Verlag, 2000.

14. J.H. Morris. Comments on "procedures and parameters". Undated and unpublished.

15. Hanne Riis Nielson and Flemming Nielson. *Semantics with Applications*. Wiley, 1992.

16. Tobias Nipkow. Winskel is (almost) right: Towards a mechanized semantics textbook. In V. Chandru and V. Vinay, editors, *Foundations of Software Technology and Theoretical Computer Science*, volume 1180 of *Lect. Notes in Comp. Sci.*, pages 180–192. Springer-Verlag, 1996.

17. Tobias Nipkow. Winskel is (almost) right: Towards a mechanized semantics textbook. *Formal Aspects of Computing*, 10:171–186, 1998.

18. Tobias Nipkow. Hoare logics for recursive procedures and unbounded nondeterminism. Draft, 2001.

19. Tobias Nipkow and David von Oheimb. Java*light* is type-safe — definitely. In *Proc. 25th ACM Symp. Principles of Programming Languages*, pages 161–170, 1998.

20. Tobias Nipkow and Lawrence Paulson. *Isabelle/HOL. The Tutorial*, 2001. http://www.in.tum.de/~nipkow/pubs/tutorial.html.

21. David von Oheimb. Hoare logic for mutual recursion and local variables. In C. Pandu Rangan, V. Raman, and R. Ramanujam, editors, *Foundations of Software Technology and Theoretical Computer Science (FST&TCS)*, volume 1738 of *Lect. Notes in Comp. Sci.*, pages 168–180. Springer-Verlag, 1999.

22. David von Oheimb. *Analyzing Java in Isabelle/HOL: Formalization, Type Safety and Hoare Logic*. PhD thesis, Technische Universität München, 2001. http://www.in.tum.de/~oheimb/diss/.

23. David von Oheimb. Hoare logic for Java in Isabelle/HOL. *Concurrency and Computation: Practice and Experience*, 13(13):1173–1214, 2001.

24. David von Oheimb and Tobias Nipkow. Hoare logic for NanoJava: Auxiliary variables, side effects and virtual methods revisited. Submitted for publication, 2001.

25. Ernst-Rüdiger Olderog. On the notion of expressiveness and the rule of adaptation. *Theoretical Computer Science*, 24:337–347, 1983.

26. Robert Pollack. *The Theory of LEGO: A Proof Checker for the Extended Calculus of Constructions*. PhD thesis, University of Edinburgh, 1994.
27. Thomas Schreiber. Auxiliary variables and recursive procedures. In *TAPSOFT'97: Theory and Practice of Software Development*, volume 1214 of *Lect. Notes in Comp. Sci.*, pages 697–711. Springer-Verlag, 1997.

PROOF THEORETIC COMPLEXITY

G. E. OSTRIN
Inst. für Informatik und angewandte Mathematik,
Universität Bern, Neubrückstrasse 10,
CH-3012 Bern, Switzerland.
(geoff@iam.unibe.ch)

S. S. WAINER
Dept. of Pure Mathematics, University of Leeds,
Leeds LS2 9JT, UK.
(s.s.wainer@leeds.ac.uk)

Abstract *A weak formal theory of arithmetic is developed, entirely analogous to classical arithmetic but with two separate kinds of variables: induction variables and quantifier variables. The point is that the provably recursive functions are now more feasibly computable than in the classical case, lying between Grzegorczyk's E^2 and E^3, and their computational complexity can be characterized in terms of the logical complexity of their termination proofs. Previous results of Leivant are reworked and extended in this new setting, with quite different proof theoretic methods.*

Keywords: proof theory; computational complexity; equation calculus; arithmetic; (sub-) elementary functions; cut elimination; ordinal bounds.

1. Introduction

The classical methods of proof theory (cut elimination or normalization) enable one to read off, from a proof that a recursively defined function terminates everywhere, a bound on its computational complexity. This was already implicit in the work of Kreisel (1951-52) fifty years ago, where the first recursion theoretic characterization of the provably terminating functions of arithmetic was given (in terms of an algebra of recursions over transfinite well orderings of order-types less than ε_0). However, although the programs defining them are quite "natural" in their structure, the functions themselves are (in general) far away from being realistically or feasibly

369

H. Schwichtenberg and R. Steinbrüggen (eds.), Proof and System-Reliability, 369–397.
© 2002 *Kluwer Academic Publishers. Printed in the Netherlands.*

computable. It took another thirty-five years before Buss (1986) had developed Bounded Arithmetic, and characterized the PTIME functions as those provably terminating in its Σ_1 fragment. More recently, through the work of Bellantoni and Cook (1992) and Leivant (1994) on their different forms of "tiered recursion", there has been a growing interest in developing theories analogous to classical arithmetic, but without the explicit quantifier-bounds of Bounded Arithmetic, whose provably terminating functions form (much) more realistic complexity classes than in the classical case. In particular, Leivant's (1995) theories based on his ramified induction schemes give proof theoretic characterizations of PTIME and the low Grzegorczyk classes E^2 and E^3. What we do here is present a simple re-formulation of arithmetic, motivated by the normal/safe variable separation of Bellantoni and Cook, in which results similar to Leivant's can be developed and extended. Our methods, however, are quite different from his, being based on traditional cut elimination with ordinal bounds. The analogies with the classical case are very strong (cf. Fairtlough and Wainer (1998)).

Bellantoni and Cook made a simple syntactic restriction on the usual schemes of primitive recursion, by inserting a semicolon to separate the variables into two distinct classes: the recursion variables, which they call "normal", and the substitution variables, which they call "safe". The effect of this is quite dramatic: every function so defined is now polynomially bounded! By working on binary number representation, they thus characterize PTIME, and if instead one works with unary representation (as we shall do) one obtains the Grzegorczyk class E^2 (see also Handley and Wainer (1999)). We shall consider the result of imposing a similar variable separation on a formal system of arithmetic. Thus "input" or "normal" variables will control the inductions, and will only occur free, whereas "output" or "safe" variables may be bound by quantifiers. The resulting theory, denoted EA(I;O) to signify this input/output separation of variables, we call Elementary Arithmetic because its provably terminating functions all lie in the Grzegorczyk class E^3 of "elementary" functions (those computable in a number of steps bounded by a finitely iterated exponential). There is a well known complexity hierarchy for E^3, according to the number of times the exponential is iterated in producing the required bound. At the bottom is E^2 with polynomial bounds, at the next level lies EXPTIME with bounds 2^p, p a polynomial, and so on. As we shall show, this hierarchy is closely related to the increasing levels of induction complexity of termination proofs.

The theory EA(I;O) will be based on minimal, not classical, logic. The reason is that from a proof in classical logic of a Σ_1 formula $\exists a A(a)$ we can, by Herbrand's Theorem, extract a finite set of terms t_i such that $\vdash A(t_1) \lor \ldots \lor A(t_n)$. Although one of the t_i will be a correct witness for A, we don't necessarily know which one. Minimal logic allows us to extract one, correct witness, so it provides more precise computational information. This will not be a restriction for us, since we will still be able to prove termination of the same functions in EA(I;O) with minimal logic, as we can using classical logic. On the other hand one needs to exercise more care in measuring the complexity of induction formulas, since minimal or intuitionistic logic is more sensitive to the precise logical structure of formulas than is classical logic (see Burr (2000)).

A very appealing feature of Leivant's "intrinsic" theories, which we too adopt, is that they are based on Kleene's equation calculus, which allows for a natural notion of provable recursiveness, completely free of any coding implicit in the more traditional definition involving the T-predicate. Thus one is allowed to introduce arbitrary partial recursive functions f by means of their equational definitions as axioms, but the logical and inductive power of the theory severely restricts one's ability to prove termination: $f(x) \downarrow$. In Leivant's theory over N (he allows for more abstract data types) this is expressed by $N(x) \to N(f(x))$. In our theory, specific to N though it could be generalised, definedness is expressed by

$$f(x) \downarrow \equiv \exists a(f(x) \simeq a).$$

This highlights the principal logical restriction which must be applied to the $\exists$-introduction and (dually) $\forall$-elimination rules of our theory EA(I;O) described below.

If arbitrary terms t were allowed as witnesses for $\exists$-introduction, then from the axiom $t \simeq t$ we could immediately deduce $\exists a(t \simeq a)$ and hence $f(x) \downarrow$ for every f ! This is clearly not what we want. In order to avoid it we make the restriction that only "basic" terms: variables or 0 or their successors or predecessors, may be used as witnesses. This is not quite so restrictive as it first appears, since from the equality rule

$$t \simeq a, \ A(t) \ \vdash \ A(a)$$

we can derive immediately

$$t \downarrow, \ A(t) \ \vdash \ \exists a A(a).$$

Thus a term may be used to witness an existential quantifier only when it has been proven to be defined. In particular, if f is introduced by a defining equation $f(x) \simeq t$ then to prove $f(x) \downarrow$ we first must prove (compute) $t \downarrow$.

For example suppose f is introduced by the defining equation:

$$f(x) \simeq g(x, h(x))$$

where g and h have been previously defined. Then the termination proof for $f(x)$ goes like this:

By the substitution axiom (which, interacting with equational cuts, constitutes the only computation rule of the equation calculus),

$$h(x) \simeq b, \ g(x, b) \simeq a \ \vdash \ g(x, h(x)) \simeq a$$

and by the defining equation for f,

$$g(x, h(x)) \simeq a \ \vdash \ f(x) \simeq a \ .$$

Therefore by an equational cut,

$$h(x) \simeq b, \ g(x, b) \simeq a \ \vdash \ f(x) \simeq a$$

and then by applying the $\exists$ rule on the right, followed by an $\exists$ rule on the left (since variable a is then not free anywhere else),

$$h(x) \simeq b, \ g(x, b) \downarrow \vdash \ f(x) \downarrow \ .$$

Now assuming that $g(x, b) \downarrow$ has already been proven, we can cut it to obtain

$$h(x) \simeq b \ \vdash \ f(x) \downarrow \ .$$

Since b does not now occur free anywhere else, we can existentially quantify it to obtain

$$h(x) \downarrow \vdash \ f(x) \downarrow$$

and then, assuming $h(x) \downarrow$ is already proven, another cut gives

$$\vdash \ f(x) \downarrow \ .$$

Note that a bottom-up (goal-directed) reading of the proof determines the order of computation of $f(x)$ thus: first find the value of $h(x)$, call it b, then find the value of $g(x, b)$, and then pass this as the value of $f(x)$.

Here we can begin to see that, provided we formulate the theory carefully enough, proofs in its Σ_1 fragment will correspond to computations in the equation calculus, and bounds on proof-size will yield complexity measures.

2. The theory EA(I;O)

There will be two kinds of variables: "input" (or "normal") variables denoted $x, y, z, \ldots$, and "output" (or "safe") variables denoted $a, b, c, \ldots$, both intended as ranging over natural numbers. Output variables may be bound by quantifiers, but input variables will always be free. The *basic terms* are: variables of either kind, the constant 0, or the result of repeated application of the successor S or predecessor P. General *terms* are built up in the usual way from 0 and variables of either kind, by application of S, P and arbitrary function symbols $f, g, h, \ldots$ denoting partial recursive functions given by sets E of Herbrand-Gödel-Kleene-style defining equations.

Atomic formulas will be equations $t_1 \simeq t_2$ between arbitrary terms, and formulas $A, B, \ldots$ are built from these by applying propositional connectives and quantifiers $\exists a$, $\forall a$ over output variables a. The negation of a formula $\neg A$ will be defined as $A \to \perp$.

It will be convenient, for later proof theoretic analysis, to work with logic in a sequent-style formalism, and the system G3 (with structural rules absorbed) as set out on page 65 of Troelstra and Schwichtenberg (1996) suits us perfectly, except that we write $\vdash$ instead of their $\Rightarrow$. However, as already mentioned above, we shall work in their system G3m of "minimal", rather than "classical", logic. This is computationally more natural, and it is not a restriction for us, since (as Leivant points out) a classical proof of $f(x) \downarrow$ can be transformed, by the double-negation interpretation, into a proof in minimal logic of

$$(\exists a((f(x) \simeq a \to \perp) \to \perp) \to \perp) \to \perp$$

and since minimal logic has no special rule for $\perp$ we could replace it throughout by the formula $f(x) \downarrow$ and hence obtain an outright proof of $f(x) \downarrow$, since the premise of the above implication becomes provable.

It is not necessary to list the propositional rules as they are quite standard, and the cut rule (with "cut formula" C) is:

$$\frac{\Gamma \vdash C \quad \Gamma, C \vdash A}{\Gamma \vdash A}$$

where, throughout, Γ is an arbitrary finite multiset of formulas. However, as stressed above, the quantifier rules need restricting. Thus the minimal left-∃ and right-∃ rules are:

$$\frac{\Gamma, A(b) \vdash B}{\Gamma, \exists a A(a) \vdash B} \qquad \frac{\Gamma \vdash A(t)}{\Gamma \vdash \exists a A(a)}$$

where, in the left-∃ rule the output variable b is not free in Γ, B, and in the right-∃ rule the witnessing term t is basic. The left-∀ and right-∀ rules are:

$$\frac{\Gamma, \forall a A(a), A(t) \vdash B}{\Gamma, \forall a A(a) \vdash B} \qquad \frac{\Gamma \vdash A(b)}{\Gamma \vdash \forall a A(a)}$$

where, in the left-hand rule the term t is basic, and in the right-hand rule the output variable b is not free in Γ.

The logical axioms are, with A atomic,

$$\Gamma, A \vdash A$$

and the equality axioms are $\Gamma \vdash t \simeq t$ and, again with $A(.)$ atomic,

$$\Gamma, t_1 \simeq t_2, A(t_1) \vdash A(t_2).$$

The logic allows these to be generalised straightforwardly to an arbitrary formula A and the quantifier rules then enable us to derive

$$\Gamma, t \downarrow, A(t) \vdash \exists a A(a)$$

$$\Gamma, t \downarrow, \forall a A(a) \vdash A(t)$$

for any terms t and formulas A.

Two further principles are needed, describing the data-type N, namely induction and cases (a number is either zero or a successor). We present these as rules rather than their equivalent axioms, since this will afford a closer match between proofs and computations. The induction rule (with "induction formula" $A(.)$) is

$$\frac{\Gamma \vdash A(0) \quad \Gamma, A(a) \vdash A(Sa)}{\Gamma \vdash A(x)}$$

where the output variable a is not free in Γ and where, in the conclusion, x is an input variable, or a basic term on an input variable.

The cases rule is
$$\frac{\Gamma \vdash A(0) \quad \Gamma \vdash A(Sa)}{\Gamma \vdash A(t)}$$

where t is any basic term. Note that with this rule it is easy to derive $\forall a(a \simeq 0 \lor a \simeq S(Pa))$ from the definition: $P(0) \simeq 0$ and $P(Sa) \simeq a$.

Definition. Our notion of Σ_1 formula will be restricted to those of the form $\exists \bar{a} A(\bar{a})$ where A is a conjunction of atomic formulas. A typical example is $f(\bar{x}) \downarrow$. Note that a conjunction of such Σ_1 formulas is provably equivalent to a single Σ_1 formula, by distributivity of $\exists$ over $\land$.

Definition. A k-ary function f is *provably recursive* in EA(I;O) if it can be defined by a system E of equations such that, with input variables $x_1, \ldots, x_k$,

$$\bar{E} \vdash f(x_1, \ldots, x_k) \downarrow$$

where $\bar{E}$ denotes the set of universal closures (over output variables) of the defining equations in E.

3. Elementary Functions are Provably Recursive

We will first look at some examples of how termination proofs work.

Examples 3.1.

1. Let E be the following program that defines the function $a + b$,

$$a + 0 \simeq a, \quad a + Sb \simeq S(a + b).$$

From an appropriate logical axiom we obtain $\bar{E} \vdash a + Sb \simeq S(a + b)$ as a consequence of left $\forall$ rules. Cutting with appropriate equality axioms we obtain,

$$\bar{E}, \ a + b \simeq c \vdash a + Sb \simeq Sc$$

which when read from left to right mirrors what we would expect the order of computation to be. As Sc is a basic term we can existentially quantify on the right. Now that c is no longer free anywhere else we can existentially quantify on the left. What remains is the step premise for an induction over b,

$$\bar{E}, \ a + b \downarrow \vdash a + Sb \downarrow.$$

An equivalent sequent for when b is 0 can be trivially obtained. Thus as an induction conclusion we obtain $\bar{E} \vdash a + x \downarrow$.

2. For multiplication we need to augment E by

$$b + c.0 \simeq b, \quad b + c.Sd \simeq (b + c.d) + c.$$

As above by starting from an appropriate axiom and applying the left $\forall$ rule three times we obtain, $\bar{E} \vdash b + x.Sd \simeq (b + x.d) + x$ and hence,

$$\bar{E}, \ b + x.d \simeq a \vdash b + x.Sd \simeq a + x.$$

Under the assumption that $a + x \downarrow$ we can existentially quantify on the right we obtain,

$$\bar{E}, \ a + x \downarrow, \ b + x.d \simeq a \vdash b + x.Sd \downarrow.$$

This assumption can be cut out as it is the result of the previous example. Now we can existentially quantify on the left over the remaining variable a. What remains is an induction step over d, from which the conclusion is $\bar{E} \vdash b + x.y \downarrow$, where y is any input variable. Note that the cut reflects that to prove termination for multiplication we need first have proved it for addition. Further note that we derive $a + x$ for addition not $a + b$. This precipitates our choice of formulating the starting axiom over an input x.

3. If we recount this last derivation whereby the starting axiom is formulated over $x.y$ instead of the input x, then after applying the left $\forall$ rules (the definedness of $x.y$ having been obtained in the previous example by setting b to be 0), we would have,

$$\bar{E} \vdash b + (x.y).Sd \simeq (b + (x.y).d) + (x.y),$$

from which we obtain,

$$\bar{E}, \ b + (x.y).d \simeq a \vdash b + (x.y).Sd \simeq a + (x.y).$$

We follow the same steps as above. Assuming $a + (x.y) \downarrow$ we can existentially quantify on the right. This assumption is then cut out as it is exactly the result obtained from the previous example. The induction step is obtained once we existentially quantify over a on the left,

$$\bar{E}, \ b + (x.y).d \downarrow \vdash b + (x.y).Sd \downarrow$$

The conclusion is $\bar{E} \vdash b + (x.y).z \downarrow$. Clearly this procedure can be repeated as many times as we would like. In particular, we therefore obtain derivations for $a + x^2 \downarrow, a + x^3 \downarrow, a + x^4 \downarrow, \ldots$, the building blocks for constructing polynomials, over inputs variables. In the following we shall generalise this argument.

Definition. Let E be a system of defining equations containing the usual primitive recursions for addition and multiplication and further equations of the forms

$$p_0 \simeq S0, \quad p_i \simeq p_{i_0} + p_{i_1}, \quad p_i \simeq p_{i_0} \cdot b$$

defining a sequence $\{p_i : i = 0, 1, 2 \ldots\}$ of polynomials in variables $\vec{b} = b_1, \ldots, b_n$. Henceforth we allow $p(\vec{b})$ to stand for any one of the polynomials so generated (clearly all polynomials can be built up in this way).

Definition. The *progressiveness* of a formula $A(a)$ with distinguished free variable a, is expressed by the formula

$$Prog_a A \equiv A(0) \wedge \forall a (A(a) \rightarrow A(Sa)).$$

Lemma 3.2. *Let $p(\vec{b})$ be any polynomial defined by a system of equations E as above. Then for every formula $A(a)$ we have, with input variables substituted for the variables of p,*

$$\bar{E}, \; Prog_a A \; \vdash \; A(p(\vec{x})).$$

Proof. Proceed by induction over the build-up of the polynomial p according to the given equations E. We argue in an informal natural deduction style, deriving the succedent of a sequent from its antecedent.

If p is the constant 1 (that is $S0$) then $A(S0)$ follows immediately from $A(0)$ and $A(0) \rightarrow A(S0)$, the latter arising from substitution of the defined, basic term 0 for the universally quantified variable a in $\forall a(A(a) \rightarrow A(Sa))$.

Suppose p is $p_0 + p_1$ where, by the induction hypothesis, the result is assumed for each of p_0 and p_1 separately. First choose $A(a)$ to be the formula $a \downarrow$ and note that in this case $Prog_a A$ is provable. Then the induction hypothesis applied to p_0 gives $p_0(\vec{x}) \downarrow$. Now again with an arbitrary formula A, we can easily derive

$$\bar{E}, \; Prog_a A, \; A(a) \; \vdash \; Prog_b(a + b \downarrow \wedge A(a + b))$$

because if $a + b$ is assumed to be defined, it can be substituted for the universally quantified a in $\forall a(A(a) \rightarrow A(Sa))$ to yield $A(a+b) \rightarrow A(a+Sb)$. Therefore by the induction hypothesis applied to p_1 we obtain

$$\bar{E}, \; Prog_a A, \; A(a) \; \vdash \; a + p_1(\vec{x}) \downarrow \wedge A(a + p_1(\vec{x}))$$

and hence

$$\bar{E}, \; Prog_a A \; \vdash \; \forall a(A(a) \rightarrow A(a + p_1(\vec{x}))).$$

Finally, substituting the defined term $p_0(\vec{x})$ for a, and using the induction hypothesis on p_0 to give $A(p_0(\vec{x}))$ we get the desired result

$$\bar{E}, Prog_a A \vdash A(p_0(\vec{x}) + p_1(\vec{x})).$$

Suppose p is $p_1.b$ where b is a fresh variable not occurring in p_1. By the induction hypothesis applied to p_1 we have as above, $p_1(\vec{x}) \downarrow$ and

$$\bar{E}, Prog_a A \vdash \forall a(A(a) \to A(a + p_1(\vec{x})))$$

for any formula A. Also, from the defining equations E and since $p_1(\vec{x}) \downarrow$, we have $p_1(\vec{x}).0 \simeq 0$ and $p_1(\vec{x}).Sb \simeq (p_1(\vec{x}).b) + p_1(\vec{x})$. Therefore we can prove

$$\bar{E}, Prog_a A \vdash Prog_b(p_1(\vec{x}).b \downarrow \wedge A(p_1(\vec{x}).b))$$

and an application of the EA(I;O)-induction principle on variable b gives, for any input variable x,

$$\bar{E}, Prog_a A \vdash p_1(\vec{x}).x \downarrow \wedge A(p_1(\vec{x}).x)$$

and hence $\bar{E}, Prog_a A \vdash A(p(\vec{x}))$ as required.

Notice that $Prog_a A \vdash A(x)$ is equivalent to the induction principle of EA(I;O). The preceding lemma extends this principle to any polynomial in $\vec{x}$ whereby, through inspection, the induction complexity is maintained as an "A" induction. What follows is the extension of this principle to any finitely iterated exponential, the cost of which being ever greater induction complexity. We therefore start by defining a measure for induction complexity.

Definition. A Σ_1 formula is said to be of "level-1" and a "level-$(n{+}1)$" formula is one of the form

$$\forall a(C(a) \to D(a))$$

where C and D are level-n. We could allow a string of universal quantifiers $\forall \vec{a}$ in the prefix, but it is not necessary for what follows. Typical examples of level-2 formulas would be $\forall a(f(a)\downarrow)$, $\forall a(f(a)\downarrow) \wedge \forall a(g(a)\downarrow)$ and $\forall c(c \leqslant d \to \forall a(f(a,c)\downarrow))$. The latter two do not fit the definition, but they are both minimally provably equivalent to level-2 formulas, namely $\forall a(f(a)\downarrow \wedge g(a)\downarrow)$ and $\forall c\forall a(c \leqslant d \to f(a,c)\downarrow)$ respectively. All these formulations play roles in the following examples.

Examples 3.3.

1. Let E contain the the following two equations,

$$a + 2^0 \simeq Sa, \quad a + 2^{Sb} \simeq (a + 2^b) + 2^b$$

from which we can now obtain,

$$\bar{E}, \; a + 2^b \simeq c \vdash a + 2^{Sb} \simeq c + 2^b.$$

Using the more general existential rule for the right we obtain,

$$\bar{E}, \; c + 2^b \downarrow, \; a + 2^b \simeq c \vdash a + 2^{Sb} \downarrow.$$

To apply the induction all that remains is to existentially bind the remaining formula over the variable c. This is inhibited since c appears free in the first formula. The solution is easy for we first universally bind this free occurrence of c, and then apply the left $\exists$ rule to obtain,

$$\bar{E}, \; \forall a(a + 2^b \downarrow), \; a + 2^b \downarrow \vdash a + 2^{Sb} \downarrow.$$

The simplicity of the solution hides the significance of its consequences; the induction complexity. The induction is going to be over b and as this variable appears throughout all the remaining formulas, we need to first "tidy up", through quantifications. Therefore our induction is no longer a level-1 induction (as for addition, multiplication and all polynomials), but of level-2,

$$\bar{E}, \; \forall a(a + 2^b \downarrow) \vdash \forall a(a + 2^{Sb} \downarrow).$$

The conclusion, $\bar{E} \vdash \forall a(a + 2^x \downarrow)$, completes the derivation. Note that the two uses of left $\forall$ rule, correspond to two applications of the induction hypothesis.

2. Let $Fib(a, b)$ be the function that gives a plus the b'th entry in the Fibonacci sequence, and let E be the following program program defining it:

$$Fib(a, 0) \simeq Sa, \quad Fib(a, 1) \simeq Sa, \quad Fib(a, SSb) \simeq Fib(Fib(a, b), Sb).$$

When unravelled the recursion step becomes,

$$\bar{E}, \; Fib(a, b) \simeq c \vdash Fib(a, SSb) \simeq Fib(c, Sb).$$

Existentially quantifying on the right, under the assumption that witness $Fib(c, Sb)$ is defined, we obtain,

$$\bar{E}, \; Fib(a, b) \simeq c, \; Fib(c, Sb) \downarrow \vdash Fib(a, SSb) \downarrow.$$

Notice that c appears twice (reflecting the substitution that increases the function's complexity) and that an existential quantifier on the left would have to bind both occurrences, something that would be undesired. Universally quantifying the righter most occurrence solves the problem and after further quantifications we now have,

$$\bar{E},\ \forall a(\,Fib(a,b)\downarrow),\ \forall a(\,Fib(a,Sb)\downarrow)\ \vdash\ \forall a(\,Fib(a,SSb)\downarrow).$$

Along with $\forall a(\,Fib(a,Sb)\downarrow)\vdash\forall a(\,Fib(a,Sb)\downarrow)$, (easily obtained from a logical axiom) we have after a right $\wedge$ rule following immediately by a left $\wedge$ rule,

$$\bar{E},\forall a(Fib(a,b)\downarrow)\wedge\forall a(Fib(a,Sb)\downarrow)\vdash\forall a(Fib(a,Sb)\downarrow)\wedge\forall a(Fib(a,SSb)\downarrow).$$

This is now an induction step, from which we could extract the result that $\forall a(\,Fib(a,x)\downarrow)$. Note that although the induction formula, $\forall a(\,Fib(a,b)\downarrow)\wedge\forall a(\,Fib(a,Sb)\downarrow)$, is not formally a level-2 formula, it is provably equivalent to one, namely $\forall a(\,Fib(a,b)\downarrow\wedge Fib(a,Sb)\downarrow)$.

3. A more complex example suggested by H. Jervell: let E contain the following three equations,

$$f(a,0,c)\simeq a+Sc,$$

$$f(a,Sb,0)\simeq f(a,b,b),$$

$$f(a,Sb,Sc)\simeq f(f(a,Sb,c),b,b).$$

describing the function $a+Fac(b).Sc$ as a double recursion, where $Fac(b)$ is the factorial of b, the function we are interested in. Following in the usual way, we obtain,

$$\bar{E},\ f(a,Sb,c)\simeq d\ \vdash\ f(a,Sb,Sc)\simeq f(d,b,b).$$

Applying a right $\exists$ rule, under the assumption that $f(d,b,b)\downarrow$, gives,

$$\bar{E},\ f(d,b,b)\downarrow,\ f(a,Sb,c)\simeq d\ \vdash\ f(a,Sb,Sc)\downarrow.$$

Again we are obliged to universally quantify on the left, to enable the appropriate left existential quantification. From this, after further $\forall$ rules, first on the left then finally on the right, we have,

$$\bar{E},\ \forall a(\,f(a,b,b)\downarrow),\ \forall a(\,f(a,Sb,c)\downarrow)\ \vdash\ \forall a(\,f(a,Sb,Sc)\downarrow).$$

We trivially obtain $\bar{E},\forall a(\,f(a,b,b)\downarrow)\ \vdash\ \forall a(\,f(a,Sb,0)\downarrow)$ from the second equation of E. Therefore we have,

$$\bar{E},\ \forall a(\,f(a,b,b)\downarrow)\ \vdash\ Prog_c(\,\forall a(\,f(a,Sb,c)\downarrow)\,).$$

Looking ahead, we know that if we were to apply the induction rule here, we will be prevented from applying the second induction at a later stage. Instead, we need a stronger form of induction, one that enables us to derive the following conclusion,

$$\bar{E}, \ \forall a(\, f(a,b,b) \downarrow\,) \ \vdash \ \forall c(\, c \leqslant x \rightarrow \forall a(\, f(a,Sb,c) \downarrow\,)\,).$$

For from this sequent after inverting the outermost universal quantifier on the right at Sb and introducing an implication on the left with a derivation of $\bar{E}, Sb \leqslant x \vdash b \leqslant x$ we arrive at,

$$\bar{E}, \ b \leqslant x \rightarrow \forall a(\, f(a,b,b) \downarrow\,) \ \vdash \ Sb \leqslant x \rightarrow \forall a(\, f(a,Sb,Sb) \downarrow\,).$$

Now we may apply an induction over b to obtain,

$$\bar{E} \ \vdash \ x \leqslant x \rightarrow \forall a(\, f(a,x,x) \downarrow\,),$$

from which we may extract the result that $a + Fac(Sx)$ is defined.

There remain two outstanding parts of the proof. The first is the stronger form of induction, that from the premise $\bar{E}, \Gamma \vdash Prog_a A(a)$ we may derive $\bar{E}, \Gamma \vdash \forall a(\, a \leqslant x \rightarrow A(a)\,)$. Thus we augment E by the following equations that define the predecessor and modified minus functions,

$$P0 \simeq 0, \quad P(Sb) \simeq b$$

$$a \doteq 0 \simeq a, \quad a \doteq Sb \simeq Pa \doteq b$$

and we define $t \leqslant r$ by $t \doteq r \simeq 0$. From these extra equations alone we derive $\bar{E}, \ a \leqslant Sb \vdash Pa \leqslant b$. Further, from the premise of progression and that $\bar{E} \vdash \forall a(\, a \simeq 0 \vee a \simeq S(Pa)\,)$ (easily obtained from cases) we derive $\bar{E}, A(Pa) \vdash A(a)$. Putting these together with the implication rules we have,

$$\bar{E}, \ Pa \leqslant b \rightarrow A(Pa) \ \vdash \ a \leqslant Sb \rightarrow A(a).$$

Universally quantifying on the left, over witness Pa, and then on the right over variable a we have the induction step. The case for when b is 0 is trivially obtained and the result then follows from an induction.

The second, and final, outstanding part of the original proof is the derivation of $\bar{E}, Sb \leqslant x \vdash b \leqslant x$. From the axiom

$$Pa \doteq Sb \simeq P(a \doteq Sb) \ \vdash \ Pa \doteq Sb \simeq P(a \doteq Sb)$$

we obtain,

$$\bar{E}, \; PPa \dot- b \simeq P(Pa \dot- b) \; \vdash \; Pa \dot- Sb \simeq P(a \dot- Sb),$$

by substitution since $Pa \dot- Sb \simeq PPa \dot- b$ and $a \dot- Sb \simeq Pa \dot- b$. Universally quantifying, first on the left with witness Pa and then on the right over variable a we obtain,

$$\bar{E}, \; \forall a(Pa \dot- b \simeq P(a \dot- b)) \; \vdash \; \forall a(Pa \dot- Sb \simeq P(a \dot- Sb)).$$

For when b is 0 we can trivially derive $\bar{E} \vdash \forall a(Pa \dot- 0 \simeq P(a \dot- 0))$, thus we obtain from an induction $\bar{E} \vdash \forall a(Pa \dot- x \simeq P(a \dot- x))$. Substituting the term Sb for the universally quantified a, we obtain

$$\bar{E} \vdash PSb \dot- x \simeq P(Sb \dot- x)$$

and from the definition of predecessor this is the same as

$$\bar{E} \vdash b \dot- x \simeq P(Sb \dot- x).$$

Thus if $Sb \dot- x \simeq 0$, (i.e. $Sb \leqslant x$), then $b \dot- x \simeq 0$, (i.e. $b \leqslant x$), since $P0 \simeq 0$, and we are therefore done.

Note that the induction takes the form $\forall c(c \leqslant d \rightarrow \forall a(f(a, Sb, c) \downarrow))$, which although is not formally a level-2 formula is indeed provably equivalent to one, namely, $\forall c \forall a(c \leqslant d \rightarrow f(a, Sb, c) \downarrow)$.

The three functions, exponential, Fibonacci and factorial, have two properties in common; firstly, they all need level-2 induction to prove their termination and secondly they are all register machine computable in an exponential number of steps. This relationship is of no coincidence, as it is a result that follows from the following more general results.

Definition. Extend the system of equations E that includes the equations for constructing polynomials, by adding the new recursive definitions:

$$f_1(a, 0) \simeq Sa, \quad f_1(a, Sb) \simeq f_1(f_1(a, b), b)$$

and for each $k = 2, 3, \ldots$,

$$f_k(a, b_1, \ldots, b_k) \simeq f_1(a, f_{k-1}(b_1, \ldots, b_k))$$

so that $f_1(a, b) = a + 2^b$ and $f_k(a, \vec{b}) = a + 2^{f_{k-1}(\vec{b})}$. Finally define

$$2_k(p(\vec{x})) \simeq f_k(0, \ldots 0, p(\vec{x}))$$

for each polynomial p given by E.

Lemma 3.4. *In EA(I;O) we can prove, for each k, for any polynomial p and any formula $A(a)$,*

$$\bar{E}, \; Prog_a A \; \vdash \; A(\, 2_k(p(\vec{x})) \,).$$

Proof. First note that by a similar argument to one used in the previous lemma (and going back all the way to Gentzen) we can prove, for any formula $A(a)$,

$$\bar{E}, \; Prog_a A \; \vdash \; Prog_b \forall a (A(a) \rightarrow f_1(a,b) \downarrow \wedge A(f_1(a,b)))$$

since the $b := 0$ case follows straight from $Prog_a A$, and the induction step from b to Sb follows by appealing to the hypothesis twice: from $A(a)$ we first obtain $A(f_1(a,b))$ with $f_1(a,b) \downarrow$, and then (by substituting the defined $f_1(a,b)$ for the universally quantified variable a) from $A(f_1(a,b))$ follows $A(f_1(a,Sb))$ with $f_1(a,Sb) \downarrow$, using the defining equations for f_1.

The result is now obtained straightforwardly by induction on k. Assuming $\bar{E}$ and $Prog_a A$ we derive

$$Prog_b \forall a (A(a) \rightarrow f_1(a,b) \downarrow \wedge A(f_1(a,b)))$$

and then by the previous lemma,

$$\forall a (A(a) \rightarrow f_1(a,p(\vec{x})) \downarrow \wedge A(f_1(a,p(\vec{x}))))$$

and then by putting a to be 0 and using $A(0)$ we have $2_1(p(\vec{x})) \downarrow$ and $A(2_1(p(\vec{x})))$, which is the case $k = 1$. For the step from k to $k+1$ do the same, but instead of the previous lemma use the induction to replace $p(\vec{x})$ by $2_k(p(\vec{x}))$.

Theorem 3.5. *Every elementary (E^3) function is provably recursive in the theory EA(I;O), and every sub-elementary (E^2) function is provably recursive in the fragment which allows induction only on Σ_1 formulas. Every function computable in number of steps bounded by an exponential function of its inputs is provably recursive in the "level-2" inductive fragment of EA(I;O).*

Proof. Any elementary function $g(\vec{x})$ is computable by a register machine M (working in unary notation with basic instructions "successor", "predecessor", "transfer" and "jump") within a number of steps bounded by $2_k(p(\vec{x}))$ for some fixed k and polynomial p. Let $r_1(c), r_2(c), \ldots, r_n(c)$ be the values held in its registers at step c of the computation, and let $i(c)$ be the number of the machine instruction to be performed next. Each of these functions depends also on the input parameters $\vec{x}$, but we suppress mention of these for brevity. The state of the computation $\langle i, r_1, r_2, \ldots, r_n \rangle$ at step $c + 1$ is obtained from the state at step c by performing the atomic act dictated by the instruction $i(c)$. Thus the values of $i, r_1, \ldots, r_n$ at step $c + 1$ can be defined from their values at step c by a simultaneous recursive definition involving only the successor S, predecessor P and definitions by cases C. So now, add these defining equations for $i, r_1, \ldots, r_n$ to the system E above, together with the equations for predecessor and cases:

$$P(0) \simeq 0, \quad P(Sa) \simeq a$$

$$C(0, a, b) \simeq a, \quad C(Sd, a, b) \simeq b$$

and notice that the cases rule built into EA(I;O) ensures that we can prove $\forall d \forall a \forall b \, C(d, a, b) \downarrow$. Since the passage from one step to the next involves only applications of C or basic terms, all of which are provably defined, it is easy to convince oneself that the Σ_1 formula

$$\exists \vec{a} \, (i(c) \simeq a_0 \wedge r_1(c) \simeq a_1 \wedge \ldots \wedge r_n(c) \simeq a_n)$$

is provably progressive in variable c. Call this formula $A(\vec{x}, c)$. Then by the second lemma above we can prove

$$\bar{E} \vdash A(\vec{x}, 2_k(p(\vec{x})))$$

and hence, with the convention that the final output is the value of r_1 when the computation terminates,

$$\bar{E} \vdash r_1(2_k(p(\vec{x}))) \downarrow.$$

Hence the function g given by $g(\vec{x}) \simeq r_1(2_k(p(\vec{x})))$ is provably recursive.

In just the same way, but using only the first lemma above, we see that any sub-elementary function (which, e.g. by Rödding (1968), is register machine computable in a number of steps bounded by just a polynomial of its inputs) is provably recursive in the Σ_1-inductive fragment. This is because the proof of $A(\vec{x}, p(\vec{x}))$ in lemma 3.2 only uses inductions on substitution instances of A, and here, A is Σ_1.

To see that a function computable in $\leqslant 2^{p(\vec{x})}$ steps is provably recursive in the level-2 inductive fragment, we need only verify that the proof of $\bar{E} \vdash A(\vec{x}, 2_1(p(\vec{x})))$ uses inductions that are at most level-2. Therefore, all we need to do is analyse the proofs of lemmas 3.2 and 3.4 a little more carefully. For any polynomial term p let $B(p)$ be the formula

$$\forall c(A(\vec{x}, c) \rightarrow p \downarrow \wedge f_1(c, p) \downarrow \wedge A(\vec{x}, f_1(c, p)))$$

and notice that although it isn't quite a level-2 formula, it is trivially and provably equivalent to one. (Recall that our notion of Σ_1 formula is restricted to an existentially quantified conjunction of equations, and the conjunction occurring after the implication inside B is equivalent to a single Σ_1 formula by distribution of $\exists$ over $\wedge$). Notice also that $B(b)$ is provably progressive since $A(\vec{x}, c)$ is. Hence by lemma 3.2 we can prove $\bar{E} \vdash B(p(\vec{x}))$ for any polynomial p, and by setting $c := 0$ we obtain the desired result. It only remains to check that this application of lemma 3.2 requires nothing more than level-2 induction. In fact the inductions required are on formulas of shape $q \downarrow \wedge B(p)$ with q other polynomial terms, but since we can prove $A(\vec{x}, 0)$ the subformulas $q \downarrow$ can also be shifted after the implication inside B, yielding provably equivalent level-2 forms. Thus level-2 induction suffices, and this completes the proof.

4. Provably Recursive Functions are Elementary

Suppose we have a derivation of $\bar{E} \vdash f(\vec{x}) \downarrow$ in EA(I;O), and suppose (arbitrary, but fixed) numerals $\bar{n}_1, \bar{n}_2, \ldots$ are substituted for the input variables $\vec{x} = x_1, x_2, \ldots$ throughout. In the resulting derivation, each application of induction takes the form:

$$\frac{\Gamma \vdash A(0) \quad \Gamma, A(a) \vdash A(Sa)}{\Gamma \vdash A(t(\bar{n}_i))}$$

where $t(x_i)$ is the basic term appearing in the conclusion of the original (unsubstituted) EA(I;O)-induction. Let m denote the value of $t(\bar{n}_i)$, so m is not greater than n_i plus the length of term t. Furthermore, let ℓ denote the length of the binary representation of m. Then, given the premises, we can unravel the induction so as to obtain a derivation of

$$\Gamma \vdash A(\bar{m})$$

by a sequence of cuts on the formula A, with proof-height $\ell + 1$. To see this we first induct on ℓ to derive

$$\Gamma, A(a) \vdash A(S^m a) \quad \text{and} \quad \Gamma, A(a) \vdash A(S^{m+1}a)$$

by sequences of A-cuts with proof-height ℓ. This is immediate when $\ell = 1$, and if $\ell > 1$ then either $m = 2m_0$ or $m = 2m_0 + 1$ where m_0 has binary length less than ℓ. So from the result for m_0 we get

$$\Gamma, A(a) \vdash A(S^{m_0}a) \quad \text{and} \quad \Gamma, A(S^{m_0}a) \vdash A(S^m a)$$

by substitution of $S^{m_0}a$ for the free variable a, and both of these derivations have proof-height $\ell - 1$. Therefore one more cut yields

$$\Gamma, A(a) \vdash A(S^m a)$$

as required. The case $A(S^{m+1}a)$ is done in just the same way.

Therefore if we now substitute 0 for variable a, and appeal to the base case of the induction, a final cut on $A(0)$ yields $\Gamma \vdash A(\bar{m})$ with height $\ell + 1$ as required.

What we have just done is unravelled the induction up to $A(\bar{m})$, replacing the single application of the induction rule by a sequence of cuts. Thus if we make any fixed numerical instantiation of the input variables, a proof in EA(I;O) can be unravelled into one in which all the inductions have been removed, and replaced by sequences of cuts on the induction formulas. The Cut-Elimination process then allows us to successively reduce the logical complexity of cut formulas, but the price paid is an exponential increase in the height of the proof. Once we get down to the level of Σ_1 cut formulas, what remains is essentially a computation (from the given system of equations E), and its complexity will correspond to the number of proof-steps. However, the size of the resulting proof (or computation) will vary with the size of numerical inputs originally substituted for the input variables. What we need is some way of giving a *uniform* bound on the proof-size, independent of the given input numbers. This is where *ordinals* enter into proof theory.

4.1. ORDINAL BOUNDS FOR RECURSION

For each fixed number k, we inductively generate an infinitary system of sequents

$$E, \, n : N, \, \Gamma \vdash^\alpha A$$

where (i) E is a (consistent) set of Herbrand-Gödel-Kleene defining equations for partial recursive functions $f, g, h, \dots$; (ii) n is a bound on the numerical inputs (or more precisely, its representing numeral); (iii) A is a closed formula, and Γ a finite multiset of closed formulas, built up from atomic equations between arbitrary terms t involving the function symbols of E; and (iv) α, β denote ordinal bounds which we shall be more specific about later (for the time being think of β as being smaller than α "modulo k").

Note that we do not explicitly display the parameter k in the sequents below, but if we later need to do this we shall insert an additional declaration $k : I$ in the antecedent thus:

$$E, \, k : I, \, n : N, \, \Gamma \vdash^\alpha A \, .$$

Intuitively, k will be a bound on the lengths of "unravelled inductions".

The first two rules are just the input and substitution axioms of the equation calculus, the next two are computation rules for N, and the rest are essentially just formalised versions of the truth definition, with Cut added.

E1 $E,\ n:N,\ \Gamma \vdash^\alpha e(\vec{n})$ where e is either one of the defining equations of E or an identity $t \simeq t$, and $e(\vec{n})$ denotes the result of substituting, for its variables, numerals for numbers $\leqslant n$.

E2 $E,\ n:N,\ \Gamma,\ t_1 \simeq t_2,\ e(t_1)\ \vdash^\alpha\ e(t_2)$ where $e(t_1)$ is an equation between terms in the language of E, with t_1 occurring as a subterm, and $e(t_2)$ is the result of replacing an occurrence of t_1 by t_2.

N1
$$E,\ n:N,\ \Gamma \vdash^\alpha\ m:N \text{ provided } m \leqslant n+1$$

N2
$$\frac{E,\ n:N,\ \Gamma \vdash^\beta n':N \quad E,\ n':N,\ \Gamma \vdash^{\beta'} A}{E,\ n:N,\ \Gamma \vdash^\alpha A}$$

Cut
$$\frac{E,\ n:N,\ \Gamma \vdash^\beta C \quad E,\ n:N,\ \Gamma,\ C \vdash^{\beta'} A}{E,\ n:N,\ \Gamma \vdash^\alpha A}$$

∃L
$$\frac{E,\ \max(n,i):N,\ \Gamma,\ B(i) \vdash^{\beta_i} A \text{ for every } i \in N}{E,\ n:N,\ \Gamma,\ \exists b B(b) \vdash^\alpha A}$$

∃R
$$\frac{E,\ n:N,\ \Gamma \vdash^\beta m:N \quad E,\ n:N,\ \Gamma \vdash^{\beta'} A(m)}{E,\ n:N,\ \Gamma \vdash^\alpha \exists a A(a)}$$

∀L
$$\frac{E,\ n:N,\ \Gamma \vdash^\beta m:N \quad E,\ n:N,\ \Gamma,\ \forall b B(b),\ B(m) \vdash^{\beta'} A}{E,\ n:N,\ \Gamma,\ \forall b B(b) \vdash^\alpha A}$$

∀R
$$\frac{E,\ \max(n,i):N,\ \Gamma \vdash^{\beta_i} A(i) \text{ for every } i \in N}{E,\ n:N,\ \Gamma \vdash^\alpha \forall a A(a)}$$

388

In addition, there are of course two rules for each propositional symbol, but it is not necessary to list them since they are quite standard. However it should be noted that the rules essentially mimic the truth definition for arithmetic, but as with EA(I;O) they are formalized in the style of **minimal**, not classical, logic. They provide a system within which the inductive proofs of EA(I;O) can be unravelled in a uniform way.

Ordinal Assignment à la Buchholz (1987).

The ordinal bounds on sequents above are intensional, "tree ordinals", generated inductively by: 0 is a tree ordinal; if α is a tree ordinal so is $\alpha+1$; and if $\lambda_0, \lambda_1, \lambda_2, \ldots$ is an ω-sequence of tree ordinals then the function $i \mapsto \lambda_i$ denoted $\lambda = \sup \lambda_i$, is itself also a tree ordinal. Thus tree ordinals carry a specific choice of fundamental sequence to each "limit" encountered in their build-up, and because of this the usual definitions of primitive recursive functions lift easily to tree ordinals. For example addition is defined by:

$$\alpha + 0 = 0, \quad \alpha + (\beta + 1) = (\alpha + \beta) + 1, \quad \alpha + \lambda = \sup(\alpha + \lambda_i)$$

and multiplication is defined by:

$$\alpha.0 = 0, \quad \alpha.(\beta + 1) = \alpha.\beta + \alpha, \quad \alpha.\lambda = \sup \alpha.\lambda_i$$

and exponentiation is defined by:

$$2^0 = 1, \quad 2^{\beta+1} = 2^\beta + 2^\beta, \quad 2^\lambda = \sup 2^{\lambda_i}.$$

For ω we choose the specific fundamental sequence $\omega = \sup(i + 1)$. For ε_0 we choose the fundamental sequence $\varepsilon_0 = \sup 2_i(\omega^2)$ where $2_i(\beta)$ is defined to be β if $i = 0$ and $2^{2_{i-1}(\beta)}$ if $i > 0$.

Definitions. For each integer i there is a predecessor function given by:

$$P_i(0) = 0, \quad P_i(\alpha + 1) = \alpha, \quad P_i(\lambda) = P_i(\lambda_i)$$

and by iterating P_i we obtain, for each non-zero tree ordinal α, the finite set $\alpha[i]$ of all its "i-predecessors" thus:

$$\alpha[i] = \{P_i(\alpha), P_i^2(\alpha), P_i^3(\alpha), \ldots, 0\}.$$

Call a tree ordinal α "structured" if every sub-tree ordinal of the form $\lambda = \sup \lambda_i$ (occurring in the build-up of α) has the property that $\lambda_i \in \lambda[i+1]$ for all i. Then if α is structured, $\alpha[i] \subset \alpha[i + 1]$ for all i, and each of its sub-tree ordinals β appears in one, and all succeeding, $\alpha[i]$. Thus we can think of a structured α as the directed union of its finite sub-orderings $\alpha[i]$. The basic example is $\omega[i] = \{0, 1, \ldots, i\}$. All tree ordinals used here will be structured ones.

Ordinal Bounds. The condition on ordinal bounds in the above sequents is to be as follows:

- In rules E1, E2, N1, the bound α is arbitrary.
- In all other rules, the ordinal bounds on the premises are governed by $\beta, \beta', \beta_i \in \alpha[k]$ where k is the fixed parameter.

Lemma 4.1. *(Weakening) If $E,\ n : N,\ \Gamma \vdash^\alpha \ A$ where $n \leqslant n'$ and $\alpha[k] \subset \alpha'[k]$ then $E,\ n' : N,\ \Gamma \vdash^{\alpha'} A$.*

4.2. BOUNDING FUNCTIONS

Definition. The bounding functions $B_\alpha(k; n)$ are given by the recursion:

$$B_0(k; n) = n + 1, \quad B_{\alpha+1}(k; n) = B_\alpha(k; B_\alpha(k; n)), \quad B_\lambda(k; n) = B_{\lambda_k}(k; n).$$

Lemma 4.2. $m \leqslant B_\alpha(k; n)$ *if and only if, using the N1, N2 rules, we can derive* $k : I,\ n : N \vdash^\alpha m : N$.

Proof. For the "only if" proceed by induction on α. If $\alpha = 0$ the result is immediate by rule N1. If $\alpha > 0$ then it's easy to see that $B_\alpha(k; n) = B_\beta(k; B_\beta(k; n))$ where $\beta = P_k(\alpha)$. So if $m \leqslant B_\alpha(k; n)$ then $m \leqslant B_\beta(k; n')$ where $n' = B_\beta(k; n)$. Therefore by the induction hypothesis we have both $k : I,\ n : N \vdash^\beta n' : N$ and $k : I,\ n' : N \vdash^\beta m : N$. Hence we obtain by the N2 rule $k : I,\ n : N \vdash^\alpha m : N$.

The "if" part again follows by induction on α. The result is immediate if the sequent $k : I,\ n : N \vdash^\alpha m : N$ comes about by the N1 rule. If it comes about by an application of N2 from premises $k : I,\ n : N \vdash^\beta n' : N$ and $k : I,\ n' : N \vdash^{\beta'} m : N$ then by the induction hypothesis $n' \leqslant B_\beta(k; n)$ and $m \leqslant B_{\beta'}(k; n')$. Thus $m \leqslant B_{\beta'}(k; B_\beta(k; n))$ and this is bounded by $B_\alpha(k; n)$ since $\beta, \beta' \in \alpha[k]$.

Lemma 4.3. $B_\alpha(k; n) = n + 2^{G(\alpha, k)}$ *where $G(\alpha, k)$ is the size of $\alpha[k]$. Furthermore, if $\alpha = 2_i(\omega.d) \prec \varepsilon_0$ then $G(\alpha, k)$ is simply obtained by substituting $k + 1$ for ω thus $G(\alpha, k) = 2_i((k + 1).d)$.*

Proof. The first part follows directly from the recursive definition of B_α, by noting also that $G(0, k) = 0$, $G(\alpha+1, k) = G(\alpha, k)+1$ and $G(\lambda, k) = G(\lambda_k, k)$.

The second part uses $G(\omega, k) = G(k+1, k) = k+1$ together with the easily verified fact that if $\varphi(\alpha, \beta)$ denotes either $\alpha + \beta$ or $\alpha.\beta$ or α^β then

$$G(\varphi(\alpha, \beta), k) = \varphi(G(\alpha, k), G(\beta, k)) \ .$$

Lemma 4.4. *If α is structured, $\beta \in \alpha[k]$, $k \leqslant k'$ and $n \leqslant n'$ then $B_\beta(k; n) < B_\alpha(k; n) \leqslant B_\alpha(k'; n')$.*

Proof. Immediate from the above.

4.3. EMBEDDING EA(I;O), AND CUT REDUCTION

Lemma 4.5. *(Embedding) If $\bar{E} \vdash f(\vec{x}) \downarrow$ in $EA(I;O)$ there is a fixed number d determined by this derivation, such that: for all inputs $\vec{n}$ of binary length $\leqslant k$, we can derive*

$$E, k : I, 0 : N \vdash^{\omega.d} f(\vec{n}) \downarrow$$

in the infinitary system. Furthermore the non-atomic cut-formulas in this derivation are the induction-formulas occurring in the $EA(I;O)$ proof.

Proof. First, by standard "free cut"-elimination arguments we can eliminate from the given EA(I;O) derivation all non-atomic cut-formulas which are not induction formulas. Then pass through the resulting free-cut-free proof, substituting the numerals for the input variables and translating each proof-rule into the corresponding infinitary rule with an appropriate ordinal bound. The non-inductive steps are quite straightforward, and each induction is unravelled in the manner described earlier. For suppose the two premises of the induction have been embedded with ordinal bound $\omega.d$. If the conclusion is $A(t(x_i))$ with t a basic term then, upon substitution of n_i for x_i, we obtain $t(n_i)$ with value $m \leqslant n_i + c$ where c measures the length of the term t (determined by the given EA(I;O) proof). The binary length of m is then $\ell < k + c$ and we thus obtain a derivation of

$$E, k : I, m : N \vdash^{\omega.d+k+c} A(m)$$

whose ordinal bound we can manipulate so that

$$E, k : I, m : N \vdash^{\omega.(d+c)+k} A(m) \ .$$

Now $m \leqslant c + 2^{k+1} = B_\omega(k; c)$ and $c \leqslant B_{\omega.c}(k; 0)$. So $m \leqslant B_{\omega.(d+c)}(k; 0)$ and by the lemma about B,

$$E, k : I, 0 : N \vdash^{\omega.(d+c)} m : N \ .$$

Therefore by applying the N2 rule,

$$E, \, k : I, \, 0 : N \, \vdash^{\omega.(d+c)+k+1} A(m)$$

and then since $\omega.(d + c + 1)[k] = \omega.(d + c) + k + 1[k]$ we have

$$E, \, k : I, \, 0 : N \, \vdash^{\omega.(d+c+1)} A(m)$$

as required.

Lemma 4.6. *(Cut Reduction) If $\bar{E} \vdash f(\vec{x}) \downarrow$ in $EA(I;O)$ then there is a tree ordinal $\alpha \prec \varepsilon_0$ such that: for all inputs $\vec{n}$ of binary length $\leqslant k$, we can derive*

$$E, \, k : I, \, 0 : N \, \vdash^{\alpha} f(\vec{n}) \downarrow$$

by an infinitary derivation in which all the cut formulas are at worst Σ_1.

Proof. The Embedding Lemma above gives a derivation of

$$E, \, k : I, \, 0 : N \, \vdash^{\omega.d} f(\vec{n}) \downarrow$$

wherein the cut formulas might be of great logical complexity, depending on the inductions used in the original $EA(I;O)$ proof. However Gentzen-style cut-reduction applies to the infinitary system (whereas it doesn't apply directly to $EA(I;O)$). Thus the "sizes" of cut formulas can be successively reduced, one level at a time, so that all that remains are cut formulas of shape Σ_1. But each time the cuts are reduced in size, there will be an exponential increase in the ordinal bound of the resulting derivation. Therefore, if the maximum size of induction formula used in the $EA(I;O)$ proof is r, then we obtain the desired derivation

$$E, \, k : I, \, 0 : N \, \vdash^{\alpha} f(\vec{n}) \downarrow$$

with at most Σ_1 cuts and ordinal bound $\alpha \prec 2_r(\omega.d) \prec \varepsilon_0$.

We now describe briefly how the cut reduction process works. First define the *size* of a formula to be 0 if it is a conjunction of atoms, 1 if it is Σ_1, and $r + 1$ otherwise, where r is the maximum size of its subformulas. Then define the *cut rank* of a derivation to be the maximum size of all cut formulas appearing in it. There are two steps to the process:

Step (i) Suppose we have two derivations in the infinitary system:

$$E, \, k : I, \, n : N, \, \Gamma \vdash^{\beta} C$$

and

$$E, \, k : I, \, n : N, \, \Gamma, \, C \vdash^{\gamma}$$

both with cut rank $\leqslant r$, and where C is of size $r + 1$. Suppose also that $\beta, \gamma \in \alpha[k]$ for some fixed α. Obviously we could apply the Cut Rule, with cut formula C, to derive

$$E, \, k : I, \, n : N, \, \Gamma \vdash^{\alpha} A$$

but the cut rank would then be $r + 1$. The point is we can do better than this, and replace this cut by another one of rank r. However the ordinal bound will increase to either $\beta + \gamma$ or $\gamma + \beta$, depending on the form of cut formula C.

As an example, suppose C is of the form $\forall a D(a)$. Then by induction on γ we show that we can remove C and derive

$$E, \, k : I, \, n : N, \, \Gamma \vdash^{\beta + \gamma} A$$

with cut rank r.

For suppose the derivation with bound γ arises by an application of the $\forall$L rule on formula C. Then, suppressing E, $k : I$ and Γ in the antecedent, we have premises $n : N, \vdash^{\gamma_0} m : N$ and $n : N, \, C, \, D(m) \vdash^{\gamma_1} A$, and by applying the induction hypothesis to remove C from the second of these, we obtain $n : N, \, D(m) \vdash^{\beta + \gamma_1} A$. Also, by inverting the derivation with bound β we obtain $\max(n, m) : N \vdash^{\beta_m} D(m)$ for some $\beta_m \in \beta[k]$ and then by the N2 rule, $n : N \vdash^{\beta + \max(\gamma_0, \gamma_1)} D(m)$ provided that γ_0 and γ_1 are not both zero. We can then derive $n : N \vdash^{\beta + \gamma} A$ by a cut on the formula $D(m)$ and the cut rank will be just r as required, since the size of D is one less than the size of C. If, on the other hand, $\gamma_0 = \gamma_1 = 0$ then $n : N, \, C, \, D(m) \vdash^{0} A$ is an axiom, and therefore so is $n : N, \, D(m) \vdash^{\beta} A$ since C is not atomic and can thus be deleted from any axiom. Hence using N2 to give $n : N \vdash^{\beta} D(m)$ we again derive $n : N \vdash^{\beta + \gamma} A$ with cut rank r by a cut on $D(m)$ as before.

If the derivation with bound γ arises by an application of any other rule, then the C will appear in the premises as an inactive "side formula" which can be removed simply by applying the induction hypothesis, and adding β to their ordinal bounds. Then that rule can be re-applied to give the desired result. This completes our example of Step (i). The other cases are handled in a similar way, though sometimes the induction must be on β instead of γ, producing an ordinal bound $\gamma + \beta$.

Step (ii) Transform any derivation with ordinal bound α and cut rank $r + 1$ into a derivation of the same sequent, but with cut rank $\leqslant r$ and ordinal bound 2^α, as follows using Step (i):

Proceed by induction on α with cases according to the last rule applied. If this last rule is a cut with cut formula C of size $r + 1$ then the premises (with all "side formulas" suppressed) will be of the form $\vdash^\beta C$ and $C \vdash^\gamma A$ with $\beta, \gamma \in \alpha[k]$. By increasing the smaller of these bounds if necessary, using weakening, we may assume that they are both the same. Applying the induction hypothesis to each premise, one obtains derivations $\vdash^{2^\beta} C$ and $C \vdash^{2^\beta} A$ both now with cut rank $\leqslant r$. Then Step (i) applies to give a derivation of A with cut rank r and ordinal bound $2^\beta + 2^\beta$, which is either equal to 2^α or can be weakened to it.

If the last rule applied is anything other than a cut of rank $r + 1$, then simply apply the induction hypothesis to the premises in order to reduce their cut rank, and then re-apply that last rule, noting that if $\beta \in \alpha[k]$ then $2^\beta \in 2^\alpha[k]$. This completes the proof of the Cut Reduction Lemma.

4.4. COMPLEXITY BOUNDS

Notation. We signify that an infinitary derivation involves only Σ_1 cut formulas C, by attaching a subscript 1 to the proof-gate thus: $E, n : N, \Gamma \vdash_1^\alpha A$. If all cut formulas are atomic equations (or possibly conjunctions of them) we attach a subscript 0 instead.

Lemma 4.7. *(Bounding) Let Γ, A consist of (conjunctions of) atomic formulas only.*

1. *If $E, k : I, n : N, \Gamma \vdash_1^\alpha m : N$ then $m \leqslant B_\alpha(k; n)$.*
2. *If $E, k : I, n : N, \Gamma \vdash_1^\alpha \exists \vec{a} A(\vec{a})$ then there are numbers $\vec{m} \leqslant B_\alpha(k; n)$ such that $E, k : I, n : N, \Gamma \vdash_0^{2 \cdot \alpha} A(\vec{m})$.*

Proof. Both parts are dealt with simultaneously by induction on α. Since only Σ_1 cuts are involved, it is only the E, N, Cut and $\exists$ rules which come into play. The E rules require no action at all, and if N1 is applied (in case 1) then we have immediately $m \leqslant n + 1 \leqslant B_\alpha(k; n)$.

(N2) Suppose the sequent in 1 or 2 comes about by the N2 rule. Then one of the premises is of exactly the same form, but with n replaced by n' and α replaced by a $\beta' \in \alpha[k]$. Therefore by the induction hypothesis, and re-application of N2 to reduce $n' : N$ to $n : N$, the desired result follows since $2.\beta' \in 2.\alpha[k]$. But the bound on existential witnesses that we have at this stage is $B_{\beta'}(k; n')$. However the other premise is a sequent of the form 1 with m replaced by n' and α replaced by a $\beta \in \alpha[k]$, so by the induction hypothesis $n' \leqslant B_\beta(k; n)$. Thus, substituting $B_\beta(k; n)$ for n' we obtain the required bound

$$B_{\beta'}(k; n') \leqslant B_{\beta'}(k; B_\beta(k; n)) \leqslant B_\alpha(k; n)$$

using the definition of B_α and its majorization properties.

($\exists$) The $\exists$L rule does not apply. If the sequent in 2 comes about by an application of $\exists$R, introducing the outermost quantifier in $\exists \vec{a} A(\vec{a})$, then the induction hypothesis applied to the first premise produces a witness $m \leqslant B_\beta(k; n)$. The induction hypothesis applied to the second premise yields the desired result, but with bound $B_{\beta'}(k; n)$ on witnesses for the remaining quantifiers. However $B_{\beta'}(k; n)$ and $B_\beta(k; n)$ are both less than the required bound $B_\alpha(k; n)$ since $\beta, \beta' \in \alpha[k]$.

(Cut) Finally, suppose the sequent in 1 or 2 arises by an application of Cut with cut formula $C \equiv \exists \vec{c} D(\vec{c})$. By applying the induction hypothesis to the first premise we obtain witnesses $\vec{\ell}$ no larger than $B_\beta(k; n)$ such that

$$E, \; k : I, \; n : N, \; \Gamma \vdash_0^{2.\beta} D(\vec{\ell})$$

and by "inverting" the $\exists \vec{c}$ in the second premise we obtain

$$E, \; k : I, \; \max(n, \vec{\ell}) : N, \; D(\vec{\ell}), \; \Gamma \vdash_1^{\beta'} F$$

where $F \equiv m : N$ or $F \equiv \exists \vec{a} A(\vec{a})$. Applying the induction hypothesis to this last sequent, in the case $F \equiv \exists \vec{a} A(\vec{a})$, we obtain numerical witnesses $\vec{m}$ bounded by:

$$B_{\beta'}(k; \max(n, \vec{\ell})) \leqslant B_{\beta'}(k; B_\beta(k; n)) \leqslant B_\alpha(k; n)$$

and

$$E, \; k : I, \; \max(n, \vec{\ell}) : N, \; D(\vec{\ell}), \; \Gamma \vdash_0^{2.\beta'} A(\vec{m}) \, .$$

Since $\vec{\ell} \leqslant B_\beta(k; n) \leqslant B_{2.\beta}(k; n)$ we have, by a lemma above,

$$E, \; k : I, \; n : N, \; D(\vec{\ell}), \; \Gamma \vdash_0^{2.\beta} \max(n, \vec{\ell}) : N$$

and so by the N2 rule, with $\gamma = \max(\beta, \beta')$, we obtain

$$E, \; k : I, \; n : N, \; D(\vec{\ell}), \; \Gamma \vdash_0^{2.\gamma+1} A(\vec{m}) \, .$$

Then by a cut on $D(\vec{\ell})$, with a weakening, we obtain the required

$$E, \; k : I, \; n : N, ; \Gamma \vdash_0^{2.\alpha} A(\vec{m})$$

since $\gamma \in \alpha[k]$ implies $2.\gamma + 1 \in 2.\alpha[k]$. The other case $F \equiv m : N$ is much simpler. This completes the proof.

Theorem 4.8. *(Complexity) Suppose f is defined by a system of equations E and $\bar{E} \vdash f(\vec{x}) \downarrow$ in EA(I;O) with induction formulas of size at most r. Then there is an $\alpha = 2_{r-1}(\omega.d)$ such that: for all inputs $\vec{n}$ of binary length at most k we have*

$$E, \, k : I, \, 0 : N \vdash_0^{2.\alpha} f(\vec{n}) \simeq m \quad where \quad m \leqslant 2_r((k+1).d) \, .$$

This is a computation of $f(\vec{n})$ from E, and the number of computation steps (nodes in the binary branching tree) is less than $4^{2_{r-1}((k+1).d)}$.

Proof. First apply the Embedding Lemma to obtain a number d such that for all inputs $\vec{n}$ of binary length $\leqslant k$,

$$E, \, k : I, \, 0 : N \vdash^{\omega.d} f(\vec{n}) \downarrow$$

with cut rank r. Then apply Cut Reduction $r - 1$ times, to bring the cut rank down to the Σ_1 level. This gives $\alpha = 2_{r-1}(\omega.d)$ such that

$$E, \, k : I, \, 0 : N \vdash_1^{\alpha} f(\vec{n}) \downarrow \, .$$

Then apply the Bounding Lemma to obtain $m \leqslant B_\alpha(k; 0) = 2^{G(\alpha, k)} = 2_r((k+1).d)$ such that

$$E, \, k : I, \, 0 : N \vdash_0^{2.\alpha} f(\vec{n}) \simeq m \, .$$

This derivation uses only the E axioms, the N rules and equational cuts, so it is a computation in the equation calculus. Since all the ordinal bounds belong to $2.\alpha[k]$, the height of the derivation tree is no greater than $G(2.\alpha, k)$ and the number of nodes (or computation steps) is therefore $\leqslant 2^{G(2.\alpha,k)} = 4^{G(\alpha,k)} = 4^{2_{r-1}((k+1).d)}$.

Theorem 4.9. *The functions provably recursive in EA(I;O) are exactly the elementary E^3 functions. The functions provably recursive in the Σ_1 inductive fragment of EA(I;O) are exactly the subelementary (E^2) functions. The functions provably recursive in the "level-2" inductive fragment of EA(I;O) are exactly those computable in a number of steps bounded by an exponential function of their inputs.*

Proof. By the above, if f is provably recursive in EA(I;O) then it is computable in a number of steps bounded by a finitely iterated exponential function of its inputs. This means it is elementary.

If f is provably recursive in the Σ_1 inductive fragment then we can take $r = 1$ in the above. So f is computable in a number of steps bounded by $4^{(k+1).d}$. But k is the maximum binary length of the inputs $\vec{n}$, so this bound is just a polynomial in max $\vec{n}$. Therefore by e.g. Rödding (1967), f is subelementary.

If f is provably recursive using "level-2" inductions then, taking $r = 2$, we obtain a computational bound $4^{2_1((k+1).d)}$ which is $\leqslant 2^{p(\vec{n})}$ for some polynomial p. (For a more detailed treatment of this case see Ostrin and Wainer (2001)).

4.5. POLYTIME FUNCTIONS

If the theory EA(I;O) were instead formulated on the basis of *binary* (rather than unary) number representation, with two successors $S_0a = 2a$, $S_1a = 2a + 1$, one predecessor $P(0) = 0$, $P(S_0a) = P(S_1a) = a$, and an induction rule of the form

$$\frac{\Gamma \vdash A(0) \quad \Gamma, A(a) \vdash A(S_0a) \quad \Gamma, A(a) \vdash A(S_1a)}{\Gamma \vdash A(x)}$$

then it only takes $\log n$ many induction steps to "climb up" to $A(n)$. Thus a similar analysis to that given here, but with n replaced by $\log n$, would show that the functions provably recursive in the Σ_1 inductive fragment of this binary theory are just those with complexity bounds polynomial in the binary lengths of their inputs, i.e. PTIME, see Leivant (1995).

References

[1] S. Bellantoni and S. Cook, *A new recursion theoretic characterization of the polytime functions*, Computational Complexity Vol. 2 (1992) pp. 97 - 110.

[2] W. Buchholz, *An independence result for $\Pi_1^1 - CA + BI$*, Annals of Pure and Applied Logic Vol. 23 (1987) pp. 131 - 155.

[3] W. Burr, *Fragments of Heyting arithmetic*, Journal of Symbolic Logic Vol. 65 (2000) pp. 1223 - 1240.

[4] S.R. Buss, *Bounded arithmetic*, Bibliopolis (1986).

[5] N. Cagman, G. E. Ostrin and S.S. Wainer, *Proof theoretic complexity of low subrecursive classes*, in F. L. Bauer and R. Steinbrueggen (Eds) Foundations of Secure Computation, NATO ASI Series F Vol. 175, IOS Press (2000) pp. 249 - 285.

[6] M. Fairtlough and S.S. Wainer, *Hierarchies of provably recursive functions*, in S. Buss (Ed) Handbook of Proof Theory, Elsevier Science BV (1998) pp. 149 - 207.

[7] W.G. Handley and S.S. Wainer, *Complexity of primitive recursion*, in U. Berger and H. Schwichtenberg (Eds) Computational Logic, Springer-Verlag ASI Series F (1999), pp. 28.

[8] G. Kreisel, *On the interpretation of non-finitist proofs, Parts I, II*, Journal of Symbolic Logic Vol. 16 (1951) pp. 241 - 267, Vol. 17 (1952) pp. 43 - 58.

[9] D. Leivant, *A foundational delineation of poly-time*, Information and Computation Vol. 110 (1994) pp. 391 - 420.

[10] D. Leivant, *Intrinsic theories and computational complexity* , in D. Leivant (Ed) Logic and Computational Complexity, Lecture Notes in Computer Science Vol. 960, Springer-Verlag (1995) pp. 177 - 194.

[11] G.E. Ostrin, *Proof theories of low subrecursive classes*, Ph.D. Leeds (1999).

[12] G.E. Ostrin and S.S. Wainer, *Elementary arithmetic*, Leeds preprint (2001).

[13] D. Rödding, *Klassen rekursiver funktionen*, in M. H. Löb (Ed) Proc. of Summer School in Logic, Leeds 1967, Lecture Notes in Mathematics Vol. 70, Springer-Verlag (1968) pp. 159 - 222.

[14] A. S. Troelstra and H. Schwichtenberg, *Basic proof theory*, Cambridge Tracts in Theoretical Computer Science Vol. 43, CUP (1996).

[15] A. Weiermann, *How to characterize provably total functions by local predicativity*, Journal of Symbolic Logic Vol. 61 (1996) pp. 52 - 69.

FEASIBLE COMPUTATION WITH HIGHER TYPES

HELMUT SCHWICHTENBERG
Mathematisches Institut der Universität München
schwicht@mathematik.uni-muenchen.de

STEPHEN J. BELLANTONI
Department of Computer Science, University of Toronto
sjb@cs.toronto.edu

Abstract. We restrict recursion in finite types so as to characterize the polynomial time computable functions. The restrictions are obtained by enriching the type structure with the formation of types $\rho \to \sigma$ and terms $\lambda \bar{x}^\rho r$ as well as $\rho \multimap \sigma$ and $\lambda x^\rho r$. Here we use two sorts of typed variables: complete ones $\bar{x}^\rho$ and incomplete ones x^ρ.

1. Introduction

Recursion in all finite types was introduced by Hilbert (1925), the system later becoming known as Gödel's system **T** (Gödel, 1958). The value computed by a higher type recursion can be any functional, which is to say a mapping that takes other mappings as arguments and produces a new mapping. Correspondingly one defines a type system of functions and functionals over some ground types.

Recursion in higher types, as in Gödel's system **T**, has long been viewed as a powerful scheme unsuitable for describing small complexity classes such as polynomial time. It is well known that ramification can be used to restrict higher type recursion. However, to characterize the very small class of polynomial-time computable functions while still admitting higher type recursion, it seems that an additional principle is required. By introducing linearity constraints in conjunction with ramified recursion, we characterize polynomial-time computability while admitting recursion in higher types.

H. Schwichtenberg and R. Steinbrüggen (eds.), Proof and System-Reliability, 399–415.
© *2002 Kluwer Academic Publishers. Printed in the Netherlands.*

We shall work with "recursion on notation", which seems appropriate in the context of poly-time computation. So we consider numbers as represented in binary notation. Recall that every positive integer can then be written uniquely as $1i_1 \ldots i_k$ with $i_\nu \in \{0, 1\}$, representing $2^k + \sum_{\nu=1}^{k} i_\nu 2^{k-\nu}$. Using the functions $s_0(x) := 2x$ and $s_1(x) := 2x + 1$, we may write $1i_1 \ldots i_k$ as $s_{i_k}(\ldots (s_{i_1} 1) \ldots)$. In our term language to be introduced below we shall denote s_i by S_i, and the number 1 by 1.

We define a restriction **LT** of Gödel's system **T** (Gödel, 1958) such that definable functions are exactly the polynomial time computable ones. To this end we combine

- a liberalized form of linearity for object and assumption variables (allowing multiple use of ground type results) with
- an extension of ramification concepts to all finite types, by allowing

$$\begin{cases} \rho \to \sigma \\ \lambda \bar{x}^\rho r \end{cases} \text{ as well as } \begin{cases} \rho \multimap \sigma \\ \lambda x^\rho r \end{cases}$$

and a corresponding syntactic distinction between incomplete and complete (typed) variables.

This paper grew out of joint work with Karl-Heinz Niggl done in 1998 and reported in (Bellantoni et al., 2000). Its aim is to simplify and clarify some of the concepts involved, with a planned extension (via the Curry-Howard correspondence) to an arithmetical system in mind. In particular, we have changed the constants, simplified the type system and the computation model, and have streamlined the treatment of ramification.

Related work has been done by Martin Hofmann (1998), who obtained similar results with a very different proof technique. Hofmann's recursive system was lifted to a polytime classical modal arithmetic by Bellantoni and Hofmann (to appear). The earlier "intrinsic theories" of Leivant (1995) followed the tradition of quantifier restrictions in induction. Ramification concepts have been considered e.g. by Simmons (1988), Bellantoni/Cook (1992) and Leivant/Marion (1993; to appear); they are extended here to all finite types. Notice also that the "tiered" typed λ-calculi of Leivant and Marion (1993) depend heavily on different representations of data (as words and as Church-like abstraction terms), which is not necessary in the approach developed here. One should also mention bounded linear logic of Girard, Scedrov and Scott (1990), and the so-called light linear logic of Girard (1998). The former differs from what we do here by requiring explicit bounds. A precise relation to the latter still needs to be clarified.

2. Motivation: examples for exponential growth

To set the stage, we discuss some examples of recursively defined functions and functionals exhibiting exponential growth. Our task will be to find appropriate restrictions on types and terms to exclude these definitions.

2.1. TWO RECURSIONS

$$d(1) := S_0(1) \qquad\qquad e(1) := 1$$
$$d(S_i(x)) := S_0(S_0(d(x))) \qquad e(S_i(x)) := d(e(x))$$

Then $|d(x)| = 2|x|$, $e(x) = d^{|x|-1}(1)$, i.e. we have exponential growth. The problem is that the previous value $e(x)$ of the second recursion is plugged into the recursion argument of d. Our cure will be to mark recursion arguments (cf. the notions of *safe* vs. *normal* arguments (Simmons, Bellantoni/Cook), and of *tiering* (Leivant)).

2.2. RECURSION WITH PARAMETER SUBSTITUTION

Consider the definition

$$e(1, y) := S_0(y) \qquad\qquad e(1) := S_0$$
$$e(S_i(x), y) := e(x, e(x, y)) \qquad\text{or}\qquad e(S_i(x)) := e(x) \circ e(x)$$

Then $e(x) = S_0^{2^{|x|-1}}$. The problem now clearly is that the previous higher type value of the recursion has been used twice. Our cure will be a linearity restriction.

A related phenomenon also involving recursion with parameter substitution occurs if we define

$$e(1, y) := y \qquad\qquad e(1) := \mathrm{id}$$
$$e(S_i(x), y) := e(x, d(y)) \qquad\text{or}\qquad e(S_i(x)) := e(x) \circ d$$

Then $e(x) = d^{(|x|-1)}$. Now the problem might be localized in the fact that we have a higher result type with argument types "marked" as recursion arguments: the type of the single (recursion) argument of d needs to be the argument type of $e(x)$. Our cure will be to exclude this.

2.3. HIGHER ARGUMENT TYPES: ITERATION

Consider the definition

$$
\begin{aligned}
I(1, f, y) &:= y \\
I(S_i(x), f, y) &:= f(I(x, f, y))
\end{aligned}
\quad \text{or} \quad
\begin{aligned}
I(1, f) &:= \mathsf{id} \\
I(S_i(x), f) &:= f \circ I(x, f)
\end{aligned}
$$

Then $I(x, f) = f^{|x|-1}$, hence $I(x, d) = d^{|x|-1}$. Now the problem lies in the substitution of the recursively defined d into a function parameter. This again will be excluded by requiring the result types to be without markers.

A related phenomenon occurs in

$$
\begin{aligned}
e(1) &:= S_0 \\
e(S_i(x)) &:= I(S_0(S_0(1)), e(x))
\end{aligned}
$$

Then: $e(x) = S_0^{2^{|x|-1}}$. Here the problem is the use of the "incomplete" higher type previous value $e(x)$, occurring as the step argument of I. This will be excluded by requiring that there are no such higher type parameters in the step terms.

3. Types, terms, and denotations

As already mentioned in the introduction, we shall work with two forms of arrow types and abstraction terms:

$$
\begin{cases} \rho \to \sigma \\ \lambda \bar{x}^\rho r \end{cases}
\quad \text{as well as} \quad
\begin{cases} \rho \multimap \sigma \\ \lambda x^\rho r \end{cases}
$$

and a corresponding syntactic distinction between incomplete and complete (typed) variables. The intuition is that a function of type $\rho \to \sigma$ may use its argument many times, whereas a function of the "linear" type $\rho \multimap \sigma$ is only allowed to use it once. As is well known, we then need a corresponding distinction for product types: the ordinary form $\times$ for $\to$, and the tensor product $\otimes$ for the linear arrow $\multimap$. Formally we proceed as follows.

3.1. TYPES

The types are:

$$
\rho, \sigma ::= \mathbf{U} \mid \mathbf{B} \mid \mathbf{L}(\rho) \mid \rho \multimap \sigma \mid \rho \to \sigma \mid \rho \otimes \sigma \mid \rho \times \sigma.
$$

The level of a type is defined by

$$
\begin{aligned}
l(\mathbf{U}) := l(\mathbf{B}) \qquad &:= 0 \\
l(\mathbf{L}(\rho)) \qquad &:= l(\rho) \\
l(\rho \multimap \sigma) := l(\rho \to \sigma) &:= \max\{l(\sigma), 1 + l(\rho)\} \\
l(\rho \otimes \sigma) \qquad &:= \max\{l(\rho), l(\sigma)\} \\
l(\rho \times \sigma) \qquad &:= \max\{l(\rho), l(\sigma), 1\}
\end{aligned}
$$

Ground types are the types of level 0, and a *higher* type is any type of level at least 1. A *function* type is a type of level at most 1. The $\to$-free types are also called *linear* types. In particular, each ground type is linear.

3.2. SET MODEL

There is an obvious set model of our type system, where we interpret every type ρ in the left column by the set $\mathbb{S}^\rho$ given in the right column:

$\mathbf{U}$	a special singleton set
$\mathbf{B}$	a special two-element set
$\mathbf{L}(\rho)$	the set of lists of elements of $\mathbb{S}^\rho$
$\rho \multimap \sigma$ and $\rho \to \sigma$	the set of total functions from $\mathbb{S}^\rho$ to $\mathbb{S}^\sigma$
$\rho \otimes \sigma$ and $\rho \times \sigma$	the cartesian product of $\mathbb{S}^\rho$ and $\mathbb{S}^\sigma$

We abbreviate $\mathbf{N} := \mathbf{L}(\mathbf{U})$ and call it the type of unary numerals; similarly $\mathbf{W} := \mathbf{L}(\mathbf{B})$ is called the type of binary numerals.

3.3. TERMS

The *constant symbols* are listed below, with their types.

$$
\begin{aligned}
&\text{⋊⋉} &&: \mathbf{U} \\
&\text{tt} &&: \mathbf{B} \\
&\text{ff} &&: \mathbf{B} \\
&\varepsilon_\rho &&: \mathbf{L}(\rho) \\
&*_\rho &&: \rho \multimap \mathbf{L}(\rho) \multimap \mathbf{L}(\rho) \\
&\text{if}_\tau &&: \mathbf{B} \multimap \tau \times \tau \multimap \tau \quad (\tau \text{ linear}) \\
&c_\tau^\rho &&: \mathbf{L}(\rho) \multimap \tau \times (\rho \multimap \mathbf{L}(\rho) \multimap \tau) \multimap \tau \quad (\tau \text{ linear}) \\
&\mathcal{R}_\tau^\rho &&: \mathbf{L}(\rho) \to (\rho \to \mathbf{L}(\rho) \to \tau \multimap \tau) \to \tau \multimap \tau \quad (\rho \text{ ground}, \tau \text{ linear})
\end{aligned}
$$

and for linear ρ, σ, τ

$$\otimes^+_{\rho\sigma} \;:\; \rho \multimap \sigma \multimap \rho \otimes \sigma$$

$$\otimes^-_{\rho\sigma\tau} \;:\; \rho \otimes \sigma \multimap (\rho \multimap \sigma \multimap \tau) \multimap \tau$$

$$\times^+_{\rho\sigma\tau} \;:\; (\tau \multimap \rho) \multimap (\tau \multimap \sigma) \multimap \tau \multimap \rho \times \sigma$$

$$\mathsf{fst}_{\rho\sigma} \;:\; \rho \times \sigma \multimap \rho$$

$$\mathsf{snd}_{\rho\sigma} \;:\; \rho \times \sigma \multimap \sigma$$

Terms are built from these constants and typed variables x^σ (incomplete variables) and $\bar{x}^\sigma$ (complete variables) by introduction and elimination rules for the two type forms $\rho \multimap \sigma$ and $\rho \to \sigma$, i.e.

c^ρ (constant) $|$

x^ρ (incomplete variable) $|$

$\bar{x}^\rho$ (complete variable) $|$

$(\lambda x^\rho r^\sigma)^{\rho \multimap \sigma}$ $|$

$(r^{\rho \multimap \sigma} s^\rho)^\sigma$ with higher type incomplete variables in r, s distinct $|$

$(\lambda \bar{x}^\rho r^\sigma)^{\rho \to \sigma}$ $|$

$(r^{\rho \to \sigma} s^\rho)^\sigma$ with s complete

We say that a term is *linear* or *ground* according as its type is. A term s is *complete* if all of its free variables are complete; otherwise it is *incomplete*. By the restriction on incomplete variables in the formation of (rs), a given higher type incomplete variable can occur at most once in a given term. We use infix notation for $*$, writing $l * r$ instead of $*rl$.

4. Conversions

The conversion rules are as expected.

$$
\begin{array}{lll}
(\lambda x r)s & \mapsto r[x := s] & \beta\text{-conversion; similar for } \bar{x} \\
\mathsf{if}_\tau \mathsf{tt}s & \mapsto \mathsf{fst}_{\tau\tau}s & \\
\mathsf{if}_\tau \mathsf{ff}s & \mapsto \mathsf{snd}_{\tau\tau}s & \\
c^\rho_\tau \varepsilon_\rho s & \mapsto \mathsf{fst}_{\tau,\sigma}s & \text{for } \sigma := \rho \multimap \mathbf{L}(\rho) \multimap \tau \\
c^\rho_\tau (l *_\rho r)s & \mapsto \mathsf{snd}_{\tau,\sigma}srl & \text{for } \sigma := \rho \multimap \mathbf{L}(\rho) \multimap \tau \\
\mathcal{R}^\rho_\tau \varepsilon_\rho st & \mapsto t & \\
\mathcal{R}^\rho_\tau (l *_\rho r)st & \mapsto srl(\mathcal{R}^\rho_\tau lst) & \\
\otimes^-_{\rho\sigma\tau}(\otimes^+_{\rho\sigma}rs)t & \mapsto trs & \\
\end{array}
$$

$$\mathrm{fst}_{\rho\sigma}(\times^{+}_{\rho\sigma\tau}rst) \mapsto rt$$

$$\mathrm{snd}_{\rho\sigma}(\times^{+}_{\rho\sigma\tau}rst) \mapsto st$$

Notice that we shall work with a representation of terms via parse dags, to be explained in Section 6. This will ensure that β-conversion of $(\lambda xr)s$ or $(\lambda \bar{x}r)s$ leads to a term within our system (cf. the proof of Lemma 6.2).

Redexes are subterms shown on the left side of conversion rules above. We assume that no two bound variables have the same name, and no bound variable has the same name as a free variable. A term is in *normal form* if it does not contain a redex. Write $r \to s$ for the one-step reduction based on the conversion rules, and $r \to^{*} s$ for its reflexive transitive closure. For every term t there is a unique normal-form term $\mathrm{nf}(t)$ such that $t \to^{*} \mathrm{nf}(t)$. Two terms are called *equivalent* if they have the same normal form.

5. Examples for exponential growth, again

We now come back to the examples form Section 2, and explain how our restrictions on the formation of types and terms make it impossible to build the corresponding terms. However, for definiteness we first have to say precisely what we mean by a numeral.

5.1. NUMERALS

Terms of the form $(\ldots(\varepsilon_\rho *_\rho r_n^\rho)\ldots *_\rho r_2^\rho) *_\rho r_1^\rho$ are called *lists*. We will make use of the following abbreviations for $\mathbf{N} := \mathbf{L}(\mathbf{U})$ and $\mathbf{W} := \mathbf{L}(\mathbf{B})$.

$$0 := \varepsilon_{\mathbf{U}}$$
$$S := \lambda l^{\mathbf{N}} l * \mathsf{x}$$
$$1 := \varepsilon_{\mathbf{B}}$$
$$S_0 := \lambda l^{\mathbf{W}} l * \mathsf{ff}$$
$$S_1 := \lambda l^{\mathbf{W}} l * \mathsf{tt}$$
$$\mathrm{id} := \lambda x.x$$

Particular lists are $S(\ldots(S0)\ldots)$ and $S_{i_1}(\ldots(S_{i_n}1)\ldots)$. The former are called *unary numerals*, and the latter *binary numerals* (or *numerals of type* $\mathbf{W}$). We denote binary numerals by ν.

5.2. TWO RECURSIONS

Recall that we considered the definition of

$$d(1) := S_0(1) \qquad\qquad e(1) := 1$$
$$d(S_i(x)) := S_0(S_0(d(x))) \qquad e(S_i(x)) := d(e(x))$$

The corresponding terms are

$$d := \lambda\bar{x}.\mathcal{R}_\mathbf{W}\bar{x}(\lambda\bar{z}\lambda\bar{l}\lambda p^\mathbf{W}.S_0(S_0 p))(S_0 1),$$
$$e := \lambda\bar{x}.\mathcal{R}_\mathbf{W}\bar{x}(\lambda\bar{z}\lambda\bar{l}\lambda p^\mathbf{W}.dp)1.$$

Here d is legal, but e is not: the application dp is not allowed.

5.3. RECURSION WITH PARAMETER SUBSTITUTION

Recall the proposed definition of

$$e(1, y) := S_0(y) \qquad\qquad e(1) := S_0$$
$$e(S_i(x), y) := e(x, e(x, y)) \quad\text{or}\quad e(S_i(x)) := e(x) \circ e(x)$$

The corresponding term

$$\lambda\bar{x}.\mathcal{R}_{\mathbf{W}\multimap\mathbf{W}}\bar{x}(\lambda\bar{z}\lambda\bar{l}\lambda p^{\mathbf{W}\multimap\mathbf{W}}\lambda y.p(py))S_0$$

does not satisfy the linearity condition: the variable p occurs twice, and p needs to be incomplete because the type of $\mathcal{R}$ uses $\cdots \to (\rho \to \mathbf{L}(\rho) \to \tau \multimap \tau)$ rather than $\ldots \multimap (\rho \to \mathbf{L}(\rho) \to \tau \multimap \tau)$. Under the term formation rules, composition using such a type can only be with a complete term.

In our second example involving recursion with parameter substitution we had

$$e(1, y) := y \qquad\qquad e(1) := \mathsf{id}$$
$$e(S_i(x), y) := e(x, d(y)) \quad\text{or}\quad e(S_i(x)) := e(x) \circ d$$

The corresponding term would be

$$\lambda\bar{x}.\mathcal{R}_{\mathbf{W}\to\mathbf{W}}\bar{x}(\lambda\bar{z}\lambda\bar{l}\lambda p^{\mathbf{W}\to\mathbf{W}}\lambda\bar{x}.p(d\bar{x}))(\lambda y y)$$

but it is not legal, since the result type is not linear.

5.4. HIGHER ARGUMENT TYPES: ITERATION

We considered

$$I(1, f, y) := y \qquad\qquad I(1, f) := \mathsf{id}$$
$$I(\mathsf{S}_i(x), f, y) := f(I(x, f, y)) \quad \text{or} \quad I(\mathsf{S}_i(x), f) := f \circ I(x, f)$$

with the corresponding term

$$I_f := \lambda \bar{x}.\mathcal{R}_{\mathbf{W} \rightarrow \mathbf{W}} \bar{x}(\lambda \bar{z} \lambda \bar{l} \lambda p^{\mathbf{W} \multimap \mathbf{W}} \lambda y.f(py))(\lambda y y)$$
$$e := \lambda x.I_d x 1$$

Here I_f is legal, but e is not: the type of d prohibits iteration. – Note that in PV^ω (Cook and Kapron (1990), Cook (1992)) I is *not* definable, for otherwise we could define $\lambda z.Idz$.

A related phenomenon occurs in

$$e(1) := \mathsf{S}_0$$
$$e(\mathsf{S}_i(x)) := I(\mathsf{S}_0(\mathsf{S}_0(1)), e(x))$$

Now the terms are

$$I_f := \lambda x.\mathcal{R}_{\mathbf{W} \multimap \mathbf{W}} \bar{x}(\lambda \bar{z} \lambda \bar{l} \lambda p^{\mathbf{W} \multimap \mathbf{W}} \lambda y.f(py))(\lambda y y)$$
$$e := \lambda \bar{x}.\mathcal{R}_{\mathbf{W} \multimap \mathbf{W}} \bar{x}(\lambda \bar{z} \lambda \bar{l} q^{\mathbf{W} \multimap \mathbf{W}}.I_q(\mathsf{S}_0(\mathsf{S}_0 1)))\mathsf{S}_0$$

Again e is not legal, this time because the free parameter f in the step term of I_f is substituted with the *incomplete* variable q. This variable needs to be complete because of the typing of the recursion operator.

5.5. POLYNOMIALS

It is high time that we give some examples of what *can* de done in our term system. It is easy to define $\oplus : \mathbf{W} \rightarrow \mathbf{W} \multimap \mathbf{W}$ such that $x \oplus y$ concatenates $|x|$ bits onto y.

$$1 \oplus y = \mathsf{S}_0 y$$
$$(\mathsf{S}_i x) \oplus y = \mathsf{S}_0(x \oplus y)$$

The representing term is $\bar{x} \oplus y := \mathcal{R}_{\mathbf{W} \multimap \mathbf{W}} \bar{x}(\lambda \bar{z} \lambda \bar{l} \lambda p^{\mathbf{W} \multimap \mathbf{W}} \lambda y.\mathsf{S}_0(py))\mathsf{S}_0$.

Similarly we define $\odot : \mathbf{W} \rightarrow \mathbf{W} \rightarrow \mathbf{W}$ such that $x \odot y$ has output length $|x| \cdot |y|$.

$$x \odot 1 = x$$
$$x \odot (\mathsf{S}_i y) = x \oplus (x \odot y)$$

The representing term is $\bar{x} \odot \bar{y} := \mathcal{R}_{\mathbf{W}}\bar{y}(\lambda\bar{z}\lambda\bar{l}\lambda p^{\mathbf{W}}\bar{x}\oplus p)\bar{x}$.

Notice that the typing $\oplus\colon \mathbf{W} \to \mathbf{W} \multimap \mathbf{W}$ is crucial: it allows using the incomplete variable p in the definition of $\odot$. If we try to go on and define exponentiation from multiplication $\odot$ just as $\odot$ was defined from $\oplus$, we find out that we cannot go ahead, because of the different typing $\odot\colon \mathbf{W} \to \mathbf{W} \to \mathbf{W}$.

6. Normalization

A *dag* is a directed acyclic graph. A *parse dag* is a structure like a parse tree but admitting in-degree greater than one. For example, a parse dag for $\lambda x r$ has a node containing λx and a pointer to a parse dag for r. A parse dag for (rs) has a node containing a pair of pointers, one to a parse dag for r and the other to a parse dag for s. Terminal nodes are labeled by constants and variables.

The *size* $|d|$ of a parse dag d is the number of nodes in it, but counting 3 for c_r nodes. Starting at any given node in the parse dag, one obtains a term by a depth-first traversal; it is the term *represented* by that node. We may refer to a node as if it were the term it represents.

A parse dag is *conformal* if (i) every node having in-degree greater than 1 is of ground type, and (ii) every maximal path to a bound variable x passes through the same binding λx node.

A parse dag is *h-affine* if each higher-type variable occurs at most once in the dag.

We adopt a model of computation over parse dags in which operations such as the following can be performed in unit time: creation of a node given its label and pointers to the sub-dags; deletion of a node; obtaining a pointer to one of the subsidiary nodes given a pointer to an interior node; conditional test on the type of node or on the constant or variable in the node. Concerning computation over terms (including numerals), we use the same model and identify each term with its parse tree. Although not all parse dags are conformal, every term is conformal (assuming a relabeling of bound variables).

A term is called *simple* if all its higher-type variables are incomplete. Obviously simple terms are closed under reductions, taking of subterms, and applications. Every simple term is h-affine, due to the linearity of incomplete higher-type variables.

Lemma 6.1 (Simplicity). *Let t be a ground type term whose free variables are of ground type. Then* $\mathsf{nf}(t)$ *contains no higher type complete variables.*

Proof. Let t be a ground type term whose free variables are of ground type, and consider $\mathsf{nf}(t)$. We must show that $\mathsf{nf}(t)$ contains no higher type complete variables. So suppose a variable $\bar{x}^\sigma$ with $l(\sigma) > 0$ occurs in $\mathrm{nf}(t)$. It must be bound in a subterm $(\lambda \bar{x}^\sigma r)^{\sigma \to \tau}$ of $\mathrm{nf}(t)$. By the well known subtype property of normal terms, the type $\sigma \to \tau$ either occurs positively in the type of $\mathrm{nf}(t)$, or else negatively in the type of one of the constants or free variables of $\mathrm{nf}(t)$. The former is impossible since t is of ground type, and the latter by inspection of the types of the constants. $\qquad \square$

A term is $\mathcal{R}$-*free* if it does not contain an occurrence of $\mathcal{R}$.

Lemma 6.2 (Sharing Normalization). *Let t be an $\mathcal{R}$-free simple term. Then a parse dag for $\mathrm{nf}(t)$, of size at most $|t|$, can be computed from t in time $O(|t|^2)$.*

Proof. Under our model of computation, the input t is a parse tree. Since t is simple, it is an h-affine conformal parse dag of size at most $|t|$. If there are no nodes which represent a redex, then we are done. Otherwise, locate a node representing a redex; this takes time at most $O(|t|)$. We show how to update the dag in time $O(|t|)$ so that the size of the dag has strictly decreased and the redex has been eliminated, while preserving conformality. Thus, after at most $|t|$ iterations the resulting dag represents the normal-form term $\mathrm{nf}(t)$. The total time therefore is $O(|t|^2)$.

Assume first that the redex in t is $(\lambda x r)s$ with x of ground type; the argument is similar for a complete variable $\bar{x}$. Replace pointers to x in r by pointers to s. Since s does not contain x, no cycles are created. Delete the λx node and the root node for $(\lambda x r)s$ which points to it. By conformality (i) no other node points to the λx node. Update any node which pointed to the deleted node for $(\lambda x r)s$, so that it now points to the revised r subdag. This completes the β reduction on the dag (one may also delete the x nodes). Conformality (ii) gives that the updated dag represents a term t' such that $t \to t'$.

One can verify that the resulting parse dag is conformal and h-affine, with conformality (i) following from the fact that s has ground type.

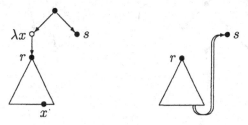

If the redex in t is $(\lambda x r)s$ with x of higher type, then x occurs at most once in r because the parse dag is h-affine. By conformality (i) there is at most one pointer to that occurrence of x. Update it to point to s instead, deleting the x node. As in the preceeding case, delete the λx and the $(\lambda x r)s$ node pointing to it, and update other nodes to point to the revised r. Again by conformality (ii) the updated dag represents t' such that $t \to t'$. Conformality and acyclicity are preserved, observing this time that conformality (i) follows because there is at most one pointer to s.

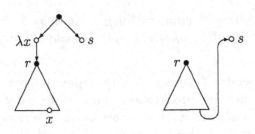

The remaining reductions are for the constant symbols.

Case $\text{if}_\tau \text{tt} s \mapsto \text{fst}_{\tau\tau}s$. Easy; similar for ff.

Case $\mathbf{c}_\tau \varepsilon_\rho s \mapsto \text{fst } s$. Easy.

Case $\mathbf{c}_\tau (l *_\rho r)s \mapsto \text{snd } srl$ with ρ a ground type.

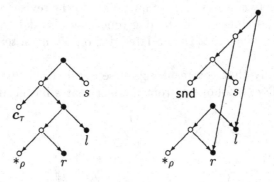

Notice that the new dag has one node more than the original one, but one $\mathbf{c}_\tau$-node less. Since we count the $\mathbf{c}_\tau$-nodes 3-fold, the total number of nodes decreases.

Case $c_\tau(l *_\rho r)s \mapsto \mathsf{snd}\, srl$ with ρ not a ground type.

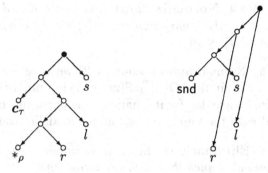

Case $\otimes^-_{\rho\sigma\tau}(\otimes^+_{\rho\sigma}rs)t \mapsto trs$ with $\rho \otimes \sigma$ a ground type.

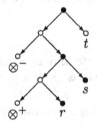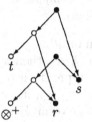

Case $\otimes^-_{\rho\sigma\tau}(\otimes^+_{\rho\sigma}rs)t \mapsto trs$ with $\rho \otimes \sigma$ not a ground type.

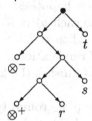

Case $\mathsf{fst}_{\rho\sigma}(\times^+_{\rho\sigma\tau}rst) \mapsto rt$. Here we need that $\rho \times \sigma$ is *never* considered as a ground type. The case of $\mathsf{snd}_{\rho\sigma}(\times^+_{\rho\sigma\tau}rst) \mapsto st$ is of course similar.

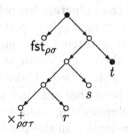

□

Corollary 6.3 (Base Normalization). *Let t be a closed $\mathcal{R}$-free simple term of type* **W**. *Then the binary numeral* $\mathsf{nf}(t)$ *can be computed from t in time* $O(|t|^2)$, *and* $|\mathsf{nf}(t)| \leq |t|$.

Proof. By Sharing Normalization (Lemma 6.2) we obtain a parse dag for $\mathsf{nf}(t)$ of size at most $|t|$, in time $O(|t|^2)$. Since $\mathsf{nf}(t)$ is a binary numeral, there is only one possible parse dag for it – namely, the parse tree of the numeral. This is identified with the numeral itself in our model of computation. □

Lemma 6.4 ($\mathcal{R}$ Elimination). *Let t be a simple term of linear type. There is a polynomial p_t such that: if $\vec{m}$ are linear type $\mathcal{R}$-free closed simple terms and the free variables of $t[\vec{x} := \vec{m}]$ are linear and incomplete, then in time $p_t(|\vec{m}|)$ one can compute an $\mathcal{R}$-free simple term $\mathsf{rf}(t; \vec{x}; \vec{m})$ such that $t[\vec{x} := \vec{m}] \to^* \mathsf{rf}(t; \vec{x}; \vec{m})$.*

Proof. By induction on $|t|$.

If t has the form $\lambda z u_1$, then z is incomplete and z, u_1 have linear type because t has linear type. If t is of the form $D\vec{u}$ with D a variable or one of the symbols xx, tt, ff, ε_ρ, $*_\rho$, if_τ, $\mathbf{c}_\tau$, $\otimes^+_{\rho\sigma}$, $\otimes^-_{\rho\sigma\tau}$, $\times^+_{\rho\sigma\tau}$, $\mathsf{fst}_{\rho\sigma}$ or $\mathsf{snd}_{\rho\sigma}$, then each u_i is a linear type term. Here (in case D is a variable x_i) we need that x_i is linear.

In all of the preceeding cases, each $u_i[\vec{x} := \vec{m}]$ has only linear and incomplete free variables. Apply the induction hypothesis as required to simple terms u_i to obtain $u_i^* := \mathsf{rf}(u_i; \vec{x}; \vec{m})$; so each u_i^* is $\mathcal{R}$-free. Let t^* be obtained from t by replacing each u_i by u_i^*. Then t^* is an $\mathcal{R}$-free simple term; here we need that $\vec{m}$ are closed, to avoid duplication of variables. The result is obtained in linear time from $\vec{u}^*$. This finishes the lemma in all of these cases.

If t is $(\lambda y r) s\vec{u}$ with an incomplete variable y of ground type, apply the induction hypothesis to yield $(r\vec{u})^* := \mathsf{rf}(r\vec{u}; \vec{x}; \vec{m})$ and $s^* := \mathsf{rf}(s; \vec{x}; \vec{m})$. Redirect the pointers to y in $(r\vec{u})^*$ to point to s^* instead. If t is $(\lambda \bar{y} r) s\vec{u}$ with a complete variable $\bar{y}$ of ground type, apply the IH to yield $s^* := \mathsf{rf}(s; \vec{x}; \vec{m})$. Notice that s^* is closed, since it is complete and the free variables of $s[\vec{x} := \vec{m}]$ are incomplete. Then apply the induction hypothesis again to obtain $\mathsf{rf}(r\vec{u}; \vec{x}, \bar{y}; \vec{m}, s^*)$. The total time is at most $q(|t|) + p_s(|\vec{m}|) + p_r(|\vec{m}| + p_s(|\vec{m}|))$, as it takes at most linear time to construct $r\vec{u}$ from $(\lambda y r) s\vec{u}$.

If t is $(\lambda y r) s\vec{u}$ with y of higher type, then y can occur at most once in r, because t is simple. Thus $|r[y := s]\vec{u}| < |(\lambda y r) s\vec{u}|$. Apply the induction hypothesis to obtain $\mathsf{rf}(r[y := s]\vec{u}; \vec{x}; \vec{m})$. Note that the time is bounded by $q(|t|) + p_{r[y:=s]\vec{u}}(|\vec{m}|)$ for a degree one polynomial q, since it takes at most linear time to make the at-most-one substitution in the parse tree.

The only remaining case is if the term is an $\mathcal{R}$ clause. Then it is of the form $\mathcal{R}ls\vec{t}$, because the term has linear type.

Since l is complete, all free variables of l are complete – they must be in $\vec{x}$ since free variables of $(\mathcal{R}ls\vec{t})[\vec{x} := \vec{m}]$ are incomplete. Then $l[\vec{x} := \vec{m}]$ is closed, implying $\mathsf{nf}(l[\vec{x} := \vec{m}])$ is a list. One obtains $\mathsf{rf}(l; \vec{x}; \vec{m})$ in time $p_l(|\vec{m}|)$ by the induction hypothesis. Then by Base Normalization (Corollary 6.3) one obtains the list $\hat{l} := \mathsf{nf}(\mathsf{rf}(l; \vec{x}; \vec{m}))$ in a further polynomial time. Let $\hat{l} = (\ldots(\varepsilon_\rho *_\rho r_N)\ldots *_\rho r_1) *_\rho r_0$ and let l_i, $0 \le i \le N$ be obtained from $\hat{l}$ by omitting the initial elements $r_0, \ldots, r_i$. Thus all $\{r_i, l_i \mid i \le N\}$ are obtained in a total time bounded by $p'_l(|\vec{m}|)$ for a polynomial p'_l.

Now consider $s\bar{y}\bar{z}$ with new variables $\bar{y}^\rho$ and $\bar{z}^{\mathbf{L}(\rho)}$. Applying the induction hypothesis to $s\bar{y}\bar{z}$ one obtains a monotone bounding polynomial $p_{s\bar{y}\bar{z}}$. One computes all $s_i := \mathsf{rf}(s\bar{y}\bar{z}; \vec{x}, \bar{y}, \bar{z}; \vec{m}, r_i, l_i)$ in a total time of at most

$$\sum_{i=1}^{N} p_{s\bar{y}\bar{z}}(|r_i| + |l_i| + |\vec{m}|) \le p'_l(|\vec{m}|) \cdot p_{s\bar{y}\bar{z}}(2p'_l(|\vec{m}|) + |\vec{m}|).$$

Each s_i is $\mathcal{R}$-free by the induction hypothesis. Furthermore, no s_i has a free incomplete variable: any such variable would also be free in s contradicting that s is complete.

Consider $\vec{t}$. The induction hypothesis gives all $\hat{t}_i := \mathsf{rf}(t_i; \vec{x}; \vec{m})$ in time $\sum_i p_{t_i}(|\vec{m}|)$. These terms are also $\mathcal{R}$-free by induction hypothesis. Clearly the t_i do not have any free (or bound) higher type incomplete variables in common. The same is true of all $\hat{t}_i$.

Using additional time bounded by a polynomial p in the lengths of these computed values, one constructs the $\mathcal{R}$-free term

$$(\lambda z.s_0(s_1 \ldots (s_N z)\ldots))\vec{\hat{t}}.$$

Defining $p_t(n) := p(\sum_i p_{t_i}(n) + p'_l(n) \cdot p_{syz}(2p'_l(n) + n))$, the total time used in this case is at most $p_t(|\vec{m}|)$. The result is a term because the $\hat{t}_i$ are terms which do not have any free higher-type incomplete variable in common and because s_i does not have any free higher-type incomplete variables at all. $\square$

7. Characterizations

Lemma 7.1 (Normalization). *Let t be a closed term of type $\vec{\rho} \multimap \sigma$, where σ and each ρ_i is a ground type. Then t denotes a polytime function.*

Proof. One must find a polynomial q_t such that for all $\mathcal{R}$-free simple closed terms $\vec{n}$ of types $\vec{\rho}$ one can compute $\mathsf{nf}(t\vec{n})$ in time $q_t(|\vec{n}|)$. Let $\vec{x}$ be new variables of types $\vec{\rho}$. The normal form of $t\vec{x}$ is computed in an amount of time that may be large, but it is still only a constant with respect to $\vec{n}$.

$\mathsf{nf}(t\vec{x})$ is simple by Lemma 6.1. By $\mathcal{R}$ elimination (Lemma 6.4) one reduces to an $\mathcal{R}$-free simple term $\mathsf{rf}(\mathsf{nf}(t\vec{x}); \vec{x}; \vec{n})$ in time $p_t(|\vec{n}|)$. Since the running time bounds the size of the produced term, $|\mathsf{rf}(\mathsf{nf}(t\vec{x}); \vec{x}; \vec{n})| \leq p_t(|\vec{n}|)$.

By Sharing Normalization (Lemma 6.2) one can compute $\mathsf{nf}(t\vec{n}) = \mathsf{nf}(\mathsf{rf}(\mathsf{nf}(t\vec{x}); \vec{x}; \vec{n}))$ in time $O(p_t(|\vec{n}|)^2)$. Let q_t be the polynomial referred to by the big-O notation. $\qquad\qquad\qquad\qquad\qquad\qquad\qquad\qquad$ □

Lemma 7.2 (Sufficiency). *Let f be a polynomial-time computable function of type $\vec{\mathbf{W}} \multimap \mathbf{W}$. Then f is denoted by a closed term t.*

Proof. In Bellantoni and Cook (1992) the polynomial time computable functions are characterized by a function algebra B based on *safe recursion* and *safe composition*. There every function is written in the form $f(\vec{x}; \vec{y})$ where $\vec{x}; \vec{y}$ denotes a bookkeeping of those variables $\vec{x}$ that are used in a recursion defining f, and those variables $\vec{y}$ that are not recursed on. We proceed by induction on the definition of $f(x_1, \ldots, x_k; y_1, \ldots, y_l)$ in B, associating to f a closed term t_f of type $\mathbf{W}^{(k)} \rightarrow \mathbf{W}^{(l)} \multimap \mathbf{W}$, such that t denotes f.

The functions in B were defined over the non-negative integers rather than the positive ones, but this clearly is a minor point. We use the bijection $x \in \mathbb{N} \Leftrightarrow (2^{|x|} + x) \in \mathbb{Z}^+$.

If f in B is an initial function 0, S_0, S_1, P, conditional C or projection $\pi_i^{m,n}$, then t_f is easily defined.

If f is defined by safe composition in system B, then

$$f(\vec{x}; \vec{y}) := g(r_1(\vec{x};), \ldots, r_m(\vec{x};); s_1(\vec{x}; \vec{y}), \ldots, s_n(\vec{x}; \vec{y})).$$

Using the induction hypothesis to obtain t_g, $t_{\vec{r}}$ and $t_{\vec{s}}$, define

$$t_f := \lambda \vec{\vec{x}} \lambda \vec{y}.t_g(t_{r_1}\vec{\vec{x}}) \ldots (t_{r_m}\vec{\vec{x}})\,(t_{s_1}\vec{\vec{x}}\vec{y}) \ldots (t_{r_m}\vec{\vec{x}}\vec{y}).$$

Finally consider f defined by safe recursion in system B.

$$f(0, \vec{x}; \vec{y}) := g(\vec{x}; \vec{y})$$
$$f(S_i n, \vec{x}; \vec{y}) := h_i(n, \vec{x}; \vec{y}, f(n, \vec{x}; \vec{y})) \quad \text{for } S_i n \neq 0.$$

One has t_g, t_{h_0} and t_{h_1} by the induction hypothesis. Let p be a variable of type $\tau := \mathbf{W}^{(\#(\vec{y}))} \multimap \mathbf{W}$; this is the linear type used in the recursion. Then define a step term by

$$s := \lambda \vec{x} \lambda \vec{l} \lambda p \lambda \vec{y}.\mathsf{if}_{\mathbf{W}}\vec{x}\big(\times^+(\lambda z.t_{h_0}\vec{l}\vec{x}\vec{y}z)(\lambda z.t_{h_1}\vec{l}\vec{x}\vec{y}z)(p\vec{y})\big).$$

Note p is used only once. Let $t_f := \lambda \vec{n} \lambda \vec{\vec{x}}.\mathcal{R}_\tau \vec{n} s(t_g \vec{\vec{x}}).$ $\qquad\qquad$ □

References

Stephen Bellantoni and Stephen Cook. A new recursion-theoretic characterization of the polytime functions. *Computational Complexity*, 2:97–110, 1992.

Stephen Bellantoni and Martin Hofmann. A New "Feasible" Arithmetic *Journal of Symbolic Logic*, to appear.

Stephen Bellantoni, Karl-Heinz Niggl, and Helmut Schwichtenberg. Higher type recursion, ramification and polynomial time. *Annals of Pure and Applied Logic*, 104:17–30, 2000.

Stephen A. Cook. Computability and complexity of higher type functions. In Y.N. Moschovakis, editor, *Logic from Computer Science, Proceedings of a Workshop held November 13–17, 1989*, number 21 in MSRI Publications, pages 51–72. Springer Verlag, Berlin, Heidelberg, New York, 1992.

Stephen A. Cook and Bruce M. Kapron. Characterizations of the basic feasible functionals of finite type. In S. Buss and P. Scott, editors, *Feasible Mathematics*, pages 71–96. Birkhäuser, 1990.

Jean-Yves Girard. Light linear logic. *Information and Computation*, 143, 1998.

Jean-Yves Girard, Andre Scedrov, and Philipp J. Scott. Bounded linear logic. In S.R. Buss and Ph.J. Scott, editors, *Feasible Mathematics*, pages 195–209. Birkhäuser, Boston, 1990.

Kurt Gödel. Über eine bisher noch nicht benützte Erweiterung des finiten Standpunkts. *Dialectica*, 12:280–287, 1958.

David Hilbert. Über das Unendliche. *Mathematische Annalen*, 95:161–190, 1925.

Martin Hofmann. Typed lambda calculi for polynomial-time computation. Habilitation thesis, Mathematisches Institut, TU Darmstadt, Germany. Available under www.dcs.ed.ac.uk/home/mxh/habil.ps.gz, 1998.

Daniel Leivant. Intrinsic theories and computational complexity. In *Logic and Computational Complexity, International Workshop LCC '94, Indianapolis* D. Leivant, ed., Springer LNCS 960, 1995, p. 177-194.

Daniel Leivant. Predicative recurrence in finite type. In A. Nerode and Y.V. Matiyasevich, editors, *Logical Foundations of Computer Science*, volume 813 of *LNCS*, pages 227–239, 1994.

Daniel Leivant and Jean-Yves Marion. Ramified recurrence and computational complexity IV: Predicative functionals and poly-space. To appear: Information and Computation.

Daniel Leivant and Jean-Yves Marion. Lambda calculus characterization of poly–time. In M. Bezem and J.F. Groote, editors, *Typed Lambda Calculi and Applications*, pages 274–288. LNCS Vol. 664, 1993.

Harold Simmons. The realm of primitive recursion. *Archive for Mathematical Logic*, 27:177–188, 1988.

9 781402 006081